T0343415

MANAGING OCEAN ENVIRONMENTS IN A CHANGING CLIMATE

MANAGING OCEAN ENVIRONMENTS IN A CHANGING CLIMATE

Sustainability and Economic Perspectives

Edited By

KEVIN J. NOONE

USSIF RASHID SUMAILA

ROBERT J. DIAZ

AMSTERDAM • BOSTON • HEIDELBERG • LONDON • NEW YORK • OXFORD
PARIS • SAN DIEGO • SAN FRANCISCO • SINGAPORE • SYDNEY • TOKYO

Elsevier
30 Corporate Drive, Suite 400, Burlington, MA 01803, USA
525 B Street, Suite 1800, San Diego, CA 92101-4495, USA

First edition **2013**

Library of Congress Cataloging-in-Publication Data
Noone, Kevin J.
Managing ocean environments in a changing climate : sustainability and economic perspectives / Kevin J. Noone, Robert J. Diaz.
pages cm
Summary: "This chapter provides background information for the scope, purpose and structure of this book. Here, we introduce the main themes that are explored in the book, and provide information that is common to all of them. We discuss the need for taking a holistic view of threats to the global ocean. We also introduce the background and boundary conditions for the economic analysis of the potential savings that could be accrued by acting now to mitigate these threats. We show how this analysis can be useful for decision support in the area of marine policy"– Provided by publisher.
Includes bibliographical references and index.
ISBN 978-0-12-407668-6 (hardback)
1. Marine ecology. 2. Marine pollution. 3. Marine resources conservation. I. Diaz, R. J. (Robert J.) II. Title.
QH541.5.S3N66 2013
577.7–dc23

2013019814

British Library Cataloguing in Publication Data
A catalogue record for this book is available from the British Library

For information on all Elsevier publications
visit our web site at store.elsevier.com

ISBN: 978-0-12-407668-6

Contents

Contributors ix
Preface xi

1. Valuing the Ocean: An Introduction

KEVIN J. NOONE, USSIF RASHID SUMAILA AND ROBERT J. DIAZ

Purpose and Scope of the Book 2
Differential Analysis for Future Scenarios 3
Expert Survey Approach 9
The Matrix: Values, Threats, and Knowledge 9
Knowledge for Decision Support 12

2. Ocean Acidification

CAROL TURLEY

Cause and Chemistry 15
Time and Space Scales 17
Future Ocean Acidification Scenarios 21
Potential Future Effects on Physiological Processes and Behavior 25
Impacts on Communities, Food Webs, and Ecosystems 32
Conclusions 35

3. Ocean Warming

KEVIN J. NOONE AND ROBERT J. DIAZ

Introduction 45
Physical Consequences of Ocean Warming 46
Biological Consequences of Ocean Warming 57
Summary and Take-Home Messages 62

4. Hypoxia

ROBERT J. DIAZ, HANNA ERIKSSON-HÄGG AND RUTGER ROSENBERG

Key Messages 67
Introduction 68
The Heart of the Problem 71
Global Patterns in Hypoxia 73

OMZs and Open Ocean Decline in Oxygen 78
Environmental Consequences of Hypoxia 81
Economic Consequences of Hypoxia 85
Global Change and Hypoxia 87
Restoration and the Future of Hypoxia 90

5. Sea-Level Rise

KEVIN J. NOONE

Introduction 97
Causes of Sea-Level Rise 97
Observations of Sea-Level Rise 109
Sea Level Rise in the Future 111
Impacts of Sea-Level Rise 115

6. Marine Pollution

DAN WILHELMSSON, RICHARD C. THOMPSON, KATRIN HOLMSTRÖM, OLOF LINDÉN AND HANNA ERIKSSON-HÄGG

Introduction 127
Chemical Pollution—POPs and Metals/Toxic Chemicals 129
Oil Pollution 135
Solid Substances 143
Radioactive Waste 150
Noise 152
Present and Forecasted Impacts of Pollution on Global Marine Ecosystems 157

7. The Potential Economic Costs of the Overuse of Marine Fish Stocks

USSIF RASHID SUMAILA, WILLIAM W.L. CHEUNG AND A.D. ROGERS

Fish and Fisheries Are Important to People 171
Evidence of Overuse of Fish Stocks 173
Climate Change Will Exacerbate the Problem of Overuse of Fish Stocks 179
Global Economic Loss due to Overfishing 181
Estimated Economic Losses 184
Conclusion 186

8. Impacts of Multiple Stressors

JULIE HALL, ROBERT J. DIAZ, NICOLAS GRUBER AND DAN WILHELMSSON

Introduction 193
Global-Scale Stressors 195
Local- and Regional-Scale Stressors 200
Feedbacks and Synergistic Effects 203
Concluding Remarks 215

9. Tipping Points, Uncertainty, and Precaution: Preparing for Surprise

KEVIN J. NOONE AND FRANK ACKERMAN

Introduction 223
Why Buy Insurance? 226
Making Decisions in the Dark 228
Peering into the Future 231
Shell Game 231
Pentagon Planning 233
Safe Standards and Planetary Boundaries 235

10. Valuing the Ocean Environment

FRANK ACKERMAN

Introduction 243
What Is Not Included 244
Classic Studies of the Value of Ocean Environments 247
Fisheries and Climate Change 252
Tourism and Climate Change 254
Costs of SLR 257
Stormy Weather 259
Shrinking the Ocean Carbon Sink 262
Valuing the Damages 265
Scenario Definitions 266
Fisheries 267
Sea-Level Rise 267
Storms 268
Tourism 268
Ocean Carbon Sink 269

11. Managing Multiple Human Stressors in the Ocean: A Case Study in the Pacific Ocean

WILLIAM W.L. CHEUNG AND USSIF RASHID SUMAILA

Introduction 277
Pattern of Biodiversity and Marine Living Resources 278
Key Human Pressures in the Pacific Ocean 282
Challenges to Sustainable Management of Fisheries Resources in the Pacific Ocean 283
Climate Change 287
Sustainable Management of the Pacific Ocean 294

12. Paths to Sustainable Ocean Resources

KATERYNA M. WOWK

Introduction 301
Implications of Major Threats and Policy Recommendations 303
Multiple Stressors: Putting the Pieces Together 323
Implications of Valuing Ocean Damages and Planning for Surprise 329
Pacific Region Case Study—Implications for Regional Ocean Governance 334
The Next Era of Global Ocean Governance: Paths to Sustainability 336
Concluding Recommendations 340

Index 349

Contributors

Frank Ackerman Synapse Energy Economics, Cambridge, Massachusetts, USA

William W.L. Cheung Changing Ocean Research Unit, Fisheries Centre, The University of British Columbia, Vancouver, British Columbia, Canada

Robert J. Diaz Department of Biological Sciences, Virginia Institute of Marine Science, College of William and Mary, Gloucester Point, Virginia, USA

Nicolas Gruber Environmental Physics, Institute of Biogeochemistry and Pollutant Dynamics, Department of Environmental Sciences, Zurich, Switzerland

Julie Hall NIWA, Kilbirnie, Wellington, New Zealand

Hanna Eriksson-Hägg Baltic Nest Institute, Stockholm Resilience Centre, Stockholm University, Kräftriket 2B, Stockholm, Sweden

Katrin Holmström Environment Department, Stockholm City, Box 8136, Sweden

Olof Lindén World Maritime University, Box 500, Malmö, Sweden

Kevin J. Noone Department of Applied Environmental Science, Stockholm University, Stockholm, Sweden

A.D. Rogers Department of Zoology, University of Oxford, Oxford, United Kingdom

Rutger Rosenberg Kristineberg Marine Research Station, University of Gothenburg, Gothenburg, Sweden

Ussif Rashid Sumaila Fisheries Economics Research Unit, Fisheries Centre, The University of British Columbia, Vancouver, British Columbia, Vancouver, Canada

Richard C. Thompson Marine Biology and Ecology Research Centre, School of Marine Science and Engineering, University of Plymouth Drake Circus, Plymouth, United Kingdom

Carol Turley Plymouth Marine Laboratory, Prospect Place, The Hoe, Plymouth PL1 3 DH, United Kingdom

Dan Wilhelmsson Swedish Secretariat for Environmental Earth System Science, Stockholm, Sweden

Kateryna M. Wowk U.S. National Oceanic and Atmospheric Administration, Washington, District of Columbia, USA

Preface

As many of the most interesting projects do, this one started in an unexpected way. A colleague from the Balaton Group mentioned that there was a German foundation that was looking for (and not finding) a "TEEB report for the oceans." This led to a conversation with Dieter Paulmann, founder of the Okeanos Foundation for the Sea. This conversation in turn led to a proposal to the foundation to do a "state-of-the-science" review of several threats to the global ocean, a review and analysis of how these threats may interact with each other, an analysis of the economic consequences and impacts of these threats, a case study for the Pacific Ocean region, and recommendations for management strategies for marine resources that cross both scales and disciplines. We wanted to do all this in a single report—not a simple task.

This required putting together an international, multidisciplinary team. Here we again ran into unexpected and very positive outcomes. For example, many of us (including the three lead editors) had not worked together before this project—which has turned out to be an enriching experience for all of us. We were delighted that so many leading marine scientists agreed to be part of the writing team, and the process of producing the book has been a truly educational and memorable experience for all involved.

Originally we anticipated a rather modest effort and a short report as the primary output from the project. After the first meeting of the lead authors, however, we decided that the subject deserved a more comprehensive effort; we hope that this book does justice to the dignity of the topic.

We would like to acknowledge some players crucial to making this book a reality, but who do not appear on the list of authors.

The Okeanos Foundation for the Sea funded all the meetings of the writing team, and without their support this book would not have happened. The Okeanos Foundation provided us with completely free hands to craft our analysis, for which all of us are extremely grateful. Beyond financial support, Dieter

Paulmann (the Okeanos Foundation founder and president) provided inspiration, encouragement, and prodding throughout the process. Dieter's passion for the sea—and the Pacific region in particular—was a reminder to all of us of how important it is to produce and convey scientific information in a way that will hopefully be useful to people connected to the sea and who are responsible for making decisions about marine resources. *Mahalo nui loa* Dieter for your commitment, enthusiasm, and patience for our many delays along the way. We hope that the result of our efforts is worth the wait.

Michael Schragger, Executive Director of the Foundation for Design and Sustainable Enterprise (www.fdse.se) was our project manager. Mike fought valiantly (and in vain...) to keep us on schedule and to make sure that we delivered on our promises. Thanks Mike for your brave efforts to keep us in line and productive.

Maria Osbeck of the Stockholm Environment Institute was the main organizer for all of the meetings of the writing team and had primary responsibility for keeping track of our budget. Maria and Mike—even though they could not keep us on schedule (despite Herculean efforts), at least kept us within our budget for the project. Maria's impressive organizational talents are nicely illustrated by the fact that she was able to stop a Stockholm commuter train for long enough to enable members of the writing team to run from a hotel and drag their bags onboard after a meeting. Stockholm commuter trains are often late, but stopping one to wait for passengers is a truly impressive feat. The project was administered through the Stockholm Environment Institute (SEI), and we are very grateful for the administrative and intellectual support we received from them throughout the project.

All three of us are grateful for the sense of community we encountered throughout our work on this project. We hope that the readers of this book will find something in it that speaks to them.

Kevin J. Noone
Ussif Rashid Sumaila
Robert J. Diaz

CHAPTER

1

Valuing the Ocean: An Introduction

Kevin J. Noone, Ussif Rashid Sumaila[§], Robert J. Diaz[†]*

*Department of Applied Environmental Science, Stockholm University, Stockholm, Sweden

[§]Fisheries Economics Research Unit, Fisheries Centre, The University of British Columbia, Vancouver, British Columbia, Vancouver, Canada

[†]Department of Biological Sciences, Virginia Institute of Marine Science, College of William and Mary, Gloucester Point, Virginia, USA

Take a deep breath. Now take one more. The oxygen in one of those breaths came from organisms in the ocean. The oceans are cornerstones of our life support system. They provide many essential ecosystem goods and services essential for humanity, including food, medicinal products, carbon storage, and roughly half the oxygen we breathe. Oceans also support many economic activities including tourism and recreation, commercial and subsistence fisheries, aquaculture, transportation, and mineral resource extraction. They contribute to local livelihoods as well as to national economies and foreign exchange receipts, government tax revenues, and employment. The global ocean is also integral to the earth's climate story, particularly since oceanic heat storage and ocean currents directly influence global climatic conditions.

Despite their central importance to the human endeavor, the oceans are essentially invisible to most of society. Even those of us fortunate enough to live near the sea seldom take the opportunity to look under the surface.

The oceans are under a number of coupled threats unprecedented in modern human history. Sea levels are rising, the oceans are warming and acidifying, oxygen is disappearing in many areas, the seas are becoming more polluted, and we are extracting many marine resources at unsustainable rates. The situation does not have to be this way; but in order to avoid further damage to the oceans, to the life they hold, and to the goods and services they provide, we must develop a holistic view of how our actions impact them. We must create a framework for ocean management in the *Anthropocene*—the current epoch in which we humans are the dominant driver of global environmental change (Crutzen, 2002).

PURPOSE AND SCOPE OF THE BOOK

The aims of this book are to:

(a) summarize the current state of the science for a number of marine-related threats (described in more detail in the following section);
(b) examine these threats to the oceans both individually and collectively;
(c) provide gross estimates of the economic and societal impacts of these threats; and
(d) deliver high-level recommendations for what is still needed to stimulate the development of policies that would help to move toward sustainable use of marine resources and services.

Threats to the Oceans: Current State of the Science

Chapters 2-7 review the state of the science for several threats to the global oceans: acidification, warming, hypoxia, sea level rise, pollution, and overuse of marine resources. Each of the chapters summarizes the latest research in each area and presents it in a way that is accessible to specialists and nonspecialists alike. There is a substantial literature on each of these issues; but thus far, these issues have been researched and reported largely separately. In this book, we want to have concise reviews of the state of the science in these areas all in one place.

A Holistic View of Threats to the Oceans

A result of the fact that research on these issues has been to a large degree done separately, we have little knowledge of the extent to which these threats interact with and feed back on each other. Chapter 8 addresses questions like:

- What are the possible feedback processes between these threats?
- Do any of the threats amplify or dampen others?
- How do local, regional, and global stressors interact?
- What sort of policy and management strategies do we need to account for multiple, interacting stressors?

Chapter 8 will show examples of how interactions between multiple stressors act across different scales and how these interactions require new approaches to marine resource management.

Chapter 9 discusses different ways to plan for the future in the context of marine resources. In many environmental areas, society has a tendency of planning for a "plain vanilla" future—a predictable, gradually changing, middle-of-the road scenario. Here, we contrast three different scenarios approaches: one used in the climate change research community, one in the private sector, and one from the military domain. Each of these approaches has its own strengths and weaknesses, but by contrasting them, we can perhaps learn how to better prepare for a future that may involve hard to predict low-probability, high impact events. From a holistic point of view, we need to develop the capability to anticipate and plan for surprises—or "unknown unknowns."

DIFFERENTIAL ANALYSIS FOR FUTURE SCENARIOS

To be relevant for policy decisions, the economic analysis will be centered on *differences between two future scenarios*. Past losses are no longer changeable with current or future decisions—they will not be included in this analysis. In order to frame our discussions of the future oceans, we will use two of the scenarios currently being employed in the 5th Assessment Report of the Intergovernmental Panel on Climate Change (IPCC).

As discussed in Chapter 9, previous assessment reports of the IPCC have used scenarios—storylines of socioeconomic and demographic development—to create estimates of human resource use and emissions into the future. As an example, the third and fourth assessment reports (AR3 and AR4, respectively) used a set of scenarios described in detail in the Special Report on Emissions Scenarios (SRES) (Nakicenovic et al., 2000); these have become known as the SRES scenarios.

A new approach will be taken for the 5th IPCC Assessment Report (AR5).

This new approach was devised to enable better integration and feedback between the impacts and climate research communities, as well as to "start in the middle" in order to minimize uncertainties (Hibbard et al., 2007; Moss et al., 2010). It also enables the research community to update the scenarios on which climate studies are based after nearly a decade of new information on economic development, technological advances, and chances in climate and environmental factors. Atmospheric carbon dioxide concentrations for the four different "representative concentration pathways" (RCPs) are shown in Figure 1.1.

Each of these four scenarios has been calculated using a different integrated assessment model, and each model has different input assumptions about, e.g., population development, demographics, energy, and resource use. For the purposes of the 5th IPCC Assessment Report, these concentration profiles are starting points with which to explore the range of different socioeconomic development pathways that are consistent with the profiles, as well as to explore the range of impacts, adaptation and mitigation options that are possible. In this sense, while the references in the previous paragraph give details about the assumptions behind the different RCPs, detailed data on the full range of parameters for these profiles are not yet available. For the purposes of this book, if the data needed for our analysis are not available from one of the sources cited above, we will use data from the SRES scenario that most closely resemble the case in question.

One example of the correspondence between atmospheric CO_2 concentrations between the new RCPs and the previous SRES scenarios is shown in Figure 1.2, which compares the concentrations between two RCPs (4.5 and 8.5) and two SRES scenarios (the A1C scenario calculated using the MESSAGE model, and the B1 scenario calculated using the IMAGE model). While not the same,

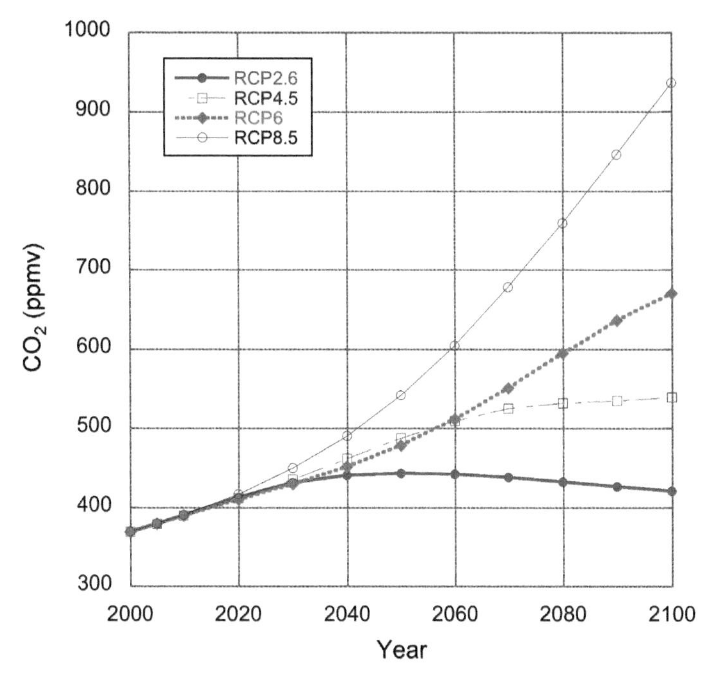

FIGURE 1.1 Atmospheric carbon dioxide concentrations for the four RCPs. IMAGE, Integrated Model to Assess the Global Environment; GCAM, Global Change Assessment Model; AIM, Asia-Pacific Integrated Model; and MESSAGE, Model for Energy Supply Strategy Alternatives and their General Environmental Impact.

the concentration pathways are similar in these two cases, meaning that the socioeconomic, technological, and demographic assumptions behind the SRES scenario calculations are roughly consistent with the RCP concentration profiles.

For the purposes of this book, we want to use two different scenarios to compare and contrast the potential impacts of following two different kinds of decision pathways in the future and to provide the basis for calculating differences in the economic consequences of taking these different decision pathways. In this regard, the scenarios should have a few important characteristics:

- One of the scenarios should reflect a decision pathway designed to reflect human activities having a relatively modest environmental impact;

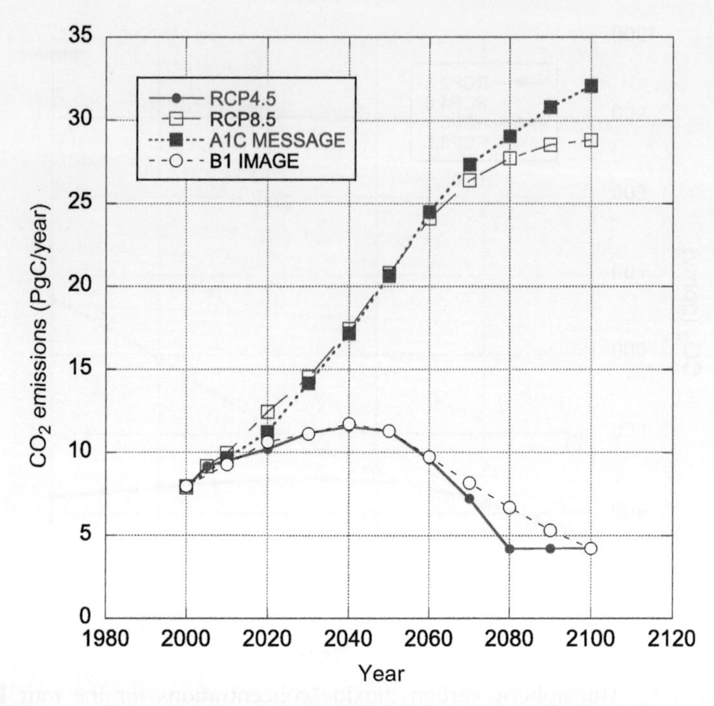

FIGURE 1.2 Comparison of CO_2 emissions for two RCPs and two SRES scenarios.

- The second scenario should be one in which the human impact on the environment is large, but not disastrously so;
- The differences between the scenarios should be sufficiently large that clear differences in impacts can be discerned, thereby providing the possibility of clearly differentiating the benefits and detriments of different policy choices.

In light of these criteria, RCP2.6 and RCP6 are the most appropriate for our analysis (van Vuuren et al., 2011). The RCP2.6 scenario is one that would require substantial economic and societal investments, but one that would likely put us on what would be a more sustainable development pathway than our current one. This scenario provides us with what can be interpreted as a desired outcome—at least in an environmental and societal sense. We chose RCP6 in order to provide a clear contrast between the end results of the different decisions pathways by the end of the century.

Global-Scale Economic Valuation: What Are the Costs of Inaction?

Chapter 10 provides a very basic attempt at a valuation of these various threats taken collectively. We recognize that there is a great deal of debate surrounding the economic and ecological concepts for valuing ecosystem goods and services. We also recognize that the literature on valuing natural resources is very heterogeneous, ranging from examinations of the value at the first point of sale for specific local or regional fisheries to valuation of global ecosystem services. The main goals of this chapter are to (a) collect illustrative examples of valuation for the various threats and present them together; (b) examine the similarities and differences between the various valuations of the different threats; (c) present valuations for the global scale; and (d) attempt a holistic evaluation of the economic impacts of the collective ocean threats, to the extent that one is scientifically sound.

More specifically, among the issues that are addressed in this valuation chapter, and paralleled in the case study, are as follows:

Monetary vs. Nonmonetary Damages

Some impacts—losses of market income, in fishing, tourism, and other ocean-dependent sectors—already have well-defined prices. Some ecosystem services (nutrient cycling, absorption of atmospheric CO_2) could in principle have market prices deduced for them. Other impacts—losses of endangered species, unique habitats/environments (e.g., the Great Barrier Reef)—are of enormous importance, but may not have any meaningful market prices. Chapter 10 attempts to tell both stories, estimating the best possible prices for the one, and at the same time honoring the nonquantified importance of the other.

Thresholds and Discontinuities

In many ecosystem problems, especially when multiple stressors are considered, there are real risks of crossing a threshold at which a species or ecosystem abruptly collapses. This is a challenge for the normal approach to pricing, which implicitly assumes constant or slowly changing marginal costs per unit (and hence constant or slowly changing prices). As a threshold approaches, how should our valuation of the critical resources change to reflect the impending discontinuity? This piece of the puzzle arises in climate change as well, where it is common to assume that there is a real risk of major, discontinuous climate catastrophe, at some unknown (perhaps unknowable) level of CO_2 or temperature rise. Chapter 10 addresses questions such as the "price" of a nonmarginal, discontinuous loss, and how far in advance should that risk be included in the current prices for the risk factors.

Major Categories for Valuation

Chapter 10 begins with two boundary conditions for the analysis: it is restricted to categories of damages that have meaningful prices, and to categories that can be affected by policy decisions today. It then reviews a few important studies that have appeared in the past, involving estimates of the value of ocean ecosystems, before turning to new calculations of these values.

Impacts are estimated for six specific categories of services provided by the oceans. The categories discussed are restricted to those subject to measurable damage with meaningful prices:

- Fishing
- Tourism and recreation
- Moderation of extreme events, including
  - sea-level rise
  - storm damages
- Carbon absorption by the oceans
- The albedo effect of Arctic sea ice

A Case Study for the Pacific: Global to Regional Aspects

Chapter 11 takes us from a global level to a regional one, focusing on the Pacific Ocean. Global issues can often seem abstract, and this chapter shows how the analyses presented in the previous chapters can be made more specific by taking a regional focus. Here, we attempt to show the potential economic and societal impacts to the Pacific Rim countries of failing to take collective action on threats to the oceans.

Economic Value and Earth System Function Value

Some things that cannot be assigned meaningful prices in markets are nonetheless important or even critical to the functioning of the Earth System. These properties of the oceans must not be forgotten in a valuation exercise, even if they cannot be assigned market values. Nutrient cycling, oxygen production, functioning ecosystems, biodiversity, and genetic resources are examples of properties of the oceans that are critical to maintaining our life support system, but to which we cannot put meaningful prices, they therefore bear enormous nonmarket values to people. We wish to highlight these kinds of *Earth System Nonmarket Values* through the use of an expert survey.

EXPERT SURVEY APPROACH

The use of expert surveys has become increasingly applied to risk assessment within the environmental sciences. It is used in the Millennium Ecosystem Assessment (MA: Reid et al., 2005) to assess the impact of various different drivers on biodiversity in a number of different ecosystem types. Figure 1.3 is an example of expert opinion results from the MA.

Expert surveys have also been used to assess the risks associated with climate change (Smith et al., 2001, 2009) and the response of the Atlantic Meridional Overturning Circulation to climate change (Zickfeld et al., 2007).

THE MATRIX: VALUES, THREATS, AND KNOWLEDGE

As mentioned previously, there are two broad categories for valuation: those elements that have identifiable and quantifiable economic values, and those elements that have value because of their importance in how the Earth System functions, but to which economic values cannot be readily assigned. The *values-threats* matrix (an example of which is shown in Figure 1.4) gathers the evidence about the risks of substantial, large-scale changes in each

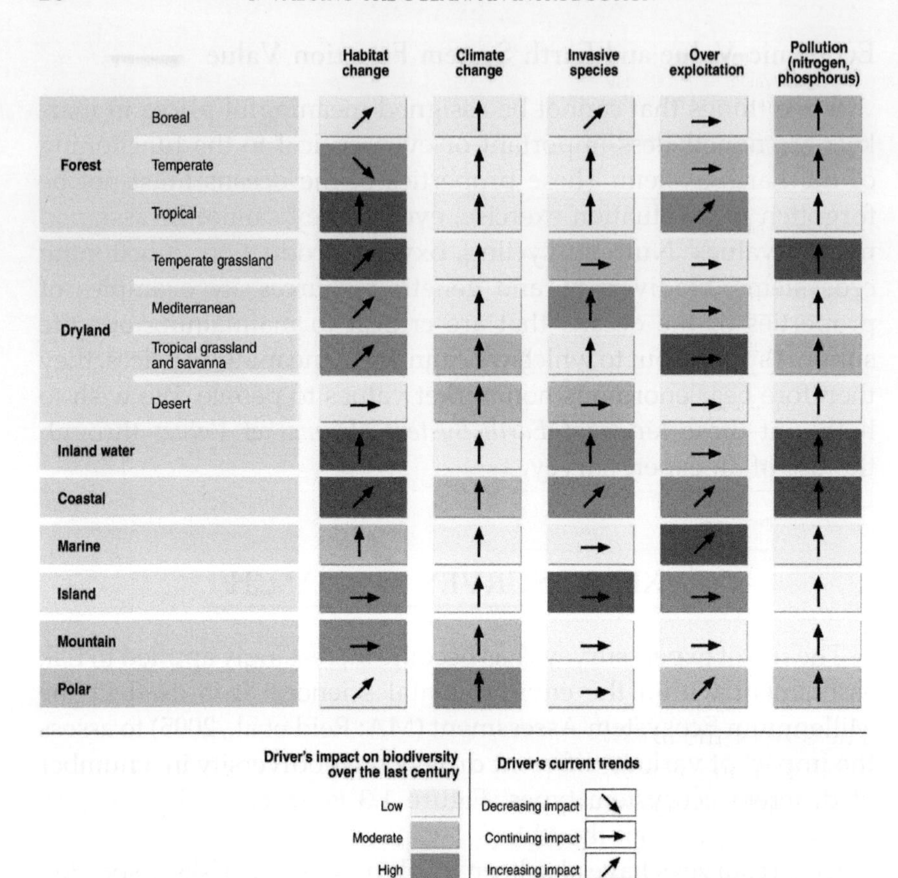

FIGURE 1.3 Example of a presentation of expert opinion results taken from the Millennium Ecosystem Assessment. *From Reid et al. (2005), figure 13. © World Resources Institute.*

threat/value area for the two scenarios (the numerical values in each matrix element) and the amount of evidence and consensus (the color coding for each element).

The matrix shown in Figure 1.4 is the result of a preliminary survey among the lead authors of the chapters in this book conducted during a writing meeting in the spring of 2011. The direction of the arrows indicates the anticipated difference between the two scenarios for each threat-value combination. An arrow pointing straight down indicates that experts thought the threat would become much worse; a 45° angle downward indicates that the situation would become worse; horizontal arrows indicate no significant change.

Threat/value category	Ocean acidification	Anoxia	Sea-level rise	Pollution	Overuse of resources	Warming
	Earth system function value					
Carbon sink	↘	?	→	→	↘	↘
Biodiversity & genetic resources	↓	↘	↘	↘	↓	↘
Oxygen production	↘	↘	→	→	→	↘
Ecosystem function	↓	↓	→	↘	↓	↘
Hydrological cycle	→	→	→	→	→	↘
DMS-cloud feedback; GHGs	↘	→	→	→	→	→
Nutrient cycling	↘	↓	→	→	→	↘

FIGURE 1.4 Example of a *values-threats* matrix.

The size of the arrows indicates the degree of consensus between experts. Long arrows indicate a high degree of consensus, short arrows little consensus.

The main idea for this *values-index* matrix is to provide stakeholders and decision makers with a tool that they can use to:

- quickly obtain an overview of the landscape of threats to the global oceans and their consequences
- easily see the similarities and differences between economic and Earth System values
- quickly identify areas in which coordinated, multinational or multi-regional efforts are necessary
- help frame and prioritize among potential management and legislative actions
- aid in strategic planning for development in individual regions.

The authors were asked for their opinion on the following questions:

1. Pick one of the matrix elements in Figure 1.4 for which you consider yourself an expert—for instance the intersection between anoxia and ecosystem function

2. Contrasting the two scenarios for future development shown in Figure 1.1 (i.e., RCP6-RCP2.6), do you think the impact of anoxia on ecosystem function would be (a) much less, (b) less, (c) not much different, (d) worse, and (e) much worse?
3. How confident are you about your opinion—(a) very certain, (b) reasonably certain, and (c) not very certain?

The categories in items 2 and 3 above are intentionally a bit vague. We are not assigning any given level of numerical certainty or uncertainty to these categories. We simply wanted to obtain expert opinion about the relative change that may occur for each threat-value intersection and roughly how confident the experts are about this relative difference.

The approach to estimating risk, the amount of evidence, and the degree of consensus is similar to the approach used to create the "burning ember" diagrams presented in the Third Assessment Report of the Intergovernmental Panel on Climate Change (IPCC AR3) and in subsequent publications (Smith et al., 2001, 2009). In our approach, however, we choose to provide separate estimates of the degree of risk and the amount of evidence and consensus.

It is important to reiterate that this matrix is intended to facilitate estimating the degree of relative risk in following the two scenarios into the future in a holistic sense, rather than for individual sectors and locations. It should not be interpreted in a predictive sense; for instance, we do not imply that global fisheries will collapse by a specific date, and that this collapse will cost us a specific amount of money. Rather, the matrix conveys experts' best estimates of the risks associated with policy decisions that cause us to follow two example scenarios into the future, to identify the greatest contributors to these risks, and hopefully allows decision makers to better plan to "avoid the unmanageable and manage the unavoidable."

KNOWLEDGE FOR DECISION SUPPORT

A key objective of this book is to expand the research frontier in marine sciences in a couple of ways. We want to improve on methodologies for holistic, cross-scale analysis of the function of the coupled human-environmental system. We also want to improve on our ability to perform global-scale economic valuation of ecosystem goods and services. Decisions in the marine domain will need to be made despite having a number of significant known and unknown unknowns. We hope that this book will help to identify some of these potential surprises and to put bounds on their significance and impact.

This approach is taken in Chapter 12, where conclusions and recommendations are discussed for moving the holistic analysis in the previous chapters into the decision support domain. Effective decision making requires a level of trust between actors—in this case between the research community and stakeholders in the policy and private sectors. Two-way communication channels are needed to establish this kind of trust. The research community needs to better understand the issues and constraints that stakeholders experience, and the stakeholder community needs to understand the ability the research community has to produce "data" on these issues as well as what the research community sees as the issues of tomorrow.

The aim of this book is not necessarily to produce specific policy advice; rather it is to derive a framework that enables more informed decision making for marine issues. We want to enable greater consistency in decision support and decision making across scales—to help see to it that local and regional decisions move us in positive directions on the global scale. We hope our efforts will be of use.

References

Crutzen, P. J. (2002), The geology of mankind, *Nature*, **415**, 23.

Hibbard, K. A., G. A. Meehl, P. M. Cox, and P. Friedlingstein (2007), A Strategy for Climate Change Stabilization Experiments, *Eos*, **88**(20), 217, 219, 221.

Moss, R. H., et al. (2010), The next generation of scenarios for climate change research and assessment, *Nature*, **463**(7282), 747–756, doi:10.1038/nature08823.

Nakicenovic, N., et al. (2000), *Special Report on Emissions Scenarios Rep.*, 612 pp., Intergovernmental Panel on Climate Change, Cambridge, U.K.

Reid, W. V., et al. (2005), Ecosystems and Human Well-being: Synthesis. In: *Millennium Ecosystem Assessment, 2005*, edited, 137 pp., Island Press, Washington, D.C.

Smith, J. B., H. J. Schellnhuber, M. M. Q. Mirza, S. Fankhauser, R. Leemans, L. Erda, L. Ogallo, B. Pittock, R. Richels, C. Rosenzweig, U. Safriel, R. S. J. Tol, J. Weyant, G. Yohe, W. Bond, T. Bruckner, A. Iglesias, A. J. McMichael, C. Parmesan, J. Price, S. Rahmstorf, T. Root, T. Wigley, and K. Zickfeld (2001), Vulnerability to Climate Change and Reasons for Concern: A Synthesis. In: *Climate Change 2001: Impacts, Adaptation and Vulnerability* [McCarthy, J.J., O. F. Canziani, N. A. Leary, D. J. Dokken, K. S. White (eds.)]. Cambridge University Press, Cambridge, U.K., 913–967.

Smith, J. B., S. H. Schneider, M. Oppenheimer, G. W. Yohe, W. Hare, M. D. Mastrandrea, A. Patwardhan, I. Burton, J. Corfee-Morlot, C. H. D. Magadza, H. -M. Füssel, A. B. Pittock, A. Rahman, A. Suarez, and J. -P. van Ypersele (2009), Assessing dangerous climate change through an update of the Intergovernmental Panel on Climate Change (IPCC) "reasons for concern", *Proc. Natl. Acad. Sci.*, **106**, 4133–4137.

van Vuuren, D., J. Edmonds, M. Kainuma, K. Riahi, A. Thomson, K. Hibbard, G. Hurtt, T. Kram, V. Krey, J. -F. Lamarque, T. Masui, M. Meinshausen, N. Nakicenovic, S. Smith, and S. Rose (2011), The representative concentration pathways: an overview, *Climatic Change*, **109**, 5–31.

Zickfeld, K., A. Levermann, M. Granger Morgan, T. Kuhlbrodt, S. Rahmstorf, and D. W. Keith (2007), Expert judgements on the response of the Atlantic meridional overturning circulation to climate change, *Climatic Change*, **82**, 235–265.

CHAPTER

2

Ocean Acidification

Carol Turley

Plymouth Marine Laboratory, Prospect Place, The Hoe, Plymouth PL1 3 DH, United Kingdom

CAUSE AND CHEMISTRY

Ocean acidification is the direct consequence of increased CO_2 emissions to the atmosphere. CO_2 emissions have increased substantially since the industrial revolution due to an increase in fossil fuel burning, cement manufacture, and land-use changes causing year-on-year increased accumulations of CO_2 in the atmosphere. Due to the increasing atmospheric CO_2 concentration over the past 200 years, the ocean takes up vast amounts of anthropogenic CO_2; currently, this is at a rate of around a million metric tons of CO_2 per hour (Brewer, 2009) and is equivalent to 25% of the accumulated CO_2 emissions (Sabine et al., 2004; Le Quéré et al., 2009). Without ocean uptake, atmospheric CO_2 would now be around 450 ppm, some 60 ppm higher than today. While this process partially buffers climate change through the removal of CO_2 from the atmosphere, there are serious consequences for the chemistry of the ocean.

Carbon dioxide (CO_2, <1%), carbonic acid (H_2CO_3, <1%), bicarbonate ions (HCO_3^-, 91%), and carbonate ions (CO_3^{2-}, 8%) comprise the forms of dissolved inorganic carbon in seawater at a mean surface seawater pH of 8.1 and salinity of 35. These occur in dynamic equilibrium, reacting with water and hydrogen ions (Equation 2.1; Zeebe and Wolf-Gladrow, 2001).

$$CO_2 + H_2O \leftrightarrow H_2CO_3 \leftrightarrow H^+ + HCO_3^- \leftrightarrow H^+ + CO_3^{2-} \quad (2.1)$$

On exchange with the atmosphere, dissolved CO_2 reacts with seawater and carbonate ions to form carbonic acid (Equation 2.1). Overall increased emissions to the atmosphere result in an *increase* in dissolved CO_2, carbonic acid, and bicarbonate, as might be expected from Equation (2.1). Hydrogen ion (H^+) concentrations also *increase*, thus pH, the measure of H^+ concentrations falls, and acidity *increases*. It should be noted that ocean pH is unlikely to become acidic (that is become lower than pH 7), and the term acidification reflects the process of becoming more acidic, just like warming reflects the process of increased temperature, although this could be from cold to not quite so cold.

Additionally, and very importantly, there is a *decrease* in the concentration of carbonate ions, as there is also a reaction between CO_2 and carbonate, further *increasing* bicarbonate levels:

$$CO_2 + CO_3^{2-} + H_2O \rightarrow 2HCO_3^- \quad (2.2)$$

Another key outcome of the *decrease* in carbonate (CO_3^{2-}) concentration is that this *increases* the rate of dissolution of calcium carbonate ($CaCO_3$) minerals in the ocean, with the net effect being the following:

$$CaCO_3 + CO_2 + H_2O \rightarrow Ca^{2+} + 2HCO_3^- \quad (2.3)$$

The saturation state (Ω) is used to express the degree of $CaCO_3$ saturation in seawater:

$$\Omega = [Ca^{2+}]\,[CO_3^{2-}]/K_{sp} \quad (2.4)$$

where $[Ca^{2+}]$ and $[CO_3^{2-}]$ are the *in situ* calcium and carbonate concentrations, respectively, and K_{sp} is the solubility product for $CaCO_3$ (concentrations when at equilibrium, neither dissolving nor forming). Values of K_{sp} depend on the crystalline form of $CaCO_3$; they also vary with temperature and pressure, with $CaCO_3$ being unusual in that it is more soluble in cold water than in warm water. Unprotected shells, skeletons, and other calcium carbonate structures start to dissolve when Ω falls below 1 for the appropriate mineral phase. Ω larger than 1 corresponds to supersaturation. The saturation horizon separates supersaturated waters (above) from undersaturated (below) waters and is projected to move upward toward the ocean surface (shoal) as a product of these changes in ocean carbonate chemistry, resulting in more organisms being exposed to the corrosive undersaturated waters (Orr et al., 2005; Turley et al., 2007).

pH describes the concentration or, more precisely, the activity of the hydrogen ion in water, a_H, by a logarithmic function:

$$pH_T = -\log_{10} a_{H^+} \quad (2.5)$$

The activity of hydrogen ions is important for all acid-base reactions. The total pH scale is used (pH_T).

These changes to ocean chemistry are a direct consequence of increased CO_2 emissions to the atmosphere and are certain, based on known chemical reactions. The degree of future ocean acidification will be dependent on CO_2 emission concentrations and rates, and if these are known, the degree of acidification is highly predicable (Caldeira and Wickett, 2003; Orr et al., 2005; Joos et al., 2011).

TIME AND SPACE SCALES

Together, atmosphere, land, and surface ocean reservoirs hold less than ~4000 PgC, while fossil fuel reserves have been estimated at ~5000 PgC (excluding hydrates). Comparing these two figures clearly shows the profound scale of the carbon perturbation the Earth will experience if fossil fuel burning is continued at the same rate over the next few hundred years and how this will overwhelm the capacity of these surface reservoirs to absorb carbon (Zeebe and Ridgwell, 2011). Model projections show that burning of these fossil fuel reserves will result in a decline in surface ocean pH of 0.77 from the preindustrial level of 8.2 (Caldeira and Wickett, 2003), with pH declines of 0.3-0.4 occurring this century (Royal Society, 2005).

It takes around a year for CO_2 emissions to be distributed throughout the atmosphere, and air-sea exchange is nearly instantaneous while equilibrium between atmosphere and the upper mixed layer of the ocean happens in under a year. Exchange between the surface mixed layer of the ocean and the massive deep ocean reservoir takes around 1000 years. On even longer timescales, the majority of the anthropogenic CO_2 will be absorbed by the oceans and neutralized by ocean carbonate sediments over

10,000-100,000 years (Archer, 2005). However, the greater the carbon perturbation, the greater the exhaustion of the oceans buffering capacity, resulting in a greater proportion of CO_2 staying in the atmosphere (Le Quéré et al., 2009). Ocean warming will affect the solubility of CO_2 and increase stratification which may slow ocean mixing, both resulting in a higher proportion of CO_2 remaining in the atmosphere.

Although there are no direct analogues for the current anthropogenic carbon perturbation, the Paleocene-Eocene Thermal Maximum (PETM) which occurred around 55 million years ago may be the most useful, although still limited, analogue as it was a transient event with a relatively rapid onset due to a large and rapid carbon input. Zeebe and Ridgwell (2011) have compared a PETM scenario of an initial carbon pulse of ~3000 PgC over ~6000 years with a Business-as-Usual (BAU) scenario of fossil fuel emissions of ~5000 PgC over ~500 year (Figure 2.1). The projected impact on aragonite saturation in surface water is substantially

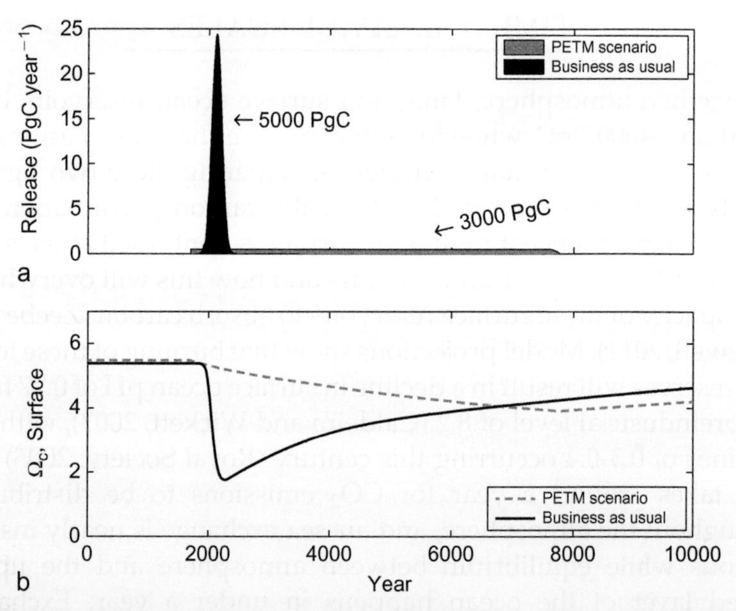

FIGURE 2.1 Model of impact of carbon perturbation scenarios for the Paleocene-Eocene Thermal Maximum (PETM) of ~3000 PgC over ~3000 years, and the current Business-as-Usual (BAU) scenario of fossil fuel emissions of ~5000 PgC over ~500 years on surface water aragonite saturation. *From Zeebe and Ridgwell (2011), figure 2.5.* © *OUP.*

different, with greater impact in the BAU scenario due to the different timescales of the carbon perturbation.

The PETM may have been the cause for an ancient reef crisis (e.g., Kiessling and Simpson, 2011) and extinction of some benthic calcareous foraminifera, with some of those surviving having thinner shell walls (Thomas, 2007). The planktonic ecosystem did not experience a comparable extinction but did however undergo migration and compositional change (see Sluijs et al., 2007 for a review). Their results indicate that the effects on surface-ocean saturation state during the PETM may have been considerably less severe than that projected for the next few centuries. These results exemplify that future rates as well as the magnitude of the carbon input are critical to changes in the chemistry of the future ocean. Such changes have probably not been experienced by the ocean for around 60 million years (Ridgwell and Schmidt, 2010) and possibly 300 million years (Hönisch et al., 2012).

Other perturbations that have disrupted the natural steady-state conditions of the oceanic carbon cycle have occurred in the past during catastrophic impact events or abrupt carbon releases from geological reservoirs. Zeebe and Ridgwell (2011) clearly rebuke those that have made comparisons between the Cretaceous and other long-term (million year), high-CO_2 steady states in the past and the current massive carbon releases that we are experiencing (over a few hundred years) to suggest that marine calcification will not be impaired in a future high-CO_2 world. Such comparisons are invalid because similar CO_2 concentrations do not imply similar carbonate chemistry conditions because of differences in, for example, the degree of $CaCO_3$ saturation or the calcium concentration (Ridgwell and Schmidt, 2010).

The direct relationship between atmospheric CO_2 emissions, ocean pH, and carbonate ion concentration can be seen from long-term time series (Figure 2.2) from the Pacific and Atlantic Oceans. The rates of change are similar although the magnitude of their seasonal variation reflects the seasonal productivity of their locations (Orr, 2011). Mean ocean pH has already decreased by 0.1, equivalent to a 30% decline in acidity (Caldeira and Wickett, 2003), as average atmospheric CO_2 has risen from 280 to 400 ppm since the industrial revolution to the present day. The rate of change in surface ocean pH is −0.019 per decade in the waters off Hawaii

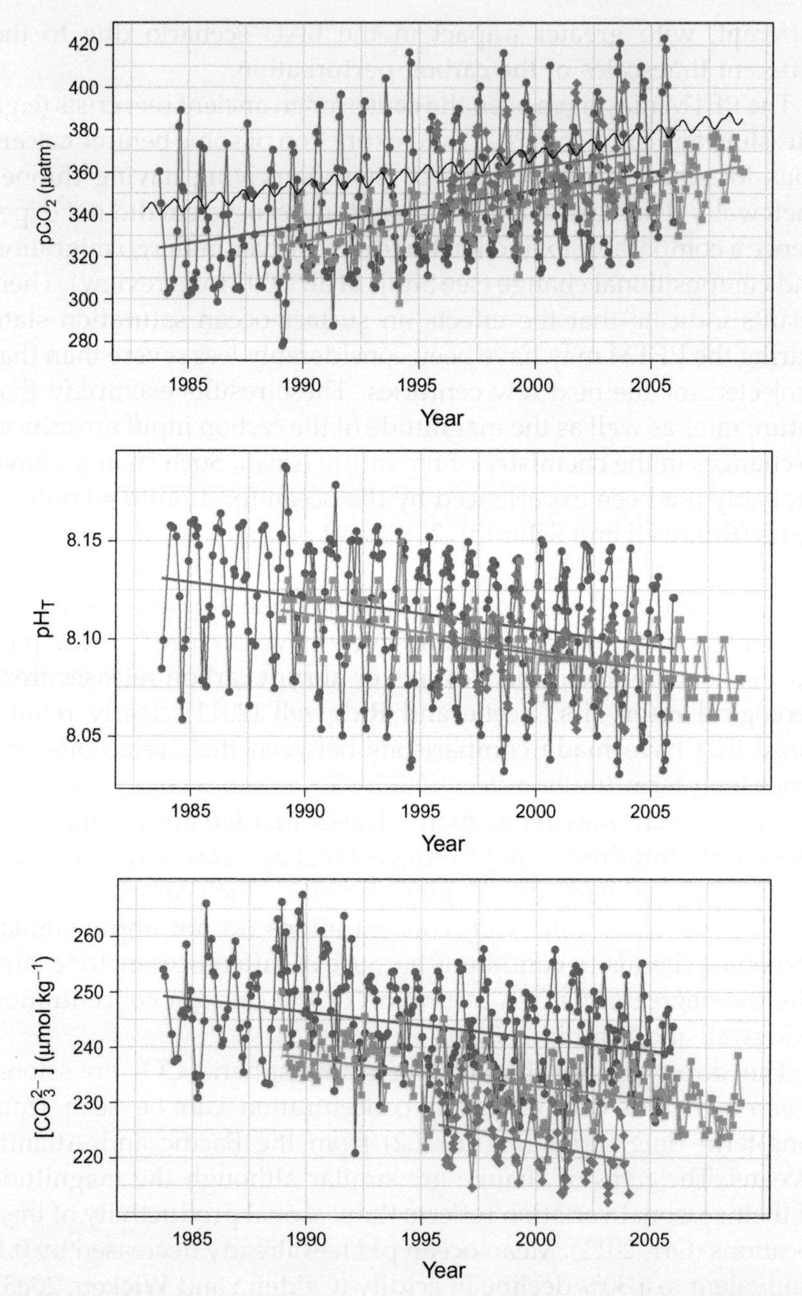

FIGURE 2.2 Time series of surface-ocean pCO_2, pH_T, and CO_3^{2-}, at three ocean time series stations: HOT (green), BATS (red), and ESTOC (blue). *From Orr (2011), figure 3.1. © OUP.*

(Figure 2.2; Doney et al., 2009), −0.012 per decade off Bermuda (Bates, 2007), −0.017 per decade off the Canary Islands (Santana-Casiano et al., 2007), and −0.020 per decade off the south coast of Honshu, Japan (Ishii et al., 2011). In the Arctic Sea, pH decline is even greater at −0.024 per decade and the aragonite saturation horizon, currently at 1750 m, is rising at 4 m per year. This means that each year about 800 km^2 of sea floor, previously bathed in saturated waters, will be exposed to undersaturated conditions and become corrosive to unprotected aragonite shells (Olafsson et al., 2009).

FUTURE OCEAN ACIDIFICATION SCENARIOS

Future global surface ocean acidification will depend on the CO_2 mitigation scenarios that humanity follows. The mean ocean pH stabilization levels of approximately 8.10, 8.01, 7.94, 7.87, 7.82, and 7.70 correspond to atmospheric CO_2 levels of 350, 450, 550, 650, 750, and 1000 ppm (Joos et al., 2011; Figure 2.3). Ocean acidification will be relatively limited in the absence of future anthropogenic emissions of CO_2, demonstrating that strong and early emissions reduction measures will make a difference to ocean chemistry and limit impacts. However, the overshoot stabilization profiles to achieve atmospheric CO_2 stabilization at 350 and 450 ppm require negative CO_2 emissions (CO_2 removal by geoengineering; Joos et al., 2011; Figure 2.3), a further argument for no delay in emission reduction.

While the above projections are of mean global surface ocean pH, it is also important to consider geographic variations across Earth's ocean. By interrogating regions within global models (Figure 2.4), it is possible to obtain projections of future changes in the carbonate chemistry for these regions.

There are areas which already naturally experience lower pH and carbonate ion concentrations than the global average—for example, the high-latitude oceans (Orr et al., 2005; Steinacher et al., 2009), upwelling zones off the west coasts of the continents (Feely et al., 2008), and the deep oceans (Orr et al., 2005; Figure 2.4). Colder waters naturally absorb more CO_2 than warmer waters, so the pH and the saturation of carbonate ions, aragonite and calcite, are lower in subpolar and polar waters than in tropical waters. If emissions of CO_2 continue at the same rate, by the end of this century, the Arctic

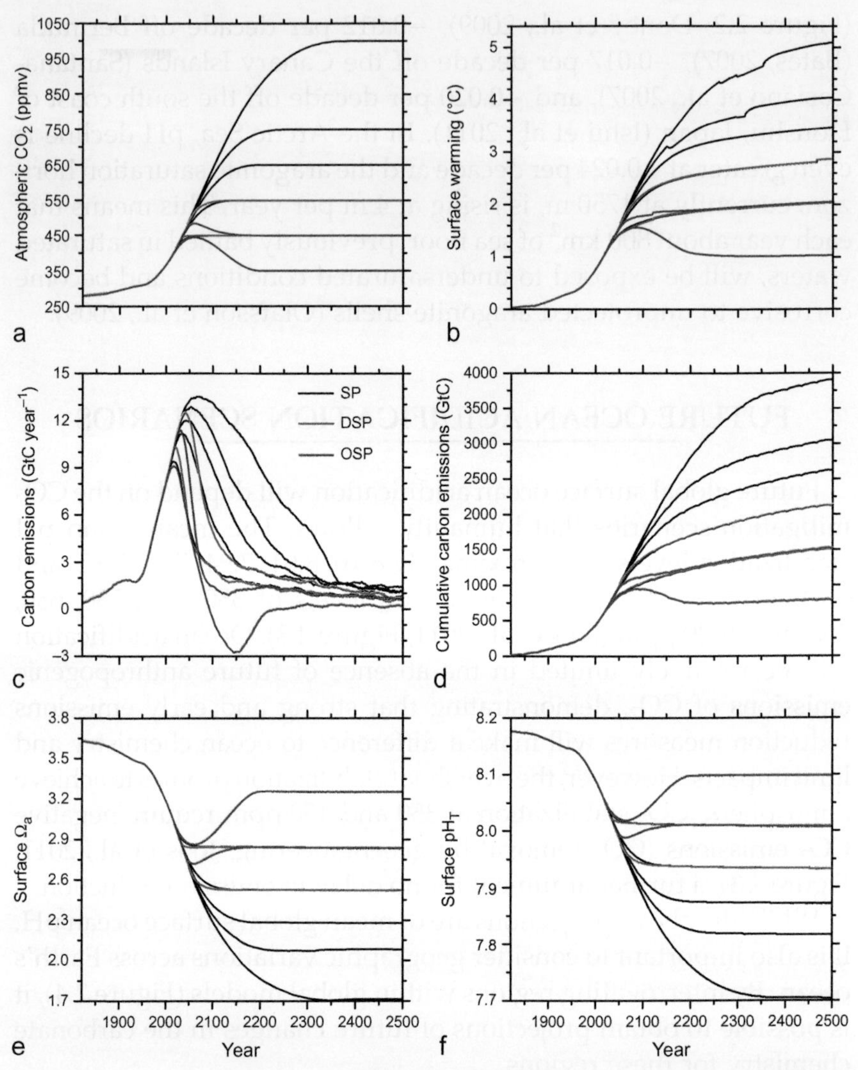

FIGURE 2.3 (a) Prescribed atmospheric CO_2 for pathways leading to stabilization and model projected (b) global-mean surface air temperature change, (c) annual and (d) cumulative carbon emissions, (e) global-mean surface saturation with respect to aragonite, and (f) global mean surface pH_T. The pH stabilization levels of 8.10, 8.01, 7.94, 7.87, 7.82, and 7.70 correspond to atmospheric CO_2 levels of 350, 450, 550, 650, 750, and 1000 ppm. Pathways where atmospheric CO_2 overshoots the stabilization concentration are shown (dotted lines) and pathways with a delayed approach to stabilization (dashed lines); the different pathways to the same stabilization target illustrate how results depend on the specifics of the stabilization pathway. The label SP refers to stabilization profile, DSP to delayed stabilization profile, and OSP to overshoot stabilization profile. See text for further explanation. *From Joos et al. (2011), figure 14.10. © OUP.*

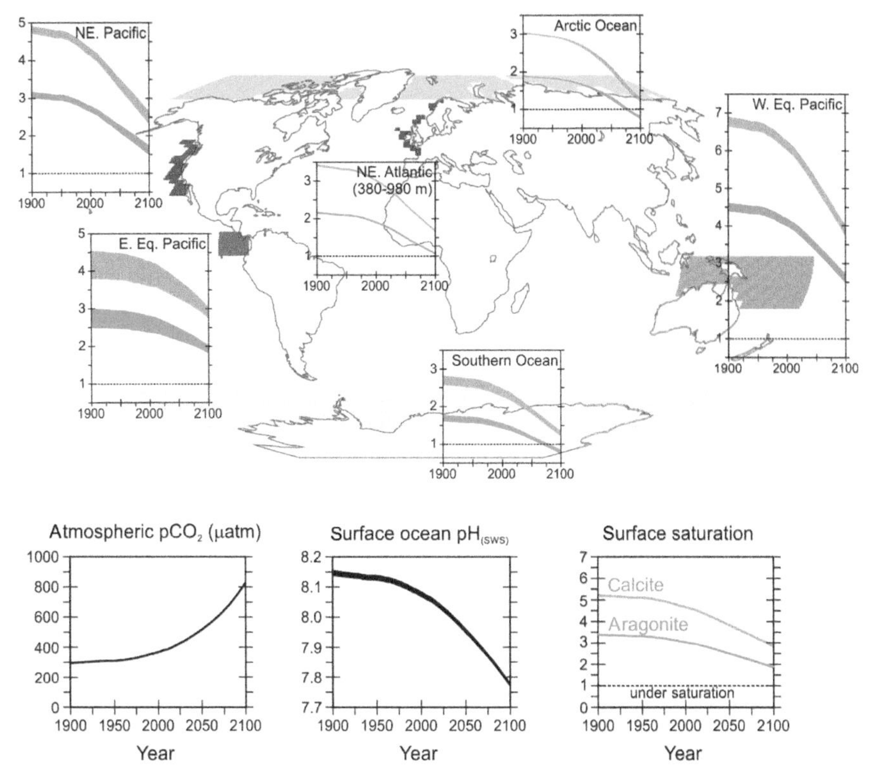

FIGURE 2.4 Projected regional changes in ocean chemistry likely to be experienced by particularly vulnerable ecosystems and compared with global-scale surface ocean changes. The transient simulation of climate and carbonate chemistry was performed with the UVic Earth System Climate Model using observed historical boundary conditions to 2006 and the SRES A2 scenario to 2100. For each of the six illustrative high-risk marine ecosystems (Arctic Ocean, Southern Ocean, NE Pacific margin, intermediate depth NE Atlantic (500-1500 m), western equatorial Pacific, eastern equatorial Pacific), the blue-shaded band indicates the annual range in ocean saturation state with respect to aragonite, while the green-shaded band indicates the range for calcite saturation. Area average surface ocean conditions are calculated for all regions with the exception of the NE Atlantic where area average benthic conditions between 380 and 980 m have been used. The thickness of the line indicates the seasonal range, with the threshold of undersaturated environmental conditions for aragonite and calcite marked as a horizontal dash line. The varying evolution in the magnitude of the seasonal range between different regions is due to the complex interplay between changes in stratification, ocean circulation, and sea-ice extent, and distorted due to the nonlinear nature of the saturation scale. The corresponding regions from which the annual ranges are calculated are shown shaded. Global ocean surface averages (bottom) are shown, from left to right: CO_2 partial pressure, $pH_{(SWS)}$, and calcite and aragonite saturation. *From Turley et al. (2010).*

and Southern Oceans will be undersaturated in these carbonate ions and therefore corrosive to unprotected shells and skeleton (Orr et al., 2005; Steinacher et al., 2009). In the Arctic Ocean, the onset of surface undersaturation shows great variability between different regions as a result of differences in climate change feedback factors such as the retreat of sea ice, changes in freshwater input, and changes in stratification (Popova et al., 2013). Deeper Arctic waters may be receiving additional CO_2 from the microbial breakdown of melting methane hydrates deposited in Arctic Ocean sediments (Biastoch et al., 2011).

Upwelling of waters naturally rich in CO_2 and with lower pH, both at the Equator and at continental margins, is a seasonal process. For example, at the NE Pacific margin, shelf waters are supersaturated with respect to aragonite during weak upwelling (Feely et al., 2008) but as soon as upwelling intensifies, the deeper undersaturated water spills over the edge of the continental margin across the gently sloping, highly productive shelf. This process may be exacerbated by ocean acidification. In addition, because calcite and aragonite saturation decreases with increasing pressure (Orr et al., 2005), organisms living at depth in areas with fast CO_2 uptake such as the North East Atlantic will on average be exposed to undersaturated waters before surface-dwelling organisms (Figure 2.4). While surface waters of tropical areas are not expected to become undersaturated with respect to aragonite (Feely et al., 2004; Orr et al., 2005; Orr, 2011), this may still impact growth rates and survival of coral reefs (Cao and Caldeira, 2008), with the natural erosion of reefs exceeding their growth due to reduced calcification rates (Kleypas et al., 2006).

Ocean acidification is not the only major stressor the ocean and its organisms are experiencing, often simultaneously. Ocean warming (Chapter 3) and deoxygenation (oxygen loss; Chapter 4) are also occurring globally, and more regionally or locally freshening of water due to sea-ice melt and changes to freshwater inflow from rivers, pollution, invasive species, and destructive fishing practices can be additional stressors, which can act synergistically (see Chapter 8).

Coastal shelf seas (Blackford and Gilbert, 2007) and estuaries zones near river mouths (Salisbury et al., 2008) are also vulnerable regions because of inputs of freshwater, which changes the carbonate chemistry. Indeed, the pH of coastal seas is more variable than

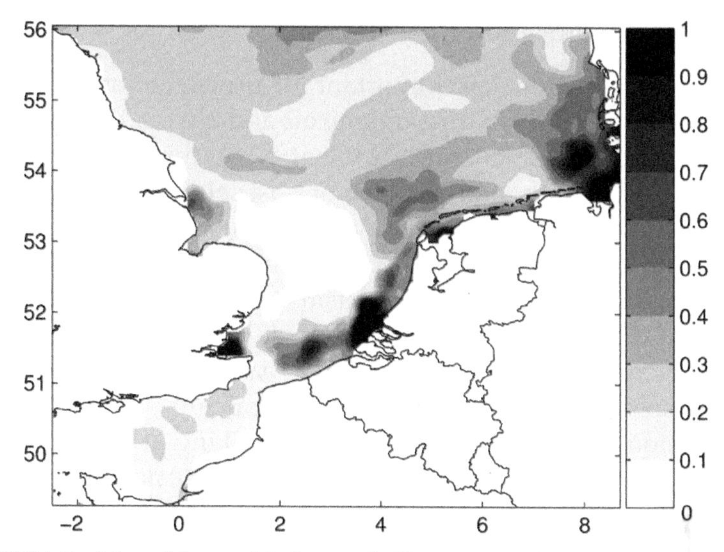

FIGURE 2.5 Map of the modeled annual pH range simulated across the southern North Sea domain. *From Blackford and Gilbert (2007).*

the open ocean (Figure 2.5) and is more difficult to predict accurately because of the interaction with sediment processes and influence from riverine input (Blackford and Gilbert, 2007). Coastal waters, in the vicinity of industry, may receive additional acidity from deposition of anthropogenic atmospheric sources of nitrogen and sulfur (Doney et al., 2007) to add to that produced by uptake of atmospheric CO_2 emissions and riverine input, although some of the acid gases may be buffered by coastal sediments (Hunter et al., 2011).

POTENTIAL FUTURE EFFECTS ON PHYSIOLOGICAL PROCESSES AND BEHAVIOR

As the science of ocean acidification has emerged, there have been several excellent in-depth reviews of the potential impacts of ocean acidification (Royal Society, 2005; Kleypas et al., 2006; Fabry et al., 2008; Turley and Findlay, 2009; Doney et al., 2009), and most recently, a 15 Chapter book is dedicated to Ocean Acidification (Gattuso and Hansson, 2011) with concluding chapter (Gattuso et al., 2011). Rather than repeat these a brief summary of their initial findings follow with an update from recent publications.

Changes in Calcification

Shells and skeletons are important for protection and/or structural support in many marine organisms and are made from different mineral forms of calcium carbonate. The three main mineral forms of calcium carbonate, in order of least soluble to most soluble, are calcite, aragonite, and magnesium-calcite. These minerals dissolve at low carbonate ion concentrations ("undersaturated" conditions), unless calcifying organisms have evolved mechanisms to prevent dissolution. Calcium carbonate becomes harder to precipitate even before undersaturation is reached and can affect the ability of some organisms to produce their calcium carbonate shells, and skeletons or liths. A meta-analysis, looking at all the experimental data available at the time, showed that calcification is the most frequently observed process impacted in experiments carried out at future seawater CO_2 concentrations (Kroeker et al., 2010). Calcification was recognized early on as being potentially vulnerable to ocean acidification (reviewed in Royal Society, 2005; Kleypas et al., 2006). Changes in calcification may impact not only individual visible shelled organisms such as the edible mussel and oyster (Gazeau et al., 2007; Miller et al., 2009), calcifying large algae (Martin et al., 2008; Martin and Gattuso, 2009), ecosystems dependent on calcification such as tropical coral reefs (Kleypas et al., 2006; Fischlin et al., 2007; Silverman et al., 2009), and cold water corals (Guinotte et al., 2006; Turley et al., 2007; Maier et al., 2009) but also microscopic planktonic calcifiers such as coccolithophores (Riebesell et al., 2000), foraminifera (Bijma et al., 2002; de Moel et al., 2009), and pteropods (Comeau et al., 2010). Through the impact on these key planktons, the removal of carbon from surface waters to deeper waters via the biological pump may be impacted (Ridgwell et al., 2009), although the extent and direction of effect is currently uncertain (for review, see Gehlen et al., 2011).

Reduced calcification has already been observed in the marine environment, at present-day CO_2 concentrations in the Great Barrier Reef (De'ath et al., 2009) and in planktonic foraminifera (de Moel et al., 2009; Moy et al., 2009). Currently, it remains uncertain whether this is due to current levels of ocean acidification. However, dissolution of living pteropod shells has very recently been found to be extensive in parts of the Southern Ocean

(Bednaršek et al., 2012), demonstrating that the impact of ocean acidification is already occurring in oceanic populations.

Such changes to calcification could have direct socioeconomic consequences. For example, the larvae of two species of commercially and ecologically valuable shellfish grown under CO_2 levels of 200 years ago displayed thicker, more robust shells, faster growth, and metamorphosis than individuals grown at present-day CO_2 concentrations, whereas when exposed to CO_2 levels expected later this century, their shells were malformed and eroded (Talmage and Gobler, 2009, 2010). The authors pose the question of whether the 30% increase in acidification that has occurred over the past 200 years may be inhibiting the development and survival of larval shellfish and contributing to global declines of some bivalve populations. Upwelling of CO_2-rich water along the western coast of North America (Feely et al., 2008) is currently being explored as the reason for failure of shellfisheries along the coast. Initial studies on potential socioeconomic impacts of ocean acidification on aquaculture, fisheries (Cooley and Doney, 2009; Cheung et al., 2011), and food quality and security (Turley and Boot, 2010, 2011; Cooley et al., 2011) have been undertaken.

Changes to Primary Production and the Microbial Loop

The main marine primary producers are microscopic-free-floating photosynthetic organisms, collectively called phytoplankton. Larger algae and seagrass can be important primary producers locally in shallower waters where they attach to sediments and rocks. All these primary producers use CO_2 and nutrients for photosynthesis. They must concentrate CO_2 to fix it, and many species within this very diverse group of organisms have evolved mechanisms of varying efficiency to do this (Royal Society, 2005). Such mechanisms require energy, so in a future ocean where CO_2 is higher, the energetic cost of concentrating it may be lower resulting in increased overall primary production. However, those species with effective CO_2 concentration mechanisms (CCMs) may lose the competitive advantage that they had over those with less effective mechanisms resulting in shifts in plankton species composition. Increased rates of photosynthesis of 10-30% have been observed in experimental seawater bioassays containing natural assemblages with raised seawater CO_2 (reviewed by Riebesell and Tortell, 2011). In contrast to these enhanced effects, there are potential pH impacts on nutrient cycles either directly on changes to the speciation of key nutrients (Royal Society, 2005) or on nitrogen fixation by the microscopic cyanobacterium *Trichodesmium* which contributes a large fraction of the new nitrogen entering the nutrient replete regions of the ocean (Hutchins et al., 2007). Elevated CO_2 could substantially increase global *Trichodesmium* dinitrogen and CO_2 fixation, fundamentally altering the current marine nitrogen and carbon cycles and potentially moving some oceanic regimes toward phosphorus limitation. In addition, the microbial oxidation of ammonia into nitrite followed by the oxidation of nitrite to nitrate (nitrification) may be particularly sensitive to lower pH (Huesemann et al., 2002). Adaptation to ocean acidification by genetic change is most likely to occur in microbes (Joint et al., 2010) with their rapid generation times of only a few days than for multicellular marine organisms, but there have been only a few studies on the effects of ocean acidification on them and other members and processes in the microbial loop (for review, see Weinbauer et al., 2011).

Changes to Behavior and Sensory Cues

In the marine environment, chemical cues are used for communication, homing ability, habitat selection, sensing predators and prey as well as courtship and mating, species recognition, and symbiotic relationships. Some of these cues are susceptible to changes in pH. In experiments with high CO_2 seawater, littorinid snails switched from thickening shells in the presence of predators to increased avoidance behavior (Bibby et al., 2008), while the resource assessment and decision-making processes are disrupted in hermit crabs (de la Haye et al., 2011). In experiments at low pH, the hermit crabs spent a longer time selecting more optimal shells and took longer transferring between shells. Larval coral reef clownfish and damselfish have shown sensitivity at CO_2 levels likely to be experienced this century through riskier behavior including losing their avoidance behavior toward a predator (Munday et al., 2010). Such changes to behavior increase mortality rates considerably and impact population recruitment. It has been proposed that disrupted internal acid-base balance caused by exposure to elevated CO_2 might potentially affect neuronal pathways that could mediate a range of functions, including olfactory discrimination, activity levels, and risk perception. The ability of juvenile clownfish to discriminate between auditory cues which help them locate reef habitats and suitable settlement sites was disrupted when larvae were reared in conditions simulating CO_2-induced ocean acidification (Munday et al., 2009; Simpson et al., 2011). CO_2 induced loss of response to predation risk ranging from 30% to 95% in four damselfish larvae (Ferrari et al., 2011). The mortality of one species of cardinal fish, another reef fish, at elevated temperatures was greater at high CO_2 (Munday et al., 2009), implying that the combination of ocean acidification and rising seawater temperature may pose a significant physiological challenge to some sea animals.

Otoliths and statoliths are sensory aragonite structures used by fish, squid, and mysids to sense three-dimensional orientation and acceleration. They are therefore very important in predator-prey interactions and consist of aragonite-protein bilayers and could be sensitive to ocean acidification. Experiments on larval white seabass indicated that in a CO_2 environment equivalent to the end of this century, otoliths are 10-14% heavier than those at current-day CO_2 concentrations, and at this stage in research, it is not known if this will impact their behavior (Checkley et al., 2009). In contrast, no changes were found in similar experiments carried out on coral reef fish (Munday et al., 2011), showing the variability between fish. The cuttlebone size and mass increase and internal structures alter in the cuttlefish at CO_2 concentrations well above those likely to be experienced (Gutowska et al., 2008), but effects of more ecologically significant CO_2 concentrations are currently unknown. Jellyfish also produce smaller statoliths when raised in high CO_2 (Winans and Purcell, 2010). The impact on the future behavior of these organisms in response to changes to these important sensory organs is currently unknown.

Changes to Reproduction, Juvenile Survival, and Recruitment

Sea urchin gametes appeared relatively robust to lowered seawater pH (Byrne, 2010; Dupont et al., 2010a; Ericson et al., 2010), although there are some indications that sperm swimming speed and motility may be significantly reduced (Havenhand et al., 2008). Early embryonic development in sea urchins is also relatively robust to reduced seawater pH (Byrne et al., 2009), but abnormalities become more apparent during the later developmental stages (Ericson et al., 2010). Furthermore, lowered pH can reduce echinoid pluteus larval growth and development rates, morphology, skeletogenesis, and calcification rates (Kurihara and Shirayama, 2004; Dupont et al., 2008; Clark et al., 2009; reviewed in Byrne, 2010) in some but not all species (Gooding et al., 2009; Dupont et al., 2010b). For example, larval development of the brittle star *Ophiothrix fragilis* was found to be very sensitive to near future ocean acidification (Dupont et al., 2008), and in contrast, larvae and juveniles of the sea star *Crossaster papposus* grow faster with no visible effects on survival or skeletogenesis when cultured at low pH (Dupont et al., 2010a). These species differences may be due to varying life history strategies, those feeding during their larval planktonic stage being less resilient to pH than those that do not feed during this stage (for review, see Dupont et al., 2010b). Similarly in bivalves, fertilization and early development in two species of oysters exhibited different tolerances to high CO_2 and temperature (Parker et al., 2010).

CO_2 concentrations expected by 2100 under business as usual significantly impaired larval development and survival by 40% in the endangered northern abalone (Crim et al., 2011). Sediment saturation state may account for up to 98% of the mortality during the settlement of juvenile bivalves (Green et al., 2009). Decreasing saturation state from ocean uptake of CO_2 may increase the already extreme mortality found in many calcifying benthic invertebrates. Coral recruitment is also critical to the persistence and resilience of coral reefs and is regulated by larval availability (e.g., gamete production, fertilization), larval settlement, postsettlement growth, and survival. Fertilization success, larval metabolic rates, larval settlement rates, and postsettlement growth rates of several common Caribbean corals are reported to be all compromised with increasing seawater CO_2 (Albright, 2011).

Changes to Hypercapnia, Respiration, Energetics, and Growth

For the normal function of an organism, internal pH must be kept within relatively narrow ranges because processes such as enzyme function, protein phosphorylation, chemical reactions, and the carrying capacity of hemoglobin for O_2 are all influenced by pH. CO_2 in seawater readily diffuses across animal tissue surfaces, lowering the pH of internal fluids, but many animals have developed compensation mechanisms to regulate their internal pH and function normally despite changes in the pH of the surrounding water. However, many marine invertebrates have poorly developed gas exchange and acid-base regulatory capacities, which may explain their apparent low tolerance for ocean acidification. The relative insensitivity of crustaceans to ocean acidification (Kurihara and Ishimatsu, 2008; Arnold et al, 2009; Ries et al., 2009) has been ascribed to well-developed ion transport regulation and high biogenic content of their exoskeletons (Kroeker et al, 2010). Nevertheless, spider crabs show a narrowing of their range of thermal tolerance by ~2 °C under high CO_2 conditions (Walther et al., 2009), and a deep-sea crab was less effective at extracellular acid-base regulation than a shallow water crab to tolerance of high CO_2 (Pane and Barry, 2007). High CO_2 substantially elevated respiration rates (by +100%) and resulted in reduced scope for growth and delayed development in sea urchin larvae (Stumpp et al., 2011). Those larvae raised in high CO_2 allocated only 39-45% of their available energy on somatic growth, while those grown under today's CO_2 concentrations spent 78-80% of the available energy on growth processes. In contrast, mammals, fish, and some molluscs have a high capacity for oxygen and CO_2 transport and exchange and appear to be tolerant of lower pH (Melzner et al., 2009). Cod (Frommel et al., 2012) and herring (Franke and Clemmesen, 2011) larvae have been found to be sensitive to high CO_2 in controlled experiments but how this impacts overall species mortality and recruitment remains uncertain. The great variations in tolerances of animals to ocean acidification may therefore be, at least in part, determined by capacity to regulating their internal acid-base balance (for reviews, see Pörtner et al., 2011 and Widdicombe et al., 2011).

However, in a future ocean with higher CO_2 and higher temperatures, there will be additional associated energetic costs of maintaining internal pH regulation to the organism's energy budget and are thus related to the rate of energy metabolism. This could lead to trade-off between physiological, metabolic, and reproductive processes. For instance, in one study investigating high

CO_2 on a brittle star, the reaction of the animal was to increase its rates of metabolism and calcification but it did this at the cost of muscle wastage (Wood et al., 2008). Such reactions to increased acidity are unlikely to be sustainable in the long term. However, many experiments have been carried out using short-term experiments that do not take account of potential acclimation that could occur over time. Acclimation or adaptation over the long term is potentially possible and explored in Pörtner et al. (2011). Adaptation potential can also be studied at natural CO_2-venting sites, where benthic marine communities have experienced low pH conditions for centuries (see below). More studies are needed to determine the physiological strain that elevated CO_2 concentrations may impose on marine organisms, their performance, and energy balance between metabolic processes and behavior activities (Pörtner, 2008; Findlay et al., 2009; Widdicombe et al., 2009). Such additional physiological strain could affect their ability to cope with other environmental stresses such as rising temperatures and decreasing oxygen.

Dissolved CO_2 increases rapidly with ocean depth, and this will be amplified in a future high CO_2 world. In some areas of the ocean, these areas are increasingly linked to areas of low oxygen, therefore placing added pressure on the respiration and metabolic partitioning of organisms that inhabit these areas (Brewer and Peltzer, 2009; Gruber, 2011; Chapter 4).

IMPACTS ON COMMUNITIES, FOOD WEBS, AND ECOSYSTEMS

Phytoplankton species with effective CCMs are likely to be less sensitive to increased CO_2 levels than those lacking efficient CCMs. These differences may alter competitive relationships among phytoplankton groups and result in shifts of phytoplankton species composition as ocean acidification increases in the future. Changes to the community structure of phytoplankton could impact the structure of the zooplankton community that depend on them for food. Additionally, zooplankton may be directly impacted by future ocean acidification, although this aspect is less well studied. For example, a decrease in the density of shelled pteropods (small planktonic snails known as sea butterflies), which already show sensitivity to ocean acidification in parts of the Southern Ocean through extensive dissolution of their shells (Bednaršek et al., 2012), may impact food webs as they are important food sources for salmon and whales in polar and subpolar waters (Orr et al., 2005; Comeau et al., 2009; Comeau et al., 2010). For in-depth reviews of the potential impact of future ocean acidification on the structure and function of pelagic and benthic communities, food webs, and ecosystems, see Riebesell and Tortell (2011) and Wicks and Roberts (2012), respectively.

Shifts in phytoplankton species composition could also have health risks. For example, a recent study raises the possibility that growth rates and toxicity of the diatom *Pseudo-nitzschia multiseries* could increase substantially in the future high-CO_2 ocean, suggesting a potentially escalating negative effect of this harmful algal bloom species on the future marine environment (Sun et al., 2011). This diatom produces the toxin domoic acid, which when transferred up the food web causes mass mortalities of wildlife, shellfish-harvesting closures, and risks to human health.

Ocean acidification will not have direct negative effects on all marine invertebrates. For example, an increase in populations of top predators which seem to exhibit no sensitivity to ocean acidification such as the sea stars *Pisaster ochraceus* and *Crossaster papposus* may occur in the future, but this will probably have significant consequences on the community structure within the ecosystem (Gooding et al., 2009; Dupont et al., 2010a). In contrast, sensitivity to ocean acidification, as exhibited by another keystone echinoderm species of brittle star which occurs in high densities throughout the shelf seas of northwestern Europe and is an important food for many common predators, including fish, and reduction in populations (Dupont et al., 2008; Wood et al., 2008), could impact food webs. Another example of potential ecosystem scale impact is that ocean acidification has the potential to negatively impact sexual reproduction and multiple early life history larval stages and processes of several common coral species and may contribute to substantial declines in sexual recruitment that are felt at the community and/or ecosystem scale (Albright, 2011; Nakamura et al., 2011).

However, taking laboratory data, often on single species in short-term experiments are open to some criticism (Riebesell et al., 2010) and could over- or underestimate potential impacts. In addition, it is difficult to upscale these experiments to understand what might happen to communities, food webs, and ecosystems. Therefore, scientists have also taken another approach to understanding the potential future impacts of ocean acidification on ecosystems by investigating submersed marine volcanic CO_2vents, where concentrations of CO_2 are naturally high. These areas provide insights into what marine ecosystems may look like in a future high CO_2 ocean. Substantial changes in shallow benthic community composition have been observed in the vicinity of CO_2 vents in the Mediterranean Sea with seagrasses doing well, a 30% decrease in the overall biodiversity, a decrease in abundance, or the loss of benthic calcifiers at lower pH (Hall-Spencer et al., 2008). Similar studies of coral reefs in the vicinity of CO_2 vents off Papua New Guinea found reductions in coral diversity, recruitment, and abundances of these structurally complex framework builders, and shifts in competitive interactions between taxa at pH 7.7 (the pH predicted for the end of this century if CO_2 emissions continue at the same rate) than found in locations away from vents, with a normal pH of 8.1 (Fabricius et al., 2011). Reef development is ceased below pH 7.7. Interestingly, the amount of hard-coral cover remained constant with increasing CO_2 concentrations between pH 8.1 and 7.8 despite low calcification rates. An explanation for this may be that the study areas were surrounded by highly diverse reefs that supply larvae from waters at normal pH and that CO_2 concentrations near volcanic vents are variable, offering periods of opportunity for larval settlement when pH levels are optimal (Hall-Spencer 2011). While these observations provide valuable insights into the responses of organisms and how they may be upscaled to ecosystems, the strong pH variation at these sites (often beyond what may be experienced in the future), the closeness of water with normal pH, and the mobility of some organisms do not make them exact analogues of a future high CO_2 ocean (Hall-Spencer, 2011).

Overall, there is mounting evidence from experiments, meta-analyses, observations of naturally high CO_2 areas and geological records, and modeling that there will be future changes to marine biodiversity and ecosystems if CO_2 emissions continue at the same rate (for reviews, see Andersson et al., 2011; Barry et al., 2011). This conclusion is supported by a survey of 59 experts on the scientific confidence relating to current understanding of ocean acidification and its impacts (Gattuso et al., 2012). The expert survey revealed a high confidence (of 90%) that "anthropogenic ocean acidification will impact ecosystems, some of them negatively (e.g., coral reefs)."

CONCLUSIONS

- Ocean acidification is happening now and will continue as more CO_2 is emitted to the atmosphere. Already, the mean ocean pH has decreased by 30% since the industrial revolution and if we keep on emitting CO_2 at the same rate pH could decrease by a further 150-200% this century.
- The rate of pH change is around 10 times faster than that of other ocean acidification events that have occurred over at least the past 60 million years.
- The only way of reducing future ocean acidification on a global scale is the rapid and substantial reduction of CO_2 emissions to the atmosphere. If CO_2 emissions are not curbed substantially and we keep on emitting CO_2 at the same rate then it will take 10,000's of years for ocean pH to return to close to today's pH.
- There is increasing concern that these rapid changes to ocean chemistry, occurring over just a few centuries will impact marine organisms, food webs, ecosystems, and biogeochemistry and through them the resources, services, and provisions that they inadvertently supply human kind.
- The growing body of scientific research finds that while there will be some animals and plants that are unaffected by, or can adapt to, ocean acidification in the coming decades, there are a growing number of organisms that show sensitivity, either in their physiology, behavior, or in their developmental stages, to ocean acidification.
- Some of these vulnerable organisms are ecosystem builders and shore protectors such as corals, which create habitats supporting high biodiversity, while others such as pteropods are key links in the food web or organisms such as shellfish, which provide a substantial and growing protein and income source for people.
- If CO_2 emissions continue at the same rate, the current evidence points toward substantial changes in this century at the species,

community, and ecosystem level from the cold polar and subpolar waters to the tropics, the deep ocean, and productive coastal and upwelling areas.

Acknowledgments

The author acknowledges support from the UK Ocean Acidification (UKOA) Research Programme funded jointly by the Natural Environment Research Council (NERC), the Department for Environment, Food, and Rural Affairs (Defra) and the Department of Energy and Climate Change (DECC), and both the European Project on Ocean Acidification (EPOCA Grant Number 211384) and the Mediterranean Sea Acidification in a Changing Climate (MedSeA Grant Number 265103) project funded by the European Community's Seventh Framework Programme (FP7/2007-2013). The assistance of colleagues in these programmes is also gratefully acknowledged. Partial support for this synthesis was provided by the Okeanos Foundation.

References

Albright, R. (2011), Effects of Ocean Acidification on Early Life History Stages of Caribbean Scleractinian Corals, *Open Access Dissertations*, **574**.

Andersson, A. J., F. T. Mackenzie, and J. -P. Gattuso (2011): Effects of Ocean Acidification on Benthic Processes, Organisms, and Ecosystems. In: *Ocean acidification* [Gattuso, J. -P. and L. Hansson (eds.)]. Oxford University Press, Oxford, Chapter 7, pp. 122–153.

Archer, D. (2005), The fate of fossil fuel CO_2 in geologic time. *J. Geophys. Res.*, **110**, C09S05, http://dx.doi.org/10.1029/2004JC002625.

Arnold, K. E., H. S. Findlay, J. I. Spicer, C. L. Daniels, and D. Boothroyd (2009), Effects of hypercapnia-related acidification on the larval development of the European lobster, *Homarus gammarus* (L.), *Biogeosciences*, **6**, 1747–1754.

Barry, J. P., S. Widdicombe, and J. M. Hall-Spencer (2011): Effects of Ocean Acidification on Marine Biodiversity and Ecosystem Function. In: *Ocean acidification* [Gattuso, J. -P. and L. Hansson (eds.)]. Oxford University Press, Oxford, Chapter 10, pp. 192–206.

Bates, N. R. (2007), Interannual variability of the oceanic CO_2 sink in the subtropical gyre of the North Atlantic Ocean over the last two decades, *J. Geophys. Res.*, **112**, C09013, http://dx.doi.org/10.1029/2006JC003759.

Bednaršek, N., G. A. Tarling, D. C. E. Bakker, S. Fielding, E. M. Jones, H. J. Venables, P. Ward, A. Kuzirian, B. Lézé, R. A. Feely, and E. J. Murphy (2012), Extensive dissolution of live pteropods in the Southern Ocean, *Nature Geoscience*, 25 November 2012, http://dx.doi.org/10.1038/NGEO1635.

Biastoch, A., T. Treude, L. H. Rüpke, U. Riebesell, C. Roth, E. B. Burwicz, W. Park, C. W. Böning, M. Latif, G. Madec, and K. Wallmann (2011), Evolution of Arctic Ocean temperatures and the fate of marine gas hydrates under global warming, *Geophys. Res. Lett.*, **38**, L08602, http://dx.doi.org/10.1029/2011GL047222.

Bibby, R., S. Widdicombe, H. Parry, J. Spicer, and R. Pipe (2008), Effects of ocean acidification on the immune response of the blue mussel *Mytilus edulis*, *Aquatic Biol.*, **2**, 67–74.

Bijma, J., B. Honisch, and R. E. Zeebe (2002), Impact of the ocean carbonate chemistry on living foraminiferal shell weight: Comment on "Carbonate ion concentration in glacial-age deep waters of the Caribbean Sea" by W.S. Broeker and E. Clark, *Geochem. Geophys. Geosyst.*, **3**, 1064, http://dx.doi.org/10.1029/2002GC0003888.

Brewer, P. G. (2009), A changing ocean seen with clarity, *Proc. Nat. Acad. Sci. USA*, **106**, 12213–12214.

Brewer, P. G., and E. T. Peltzer (2009), Limits to marine life, *Science*, **324**, 347–348.

Byrne, M. (2010), Impact of climate change stressors on marine invertebrate life histories with a focus on the Mollusca and Echinodermata, In: *Climate alert: climate change monitoring and strategy* [You Y. and A. Henderson- Sellers (eds.)]. University of Sydney Press, Sydney, 142–185.

Byrne, M., M. Ho, P. Selvakumaraswamy, H. D. Nguyen, S. A. Dworjanyn, and A. R. Davis (2009), Temperature, but not pH, compromises sea urchin fertilization and early development under near-future climate change scenarios, *Proc. Roy. Soc. B*, **276**, 1883–1888.

Blackford, J. C., and F. J. Gilbert (2007), pH variability and CO_2 induced acidification in the North Sea, *J. Mar. Systems*, **64**, 229–241.

Caldeira, K., and M. E. Wickett (2003), Anthropogenic carbon and ocean pH, *Nature*, **425**, 365.

Cao, L., and K. Caldeira (2008), Atmospheric CO_2 stabilization and ocean acidification, *Geophys. Res. Lett.*, **35**, L19609, http://dx.doi.org/10.1029/2008GL035072.

Checkley, D. M., A. G. Dickson, M. Takahashi, J. A. Radish, N. Eisenkolb, and R. Asch (2009), Elevated CO_2 enhances otolith growth in young fish, *Science*, **324**, 1683. http://dx.doi.org/10.1126/science.1169806.

Cheung, W. W. L., J. Dunne, J. L. Sarmiento, and D. Pauly (2011), Integrating ecophysiology and plankton dynamics into projected maximum fisheries catch potential under climate change in the Northeast Atlantic, *ICES J. Mar. Sci.*, **68**(8), 1008–1018. http://dx.doi.org/10.1093/icesjms/fsr012.

Clark, D., M. Lamare, and M. Barker (2009), Response of sea urchin pluteus larvae (Echinodermata: Echinoidea) to reduced seawater pH: a comparison among a tropical, temperate, and a polar species, *Mar. Biol.*, **156**, 1125–1137.

Comeau, S., G. Gorsky, R. Jeffree, J. L. Teyssié, and J. -P. Gattuso (2009), Impact of ocean acidification on a key Arctic pelagic mollusc (*Limacina helicina*), *Biogeosciences*, **6**, 1877–1882.

Comeau, S., R. Jeffree, J. -L. Teyssié, and J. -P. Gattuso (2010), Response of the Arctic pteropod *Limacina helicina* to projected future environmental conditions, *PLoS ONE*, **5**, e11362.

Cooley, S., and S. C. Doney (2009), Anticipating ocean acidification's economic consequences for commercial fisheries, *Environ. Res. Letters*, **4**, http://dx.doi.org/10.1088/1748-9326/4/2/024007.

Cooley, S. R., N. Lucey, H. Kite-Powell, and S. C. Doney (2011), Nutrition and income from molluscs today imply vulnerability to ocean acidification tomorrow, *Fish Fish.*, http://dx.doi.org/10.1111/j.1467-2979.2011.00424.x.

Crim, R. N., J. M. Sunday, and C. D. G. Harley (2011), Elevated seawater CO_2 concentrations impair larval development and reduce larval survival in endangered northern abalone (Haliotis kamtschatkana), *J. Exp. Mar. Biol. Ecol.*, **400**, 272–277.

de la Haye, K. L., J. I. Spicer, S. Widdicombe, and M. Briffa (2011), Reduced sea water pH disrupts resource assessment and decision making in the hermit crab *Pagurus bernhardus*, *Anim. Behav.*, http://dx.doi.org/10.1016/j.anbehav.2011.05.030.

de Moel, H., G. M. Ganssen, F. J. C. Peeters, S. J. A. Jung, G. J. A. Brummer, D. Kroon, and R. E. Zeebe (2009), Planktic foraminiferal shell thinning in the Arabian Sea due to anthropogenic ocean acidification? *Biogeosciences*, **6**, 1917–1925.

De'ath, G., J. M. Lough, and K. E. Fabricius (2009), Declining coral calcification on the Great Barrier Reef, *Science*, **323**, 116–119.

Doney, S. C., N. Mahowald, I. Lima, et al. (2007), Impact of anthropogenic atmospheric nitrogen and sulfur deposition on ocean acidification and the inorganic carbon system, *PNAS*, **104**, 14580–14585.

Doney, S. C., V. J. Fabry, R. A. Feely, and J. A. Kleypas (2009), Ocean Acidification: The Other CO_2 Problem, *Annu. Rev. Marine. Sci.*, **1**, 169–192.

Dupont, S., J. Havenhand, W. Thorndyke, L. Peck, and M. Thorndyke (2008), CO_2-driven ocean acidification radically affects larval survival and development in the brittlestar *Ophiothrix fragilis*, *Mar. Ecol. Prog. Ser.*, **373**, 285–294.

Dupont, S., B. Lundve, and M. Thorndyke (2010a), Near future ocean acidification increases growth rate of the lecithotrophic larvae and juveniles of the sea star *Crossaster papposus*, *J. Exp. Zool. B*, **314B**, 382–389.

Dupont, S., O. Ortega-Martínez, and M. C. Thorndyke (2010b), Impact of near-future ocean acidification on echinoderms, *Ecotoxicology*, **19**, 449–462.

Ericson, J. A., M. D. Lamare, S. A. Morley, and M. F. Barker (2010), The response of two ecologically important Antarctic invertebrates (*Sterechinus neumayeri* and *Parborlasia corrugatus*) to reduced seawater pH: Effects on fertilisation and embryonic development, *Mar. Biol.*, **157**, 2689–2702, http://dx.doi.org/10.1007/s00227-010-1529-y.

Fabricius, K. E., C. Langdon, S. Uthicke, C. Humphrey, S. Noonan, G. De'ath, R. Okazaki, N. Muehllehner, M. S. Glas, and J. M. Lough (2011), Losers and winners in coral reefs acclimatized to elevated carbon dioxide concentrations, *Nature Clim. Change*, **1**, 165–169.

Fabry, V. J., B. A. Seibel, R. A. Feely, and J. C. Orr (2008), Impacts of ocean acidification on marine fauna and ecosystem processes, *J. Mar. Sci.*, **65**, 414–432.

Feely, R. A., C. L. Sabine, K. Lee, W. Berelson, J. Kleypas, V. J. Fabry, and F. J. Millero (2004), Impact of anthropogenic CO_2 on the $CaCO_2$ system in the ocean, *Science*, **305**, 362–366.

Feely, R. A., C. L. Sabine, J. M. Hernandez-Ayon, D. Lanson, and B. Hales (2008), Evidence for upwelling of corrosive "acidified" water onto the continental shelf, *Science*, **320**, 1490–1492, http://dx.doi.org/10.1126/science.1155676.

Ferrari, M. C. O., D. L. Dixon, P. L. Munday, M. I. McCormick, M. G. Meekan, A. Sih, and D. P. Chivers (2011), Intrageneric variation in antipredator responses of coral reef fishes affected by ocean acidification: implications for climate change projections on marine communities, *Global Change Biology*, http://dx.doi.org/10.1111/j.1365-2486.2011.02439.x.

Findlay, H. S., H. L. Wood, M. A. Kendall, J. I. Spicer, R. J. Twitchett, and S. Widdicombe (2009), Calcification, a physiological process to be considered in the context of the whole organism, *Biogeosciences Discuss.*, **6**, 2267–2284.

Fischlin, A., G. F. Midgley, J. Price, R. Leemans, B. Gopal, C. Turley, M. Rounsevell, P. Dube, J. Tarazona, and A. Velichko (2007): Ecosystems, their properties, goods and services. In: *Climate Change 2007: Climate Change Impacts, Adaptation and vulnerability*, Fourth Assessment Report of the Intergovernmental Panel on Climate Change, Cambridge University Press, Cambridge, 211–272.

Franke, A., and C. Clemmesen (2011), Effect of ocean acidification on early life stages of Atlantic herring (*Clupea harengus* L.), *Biogeosciences Discuss.*, **8**, 7097–7126. http://dx.doi.org/10.5194/bgd-8-7097-2011.

Frommel, A. Y., R. Maneja, D. Lowe, A. M. Malzahn, A. J. Geffen, A. Folkvord, U. Piatkowski, T. B. H. Reusch, and C. Clemmesen (2012), Severe tissue damage in Atlantic cod larvae under increasing ocean acidification, *Nature Climate Change*, **2**, 42–46, http://dx.doi.org/10.1038/NCLIMATE1324.

Gattuso, J. -P., and L. Hansson (2011), *Ocean Acidification,* xix, 326 pp., Oxford University Press, Oxford.

Gattuso, J. -P., J. Bijma, M. Gehlen, U. Riebesell, and C. Turley (2011): Ocean Acidification: Knowns, Unknowns, and Perspectives. In: *Ocean acidification* [Gattuso J. -P. and L. Hansson (eds.)]. Oxford University Press, Oxford, Chapter 15, 291–312.

Gattuso, J. -P., K. J. Mach, and G. Morgan (2012), Ocean acidification and its impacts: an expert survey, *Climatic Change (Online)*, http://dx.doi.org/10.1007/s10584-012-0591-5.

Gazeau, F., C. Quiblier, J. M. Jansen, J. -P. Gattuso, J. J. Middelburg, and C. H. R. Heip (2007), Impact of elevated CO_2 on shellfish calcification, *Geophys. Res. Lett.*, **34**, L07603, http://dx.doi.org/10.1029/2006GL028554.

Gehlen, M. , N. Gruber, R. Gangst, L. Bopp, and A. Oschlies (2011): Biogeochemical Consequences of Ocean Acidification and Feedback to the Earth System. In: *Ocean acidification* [Gattuso. J. -P. and L. Hansson (eds.)]. Oxford University Press, Oxford, Chapter 12, 230–248.

Gooding, R. A., C. D. G. Harley, and E. Tang (2009), Elevated water temperature and carbon dioxide concentration increase the growth of a keystone echinoderm, *PNAS*, **106**, 9316–9321.

Green, M. A., G. G. Waldbusser, S. L. Reilly, K. Emerson, and S. O'Donnell (2009), Death by dissolution: sediment saturation state as a mortality factor for juvenile bivalves, *Limnol. Oceanogr.*, **54**, 1037–1047.

Gruber, N. (2011), Warming up, turning sour, losing breath: ocean biogeochemistry under global change, *Phil. Trans. R. Soc. A*, **369**, 1980–1996.

Guinotte, J. M., J. Orr, S. Cairns, A. Freiwald, L. Morgan, and R. George (2006), Will human induced changes in seawater chemistry alter the distribution of deep-sea scleractinian corals? *Frontiers Ecol. Environ.*, **4**, 141–146.

Gutowska, M. A., H. O. Pörtner, and F. Melzner (2008), Growth and calcification in the cephalopod Sepia officinalis under elevated seawater pCO_2, *Mar. Ecol. Prog. Ser.*, **373**, 303–309, http://dx.doi.org/10.3354/meps07782.

Hall-Spencer, J. M. (2011), No reason for complacency, *Nature Clim. Change*, **1**, 174 http://dx.doi.org/10.1038/nclimate1159.

Hall-Spencer, J. M., R. Rodolfo-Metalpa, S. Martin, E. Ransome, M. Fine, S. M. Turner, S. J. Rowley, D. Tedesco, and M. -C. Buia (2008), Volcanic carbon dioxide vents show ecosystem effects of ocean acidification, *Nature*, **454**, 96–99.

Havenhand, J., F. R. Buttler, M. C. Thorndyke, and J. E. Williamson (2008), Near-future levels of ocean acidification reduce fertilization success in a sea urchin, *Curr. Biol.*, **18**, R651–R652, http://dx.doi.org/10.1016/j.cub.2008.06.015.

Hönisch, B., A. Ridwell, D. N. Schmidt, E. Thomas, S. J. Gibbs, A. Sluijs, R. Zeebe, L. Kump, R. C. Martindale, S. E. Greene, W. Kiessling, J. Ries, J. C. Zachos, D. L. Royer, S. Barker, T. M. Marchitto, R. Moyer, C. Pelejero, P. Ziveri, G. L. Foster, and B. Williams (2012), The geological record of ocean acidification, *Science*, **335**, 1058–1063, http://dx.doi.org/10.1126/science.1208277.

Huesemann, M. H., A. D. Skillman, and E. A. Crecelius (2002), The inhibition of marine nitrification by ocean disposal of carbon dioxide, *Mar. Pollut. Bull.*, **44**, 142–148.

Hunter, K. A., P. S. Liss, V. Surapipith, F. Dentener, R. Duce, M. Kanakidou, N. Kubilay, N. Mahowald, G. Okin, M. Sarin, M. Uematsu, and T. Zhu (2011), Impacts of anthropogenic SOx, NOx and NH_3 on acidification of coastal waters and shipping lanes, *Geophys. Res. Lett.*, **38**, L13602, http://dx.doi.org/10.1029/2011GL047720.

Hutchins, D. A., F. X. Fu, Y. Zhang, M. E. Warner, Y. Feng, K. Portune, P. W. Bernhardt, and M. R. Mulholland (2007), CO_2 control of *Trichodesmium* N_2 fixation, photosynthesis, growth rates, and elemental ratios: implications for past, present, and future ocean biogeochemistry, *Limnol. Oceanogr.*, **52**, 1293–1304.

Ishii, M., N. Kosugi, D. Sasano, S. Saito, T. Midorikawa, and H. Y. Inoue (2011), Ocean acidification off the south coast of Japan: A result from time series observations of CO_2 parameters from 1994 to 2008, *J. Geophys. Res.*, **116**, C06022, http://dx.doi.org/10.1029/2010JC006831.

Joint, I., S. C. Doney, and D. M. Karl (2010), Will ocean acidification affect marine microbes? *The ISME J.*, http://dx.doi.org/10.1038/ismej.2010.79.

Joos, F., T. L. Frölicher, M. Steinacher, and G. -K. Plattner (2011): Impact of climate change mitigation on ocean acidification projections. In: *Ocean acidification* [Gattuso J. -P. and L. Hansson (eds.)]. Oxford University Press, Oxford, Chapter 14, 272–290.

Kiessling, W., and C. Simpson (2011), On the potential for ocean acidification to be a general cause of ancient reef crises, *Global Change Biology*, **17**, 56–67, http://dx.doi.org/10.1111/j.1365-2486.2010.02204.x.

Kleypas, J. A., R. A. Feely, V. J. Fabry, C. Langdon, C. L. Sabine, and L. L. Robbins (2006), *Impacts of Ocean Acidification on Coral Reefs and other Marine Calcifiers: a Guide for Future Research*, 88 pp., NSF, NOAA, and the US Geological Survey, St Petersburg, Florida.

Kroeker, K. J., R. L. Kordas, R. N. Crim, and G. G. Singh (2010), Meta-analysis reveals negative yet variable effects of ocean acidification on marine organisms, *Ecol. Lett.*, **86**, 157–164.

Kurihara, H., and A. Ishimatsu (2008), Effects of high CO_2 seawater on the copepod (*Acartia tsuensis*) through all life stages and subsequent generations, *Mar. Pollut. Bull.*, **56**, 1086–1090.

Kurihara, H., and Y. Shirayama (2004), Effects of increased atmospheric CO_2 on sea urchin early development, *Mar. Ecol. Prog. Ser.*, **274**, 161–169.

Le Quéré, C., M. R. Raupach, J. G. Canadell, G. Gregg Marland, et al. (2009), Trends in the sources and sinks of carbon dioxide, *Nature Geosci*, **2**, 831–836, http://dx.doi.org/10.1038/ngeo689.

Maier, C., J. Hegeman, M. G. Weinbauer, and J. -P. Gattuso (2009), Calcification of the cold- water coral *Lophelia pertusa* under ambient and reduced pH, *Biogeosc.*, **6**, 1671–1680.

Martin, S., and J. -P. Gattuso (2009), Response of Mediterranean coralline algae to ocean acidification and elevated temperature, *Global Change Biology*, **15**, 2089–2100, http://dx.doi.org/10.1111/j.1365-2486.2009.01874.x.

Martin, S., R. Rodolfo-Metalpa, E. Ransome, S. Rowley, M. -C. Buia, J. -P. Gattuso, and J. Hall-Spencer (2008), Effects of naturally acidified seawater on seagrass calcareous epibionts, *Biol. Lett.*, **4**, 689–692, http://dx.doi.org/10.1098/rsbl.2008.0412.

Melzner, F., M. A. Gutowska, M. Langenbuch, S. Dupont, M. Lucassen, M. C. Thorndyke, M. Bleich, and H. -O. Pörtner (2009), Physiological basis for high CO_2 tolerance in marine ectothermic animals: pre-adaptation through lifestyle and ontogeny? *Biogeosciences*, **6**, 2313–2331.

Miller, A. W., A. C. Reynolds, C. Sobrino, and G. F. Riedel (2009), Shellfish face uncertain future in high CO_2 world: influence of acidification on oyster larvae calcification and growth in estuaries, *PLoS One*, **4**, e5661, http://dx.doi.org/10.1371/journal.pone.0005661.

Moy, A. D., W. R. Howard, S. G. Bray, and T. W. Trull (2009), Reduced calcification in modern Southern Ocean planktonic foraminifera, *Nature Geoscience*, **2**, 276–280.

Munday, P. L., D. L. Dixson, J. M. Donelson, G. P. Jones, M. S. Pratchett, G. V. Devitsina, and K. B. Dving (2009), Ocean acidification impairs olfactory discrimination and homing ability of a marine fish, *PNAS*, **106**, 1848–1852, http://dx.doi.org/10.1073/pnas.0809996106.

Munday, P. L., D. L. Dixson, M. I. McCormick, M. Meekan, M. C. O. Ferrari, and D. P. Chivers (2010), Replenishment of fish populations is threatened by ocean acidification, *PNAS*, 12930–12934. http://dx.doi.org/10.1073/pnas.1004519107.

Munday, P. L., V. Hernaman, D. L. Dixson, and S. R. Thorrold (2011), Effect of ocean acidification on otolith development in larvae of a tropical marine fish, *Biogeosciences*, **8**, 1631–1641.

Nakamura, M., S. Ohki, A. Suzuki, and K. Sakai (2011), Coral Larvae under Ocean Acidification: Survival, Metabolism, and Metamorphosis, *PLoS One*, **6**, e14521, http://dx.doi.org/10.1371/journal.pone.0014521.

Olafsson, J., J. Olafsdottir, A. Benoit-Cattin, M. Danielsen, T. S. Arnarson, and T. Takahashi (2009), Rate of Iceland Sea acidification from time series measurements, *Biogeosciences*, **6**, 2661–2668.

Orr, J. C., V. J. Fabry, O. Aumont, and 24 others (2005), Anthropogenic ocean acidification over the twenty-first century and its impact on calcifying organisms, *Nature*, **437**, 681–686.

Orr, J. C. (2011): Recent and Future Changes in Ocean Carbonate Chemistry. In: *Ocean acidification* [Gattuso J. -P. and L. Hansson (eds.)]. Oxford University Press, Oxford, Chapter 3, pp. 41–66.

Pane, E. F., and J. P. Barry (2007), Extracellular acid–base regulation during short-term hypercapnia is effective in a shallow water crab, but ineffective in a deep-sea crab, *Mar. Ecol. Prog. Ser.*, **334**, 1–9.

Parker, L. M., P. M. Ross, and W. O. O'Connor (2010), Comparing the effect of elevated pCO_2 and temperature on the fertilization and early development of two species of oysters, *Mar. Biol.*, **157**, 2435–2452, http://dx.doi.org/10.1007/s00227-010-1508-3.

Popova, E. E., A. Yool, A. C. Coward, and T. R. Anderson (2013), Regional variability of acidification in the Arctic: a sea of contrasts, *Biogeosciences Discussions*, **10**, 2937–2965.

Pörtner, H. O. (2008), Ecosystem effects of ocean acidification in times of ocean warming: a physiologist's view, *Mar. Ecol. Prog. Ser.*, **373**, 203–217.

Pörtner, H. -O., M. Gutowska, A. Ishimatsu, M. Lucassen, F. Melzner, and B. Seibel (2011), Effects of Ocean Acidification on Nektonic Organisms. In: *Ocean acidification* [Gattuso, J. -P. and L. Hansson (eds.)]. Oxford University Press, Oxford, Chapter 8, 154–169.

Ridgwell, A., and D. N. Schmidt (2010), Past constraints on the vulnerability of marine calcifiers to massive carbon dioxide release, *Nature Geoscience*, **3**, 196–200.

Ridgwell, A., D. N. Schmidt, C. Turley, C. Brownlee, M. T. Maldonado, P. Tortell, and J. R. Young (2009), From laboratory manipulations to Earth system models: scaling calcification impacts of ocean acidification, *Biogeosciences*, **6**, 2611–2623.

Riebesell, U., I. Zondervan, B. Rost, P. D. Tortell, R. E. Zeebe, and F. M. M. Morel (2000), Reduced calcification of marine plankton in response to increased atmospheric CO_2, *Nature*, **407**, 364–367.

Riebesell, U., V. J. Fabry, L. Hansson, and J. -P. Gattuso (2010), *Guide to Best Practices for Ocean Acidification Research and Data Reporting*, 260 pp., EUR 24328, EUR 24328, European Union, Luxembourg.

Riebesell, U., and P. D. Tortell (2011): Effects of Ocean Acidification on Pelagic Organisms and Ecosystems. In: *Ocean acidification* [Gattuso, J. -P. and L. Hansson (eds.)]. Oxford University Press, Oxford, Chapter 6, 99–121.

Ries, J. B., A. L. Cohen, and D. C. McCorkle (2009), Marine calcifiers exhibit mixed responses to CO_2-induced ocean acidification, *Geology*, **37**, 1131–1134, http://dx.doi.org/10.1130/G30210A.1.

Royal Society (2005), *Ocean Acidification due to Increasing Atmospheric Carbon Dioxide*, 68 pp., Policy document 12/05. The Royal Society, London.

Sabine, C. L., R. A. Feely, N. Gruber, R. M. Key, K. Lee, J. L. Bullister, R. Wanninkhof, C. S. Wong, D. W. R. Wallace, B. Tilbrook, F. J. Millero, T. H. Peng, A. Kozyr, T. Ono, and A. F. Rios (2004), The oceanic sink for anthropogenic CO_2, *Science*, **305**, 367–371.

Salisbury, J., M. Green, C. Hunt, and J. Campbell (2008), Coastal acidification by Rivers: a threat to shellfish? *Eos, Transactions American Geophysical Union*, **89**, 513.

Santana-Casiano, J. M., M. González-Dávila, M. -J. Rueda, O. Llinás, and E. -F. González-Dávila (2007), The interannual variability of oceanic CO_2 parameters in the northeast Atlantic subtropical gyre at the ESTOC site, *Global Biogeochem. Cycles*, **21**, GB1015, http://dx.doi.org/10.1029/2006GB002788.

Silverman, J., B. Lazar, L. Cao, K. Caldeira, and J. Erez (2009), Coral reefs may start dissolving when atmospheric CO_2 doubles, *Geophys. Res. Lett.*, **36**, L05606, http://dx.doi.org/10.1029/2008GL036282.

Sluijs, A., G. J. Bowen, H. Brinkhuis, L. J. Lourens, and E. Thomas (2007): The Palaeocene-Eocene Thermal Maximum super greenhouse: biotic and geochemical signatures, age models and mechanisms of global change. In: *Deep time perspectives on climate change—marrying the signal from computer models and biological proxies* [Williams, M., M., A. M. Haywood, F. J. Gregory and D. N. Schmidt (eds.)]. TMS Special Publication 2, The Geological Society of London, London.

Steinacher, M., F. Joos, T. L. Frölicher, G. -K. Plattner, and S. C. Doney (2009), Imminent ocean acidification in the Arctic projected with the NCAR global coupled carbon cycle-climate model, *Biogeosciences*, **6**, 515–533.

Simpson, S. D., P. L. Munday, M. L. Wittenrich, R. Manassa, D. L. Dixson, M. Gagliano, and H. Y. Yan (2011), Ocean acidification erodes crucial auditory behaviour in a marine fish. *Biol. Lett.*, http://dx.doi.org/10.1098/rsbl.2011.0293R.

Stumpp, M., J. Wren, F. Melzner, M. C. Thorndyke, and S. Dupont (2011), CO_2 induced seawater acidification impacts sea urchin larval development I: elevated metabolic rates decrease scope for growth and induce developmental delay, *Comp. Biochem. Physiol. A*, http://dx.doi.org/10.1016/j.cbpa.2011.06.022.

Sun, J., D. A. Hutchins, Y. Feng, E. L. Seubert, D. A. Caron, and F. -X. Fu (2011), Effects of changing pCO_2 and phosphate availability on domoic acid production and physiology of the marine harmful bloom diatom *Pseudo-nitzschia multiseries*, *Limnol. Oceanogr.*, **56**, 829–840.

Talmage, S. C., and C. J. Gobler (2009), The effects of elevated carbon dioxide concentrations on the metamorphosis, size, and survival of larval hard clams (*Mercenaria mercenaria*), bay scallops (*Argopecten irradians*), and Eastern oysters (*Crassostrea virginica*), *Limnol. Oceanogr.*, **54**, 2072–2080.

Talmage, S. C., and C. J. Gobler (2010), Effects of past, present, and future ocean carbon dioxide concentrations on the growth and survival of larval shellfish. *PNAS*, **107**, 17246–17251.

Thomas, E. (2007): Cenozoic mass extinctions in the deep sea: what perturbs the largest habitat on earth? In: *Large Ecosystem Perturbations: Causes and Consequences* [S. Monechi, et al. (eds.)]. Geological Society of America Special Paper, pp. 1–23.

Turley, C., and K. Boot (2010), Environmental consequence of ocean acidification: a threat to food security, *UNEP Emerging Issues Bulletin*, **9**.

Turley, C., and K. Boot (2011), The ocean acidification challenges facing science and society. In: *Ocean acidification* [Gattuso, J. -P. and L. Hansson (eds.)] Oxford University Press, Oxford, Chapter 13, pp. 249–271.

Turley, C., M. Eby, A. J. Ridgwell, D. N. Schmidt, H. S. Findlay, C. Brownlee, U. Riebesell, J. -P. Gattuso, V. J. Fabry, and R. A. Feely (2010), The societal challenge of ocean acidification, *Mar. Poll. Bull.*, **60**, 787–792.

Turley, C. M., and H. S. Findlay (2009): Ocean acidification as an indicator for climate change. In: *Climate and Global Change: Observed Impacts on Planet Earth* [Letcher, T. M. (ed.)]. Elsevier, Amsterdam, pp. 367–390.

Turley, C. M., J. M. Roberts, and J. M. Guinotte (2007), Corals in deep-water: will the unseen hand of ocean acidification destroy cold-water ecosystems? *Coral Reefs*, **26**, 445–448.

Walther, K., F. J. Sartoris, C. Bock, and H. -O. Pörtner (2009), Impact of anthropogenic ocean acidification on thermal tolerance of the spider crab *Hyas araneus*. *Biogeosciences*, **6**, 2207–2215.

Weinbauer, M. G., X. Mari, and J. -P. Gattuso (2011): Effect of ocean acidification on the diversity and activity of heterotrophicmarine microorganisms. In: *Ocean acidification* [Gattuso J. -P. and L. Hansson (eds.)]. Oxford University Press, Oxford, Chapter 5, 83–98.

Wicks, L. C., and J. M. Roberts (2012), Benthic invertebrates in a high CO_2 world. *Oceanogr. Mar Biol. Ann. Rev.*, **50**, 127–188.

Widdicombe, S., S. L. Dashfield, C. L. McNeill, et al. (2009), Effects of CO_2 induced seawater acidification on infaunal diversity and sediment nutrient fluxes, *Mar. Ecol. Prog. Ser.*, **379**, 59–75.

Widdicombe, S., J. I. Spicer, and V. Kitidis (2011): Effects of Ocean Acidification on Sediment Fauna. In: *Ocean acidification* [Gattuso J. -P. and L. Hansson (eds.)]. Oxford University Press, Oxford, Chapter 10, 176–191.

Winans, A. K., and J. E. Purcell (2010), Effects of pH on asexual reproduction and statolith formation of the scyphozoan, *Aurelia labiata*, *Hydrobiol.*, **645**, 39–52.

Wood, H. L., J. I. Spicer, and S. Widdicombe (2008), Ocean acidification may increase calcification rates, but at a cost, *Proc. R. Soc. B: Biol. Sci.*, **275**, 1767–1773.

Zeebe, R. E., and D. A. Wolf-Gladrow (2001), *CO_2 in Seawater: Equilibrium, Kinetics and Isotopes*, 346 pp., Elsevier Oceanography Series, 65.

Zeebe, R. E., and A. Ridgwell (2011): Past Changes of Ocean Carbonate Chemistry. In: *Ocean acidification* [Gattuso, J. -P. and L. Hansson (eds.)]. Oxford University Press, Oxford, Chapter 2, 21–40.

CHAPTER

3

Ocean Warming

Kevin J. Noone, Robert J. Diaz*[†]

*Department of Applied Environmental Science, Stockholm University, Stockholm, Sweden

[†]Department of Biological Sciences, Virginia Institute of Marine Science, College of William and Mary, Gloucester Point, Virginia, USA

INTRODUCTION

Our oceans have been warming as a result of our changing climate. Over the past 50 years, most of the added heat from anthropogenic climate change has been absorbed by the oceans. Estimates are that from the surface down to 700 m ocean temperature has increased by 0.2 °C (Bindoff et al., 2007). In addition, there are now indications that deep oceanic waters are warming (Roemmich et al., 2012).

Ocean warming is an important aspect of sea-level rise, observations of which are discussed in detail in Chapter 5. In this chapter, we treat the physical and biological aspects of ocean warming that are not related to sea-level rise, but which will have important impacts on society; we will concentrate on the consequences of a warming ocean.

Some of these impacts will be felt in the oceans themselves, such as damage to or even the complete destruction of coral reefs, or alterations to the patterns of biological production and fisheries landings (Sumaila et al., 2011; Chapters 8, 10, and 11). Others will also be felt far away from the ocean domain, such as changes in severe weather or hurricane intensity. Changes of this kind are driven by ocean warming, either almost entirely (as in the case of hurricane strength) or partially (such as increases in the intensity of precipitation over land).

This chapter discusses the major causes and consequences of ocean warming. We begin by looking at the physical consequences of ocean warming, and how a warmer ocean influences "extreme" weather such as intense precipitation and flooding, and tropical cyclones. We then examine some of the basic biological consequences of ocean warming, including how ocean warming will influence the availability of dissolved oxygen, and the metabolism and distribution of marine organisms. We conclude with a summary and take-home messages about the consequences of a warming ocean.

PHYSICAL CONSEQUENCES OF OCEAN WARMING

"Extreme Weather": Precipitation and Flooding

In a very simplified sense, as the oceans and atmosphere warm, the amount of water vapor in the atmosphere at a given relative humidity increases. The increase is not linear, but rather exponential. As temperature increases, the amount of water at a given relative humidity increases very rapidly. Observations of significant increases in specific humidity (the actual amount of water vapor in the atmosphere) in the Northern Hemisphere confirm this simplified picture in an overall sense (Allen and Ingram, 2002).

These changes in the amount of water vapor are expected to lead to concomitant changes in the hydrological cycle. One such expected change is an increase in the intensity of precipitation. Climate extremes such as extreme precipitation or droughts are not driven exclusively or sometimes even dominantly by ocean temperature increases. A detailed discussion of these kinds of climate extremes is beyond the scope of this chapter, and even of this book. Here, we present a short discussion of precipitation extremes as an example of the kinds of changes in global climate that are due at least in part to ocean warming, and that will be felt in areas well removed from the ocean basins.

Min et al. (2011) examined both observations and model simulations of trends in extreme precipitation in the Northern Hemisphere. They looked at changes based on the annual maxima of daily (RX1D) and 5-day consecutive (RX5D) precipitation amounts for the latter half of the twentieth century. These two indices were chosen because they relate closely to the kinds of extreme events that impact society through, e.g., flood damage, crop losses, and damage to built infrastructure. They used the Hadley Centre global climate extremes data set (Alexander et al., 2006), which is based on daily observations from 6000 stations. They examined the period between 1951 and 1999. Model results were taken from the Coupled Model Intercomparison Project Phase 3 (CMIP3) archive (http://esg.llnl.gov:8080/index.jsp). Their results are shown in Figure 3.1.

Panels (a) and (b) show observations of changes in 1-day (RX1D) and 5-day (RX5D) extreme precipitation values for the period 1951-1999. Blue colors indicate areas where extreme precipitation increased; orange/red colors indicate areas where it decreased. Observations show an increase in extreme precipitation over much of the United States, Mexico, and Central America, as well as over central and eastern Europe. Observations show both areas of increased and decreased extreme precipitation over much of east and south Asia, with decreased extreme precipitation (associated with warming and drying) over the Iberian Peninsula in Europe. Overall, the observations show a 65% increase in 1-day extreme precipitation over the period and a 61% increase in 5-day extreme precipitation. The CMIP3 models tend to underestimate the changes in extreme precipitation. However, the spatial patterns of the changes were sufficiently robust for "fingerprinting" techniques to be used to detect a clear signal of anthropogenic influence on the extreme values.

This clear relationship may not exist for the entire planet (Peixoto and Oort, 1996). Changes in atmospheric circulation can also affect extreme precipitation. The tropics in particular is an area where detecting changes in extreme precipitation and attributing the changes to a particular cause has proven difficult (O'Gorman and Schneider, 2009). Current climate models have difficulty in resolving the kinds of tropical convective systems that are responsible for heavy precipitation in the region, and the precipitation itself is much more intense and variable compared with midlatitude regions.

Extreme precipitation impacts a number of areas of importance to society. Table 3.1 lists examples of impacts and potential mitigation options. The economic impacts of extreme weather are addressed in

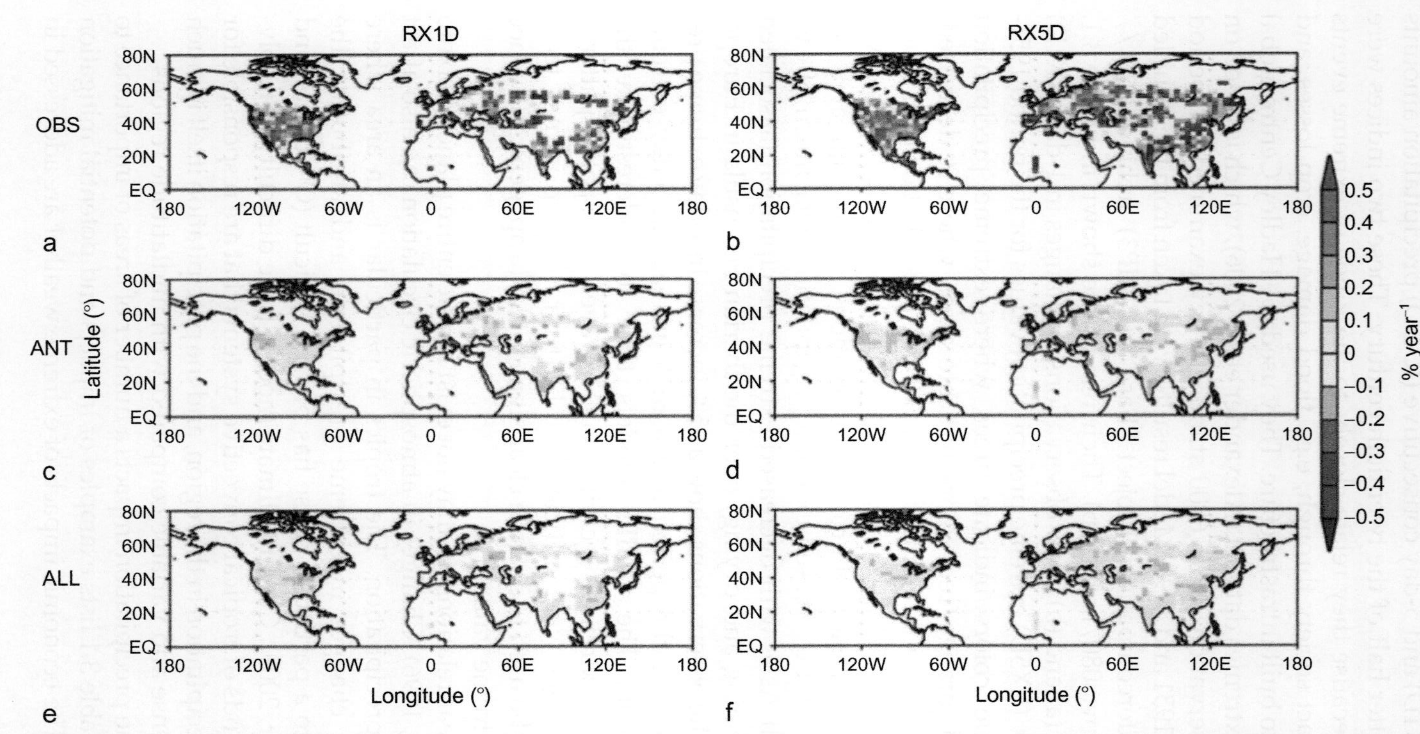

FIGURE 3.1 Changes in extreme precipitation for the period 1951-1999. Panels (a and b) are observations; panels (c) and (d) are model simulations with only anthropogenic forcing; panels (e) and (f) are model calculations with anthropogenic plus natural forcing. Units: percent probability per year.

TABLE 3.1 Examples of Impacts of and Potential Mitigation Options for Extreme Precipitation

Category	Food, Fiber, Forestry, Ecosystems	Water Resources	Human Health	Infrastructure and Society
Impacts	Damage to crops; waterlogged soil	Decrease in water quality; contamination of water supply; relief of water stress	Increase risk of infectious diseases	Infrastructure damage due to flooding; loss of property; soil erosion and landslides
Mitigation options	Improved drainage; adjustment of planting/ harvesting times; alternative crops	Better forecasting; improved watershed management	Early warning systems; more rapid postevent relief	Improved flood forecasting, improved zoning laws

more detail in Chapter 10. Many of the impacts of extreme precipitation revolve around damage due to flooding and erosion.

In a recent study, Pall et al. (2011) showed that anthropogenic greenhouse gas emissions (and the warming and changes in the hydrological cycle that they cause) had more than doubled the risk of serious flooding in England and Wales for the year 2000. Between September and November 2000, more than 10,000 homes were flooded in the region, and roughly 3.5B GBP of insurance claims were submitted (http://www.guardian.co.uk/environment/2011/feb/16/climate-change-risk-uk-floods). While the quantification of a doubling of risk of flooding due to our emissions of greenhouse gases strictly holds only for England and Wales (where the study was focused), it is clear that human-induced changes in extreme precipitation will have large societal and economic consequences.

Tropical Cyclones

The changes in extreme precipitation discussed in the previous section were only indirectly linked to ocean warming. The life cycles of tropical cyclones, on the other hand, are intimately linked to sea surface temperatures. Tropical cyclone is a generic term for storm systems with sustained winds of more than 33 m/s. In the northwest Pacific region, they are called typhoons, and in the north Atlantic and northeast Pacific regions, they are referred to as hurricanes.

A number of factors are necessary for a tropical cyclone to form. These include the following:

- Seawater temperatures (down to at least roughly 50 m) of more than 26 °C
- High humidity in the atmosphere
- Strong convection and cloud formation (which releases latent heat and further strengthens convection)
- Little change in wind speed or direction with height (low wind shear)
- Some disturbance in weather patterns to start the cyclone formation process

These conditions are met most often between roughly 10-30° away from the Equator. Figure 3.2 (Lloyd and Vecchi, 2011) shows a map of global tropical cyclones during the period 1998-2007.

Signal or Noise?

It is very difficult to extract a signal for changes in the number and properties of tropical cyclones from the very large interannual and decadal-scale variability that exists in their life cycles. The observational time series are often too short, since several decades

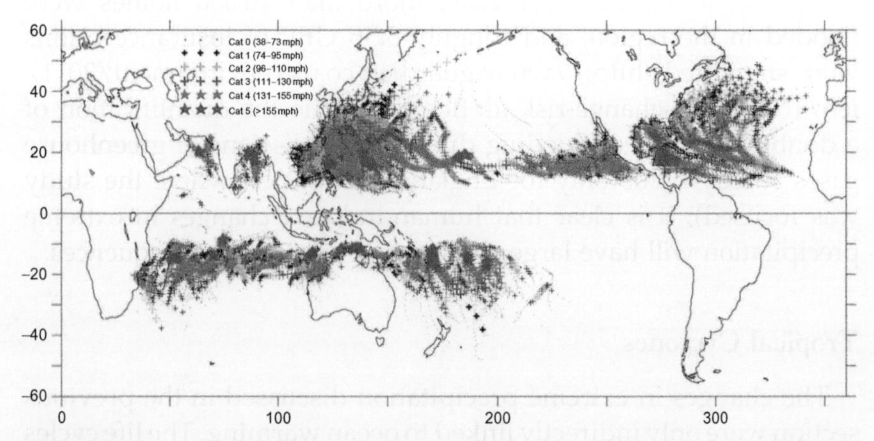

FIGURE 3.2 Map of global tropical cyclones in different Saffir-Simpson categories.

of continuous and consistent observations are often needed to reveal a statistically robust signal. In addition, most current global climate models do not sufficiently resolve tropical cyclones (if they do at all). There has been intense research activity in the past decade or so examining observations of tropical cyclones and projected future changes in their properties. Most studies have concentrated on trying to detect a signal in the frequency or in the strength of tropical cyclones and to assess to what extent any observed or modeled changes can be linked to human activities.

CHANGES IN FREQUENCY

To date, there is no robust observational evidence that changes in tropical cyclone frequency have exceeded natural variability (Knutson et al., 2010; Webster et al., 2005). Webster et al. (2005) used satellite data from the period 1970 to 2004 to examine global tropical cyclone frequency. Their results are shown in Figure 3.3.

There is a good deal of variability in the number of storms during this period, with no clear trend. Knutson et al. (2010) looked at a

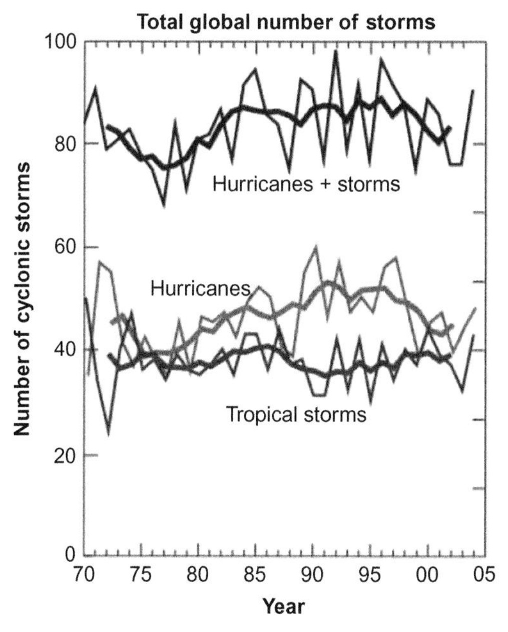

FIGURE 3.3 Time series of the number of tropical cyclones. *From Lloyd and Vecchi (2011), figure 1a. © American Meteorological Society. Used with permission.*

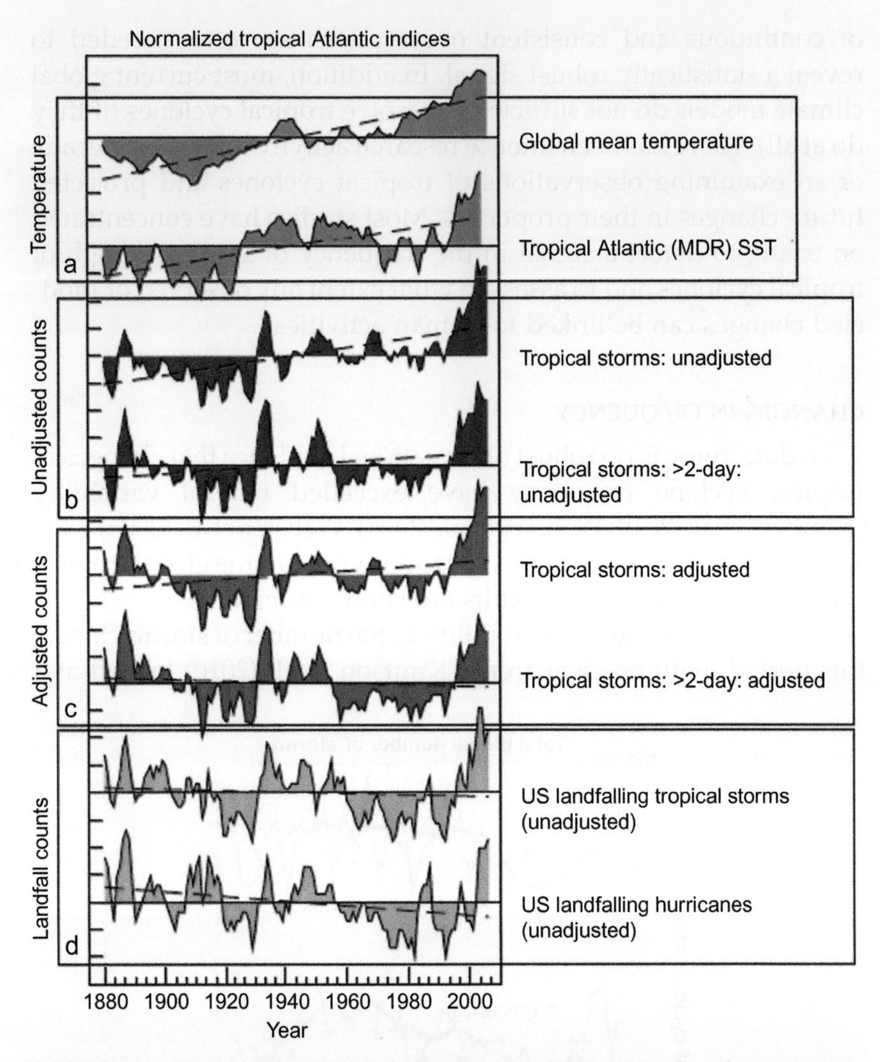

FIGURE 3.4 Time series of (a) global and tropical Atlantic sea surface temperature; (b) the number of all tropical storms in the Atlantic, and those with a >2-day duration; (c) number of Atlantic tropical storms adjusted for missing observations; (d) number of storms and hurricanes making landfall in the United States. *From Webster et al. (2005), figure 2a. © AAAS.*

longer time series of storms in the tropical Atlantic region. They looked at observations from the mid-1880s to past 2000 (shown in Figure 3.4).

As in the Webster et al. (2005) data, Knutson et al. (2010) concluded that there is no clear trend in the number of tropical storms over this longer time period.

CHANGES IN INTENSITY, GENESIS, AND STORM TRACKS

OBSERVATIONAL EVIDENCE OF AN INCREASE IN THE INTENSITIES OF THE STRONGEST TROPICAL CYCLONES In contrast to the number of tropical cyclones, an increase in the intensity of tropical cyclones has been observed—at least over the period of satellite observations. Webster et al. (2005) showed that while the number of tropical cyclones showed no trend, the percentage of the strongest storms (categories 4 and 5 on the Saffir-Simpson scale) increased from less than 20% in 1970-1974 to about 35% in 2000-2004 (panel a of Figure 3.5).

They found that this increase held true across all the major ocean basins (shown in panel b of Figure 3.5). Similar results were found by Emanuel (2005a). Kossin et al. (2007) used a different data processing technique and found an increasing trend in hurricane intensity in the Atlantic region, but did not find similar trends in other ocean basins. The same conclusion was drawn by Elsner et al. (2008), who found that Atlantic tropical cyclones were getting stronger, but that no trend was discernable for other tropical ocean basins. Saunders and Lea (2008) used data from the Atlantic from the period 1965 to 2005 to deduce the increase in hurricane frequency/activity for a given change in sea surface temperature. They employ a model that uses two environmental variables (sea surface temperature and the atmospheric wind field). These two

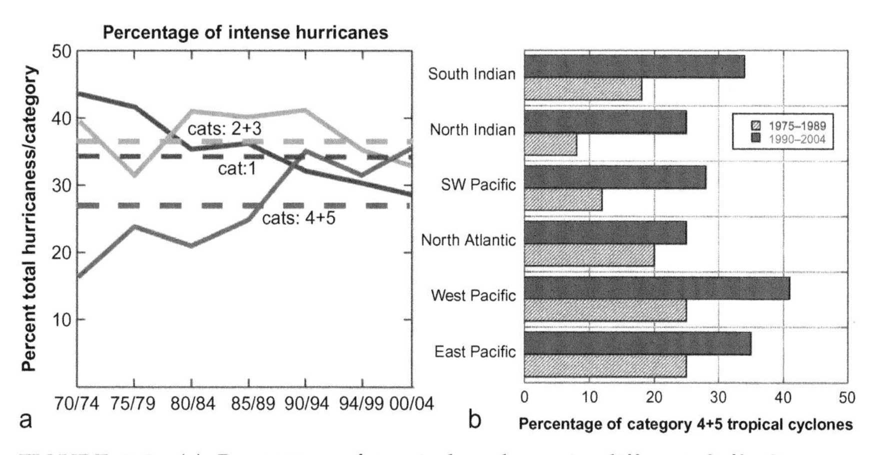

FIGURE 3.5 (a) Percentage of tropical cyclones in different Saffir-Simpson hurricane categories (1, weakest; 5, strongest); (b) increase in percentage of the strongest storms across ocean basins.

variables capture 75-80% of the variance in Atlantic tropical cyclone activity between 1965 and 2005. They use the period 1950-2000 to establish a baseline and then looked at increases in hurricane frequency and intensity for the period 1996-2005 relative to this baseline. There was an observed temperature increase of 0.27 °C in the main development region for Atlantic hurricanes during 1996-2005 compared with the 1950-2000 baseline period. Their results are shown in Figure 3.6.

The Accumulated Cyclone Energy index is a measure of hurricane strength. It should be noted that the Saunders and Lea (2008) results refer to a shorter and more recent period than the other references cited here.

Given the variability of tropical cyclones, the number of factors that influence their formation and life cycle, and the limited length of consistent global observations (especially from satellites), it is perhaps not surprising that there is continued debate about detecting and attributing trends in tropical cyclone intensity and frequency (e.g., Emanuel, 2005b; Landsea, 2005). Though there is increasing evidence that Atlantic tropical cyclones have increased in intensity in recent decades, extending this observed trend beyond the Atlantic ocean basin and farther back in time has not been possible (Field et al., 2012). It is important to note that while

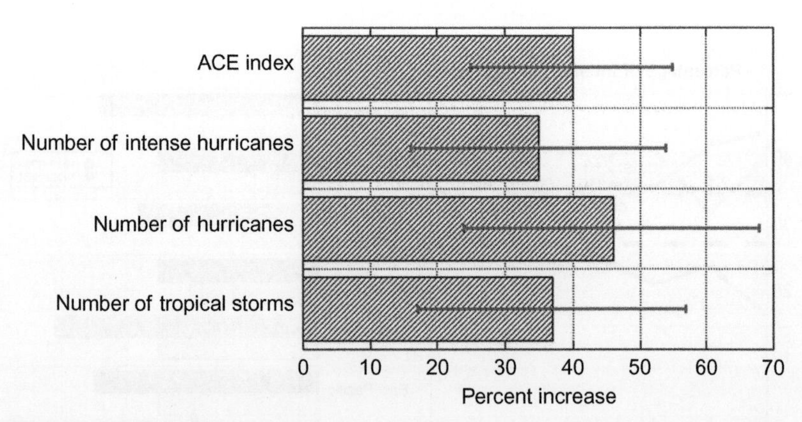

FIGURE 3.6 Increase in Atlantic hurricane frequency/activity between 1996 and 2005 (compared with the period 1950-2000) linked to the observed increase in August-September sea surface temperature of 0.27 °C in the main development region for Atlantic hurricanes. Error bars reflect the 95% confidence interval. *From Webster et al. (2005), figure 4b. © AAAS.*

the Field et al. (2012) review concluded that "there is low confidence that any observed long-term (i.e., 40 years or more) increases in tropical cyclone activity are robust, after accounting for past changes in observing capabilities," they did not dispute the significant upward trend observed in the intensity of the strongest tropical cyclones using a homogeneous satellite record. The scientific debate on this issue does not revolve around whether the satellite observations of an increase in tropical cyclone intensity are correct, rather it is centered on whether the period over which a homogeneous satellite record is available (about 30 years) is sufficiently long to declare an observed upward trend since other observations further back in time do not support this conclusion.

PROJECTIONS OF TROPICAL CYCLONE ACTIVITY IN THE FUTURE While observations generally show no discernable change in the frequency of tropical storms on the global scale, most model projections into the future indicate that there will be a general reduction in frequency toward the end of the century (Bengtsson et al., 2007; Gualdi et al., 2008; Knutson et al., 2010; Murakami et al., 2011; Zhao et al., 2009). Many of the studies show a good deal of variability between ocean basins. For example, Zhao et al. (2009) project reductions in frequency in the Pacific basin, and increases in the Atlantic. Bengtsson et al. (2007) and Murakami et al. (2011) project general reductions in frequency and shifts in storm track locations. The decrease in number of tropical cyclones is generally driven by factors related to changes in atmospheric circulation. For example, many models predict an increase in stability for the lower atmosphere, which would reduce the strength of convection. This could act to decrease tropical cyclone formation.

Bender et al. (2010) noted that while many models predict a decrease in tropical cyclone frequency, some models are not able to reproduce storms of category 3 or higher. They used a downscaling technique to look at the impact of anthropogenic warming on the frequency of intense Atlantic hurricanes. They too projected a decrease in the total number of storms, but nearly a doubling in frequency of category 4 and 5 storms by the end of the twenty-first century.

Overall, a literature review by Knutson et al. (2010) with a global perspective bracketed projections of a decrease in frequency of tropical cyclones by 6-34%, with an increase in intensity of 2-11% by 2100.

POTENTIAL IMPACTS OF FEWER BUT STRONGER TROPICAL CYCLONES

Why is it so important to pay attention to the most intense tropical cyclones, if the total number is expected to decline? Karl et al. (2008) looked at the mean damage ratio (MDR) due to hurricanes in the United States. The MDR is the average expected economic loss as a percent of the total insured value. For reference, Table 3.2 shows the sustained wind speeds that correspond to the various Saffir-Simpson tropical cyclone categories.

Category 1-3 hurricanes do not usually result in mean damage ratios above a few percent. The mean damage ratio increases rapidly once storms reach the category 4 and 5 level, with ratios of up to 80% for the strongest storms (shown in Figure 3.7, from Karl et al., 2008).

TABLE 3.2 Wind Speeds for the Different Saffir-Simpson Tropical Cyclone Categories

Category	1	2	3	4	5
Wind speed (km/h)	119-153	154-177	178-209	210-249	>249

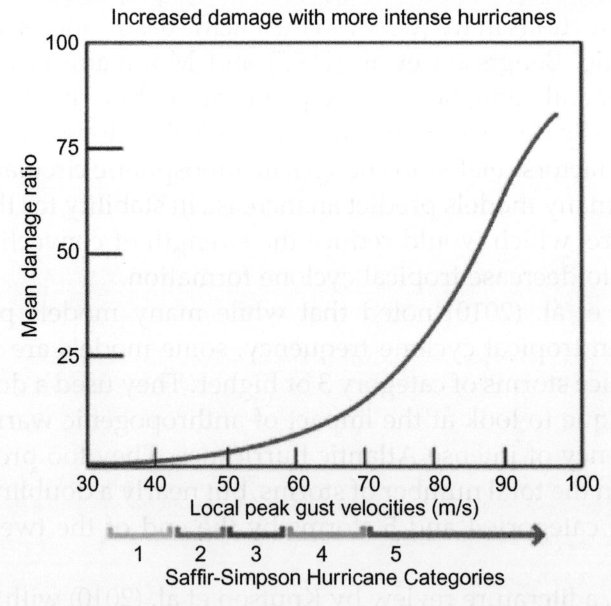

FIGURE 3.7 Increase in mean damage ratio as a function of tropical cyclone strength.

Pielke et al. (2008) point out that economic damage is a function not only of hurricane strength but also of demographic changes such as population and housing in coastal areas subject to tropical cyclone landfall. They conclude that unless action is taken to address the growing concentration of people and property in coastal areas where tropical cyclones reach land, damage will increase, and by a great deal, as more and wealthier people increasingly inhabit these coastal locations.

While the economic losses due to tropical cyclones may be skewed toward wealthy coastal settlements, human impacts will not be limited to areas where rich people live. Figure 3.2 clearly shows that intense tropical cyclones make landfall in many areas of the Americas, East and South Asia, the Pacific, and even some parts of southern Africa. Less affluent people in these areas are even more vulnerable to the impacts of intense tropical cyclones, since their built infrastructure is generally worse and less insured than in more affluent areas, and the disaster response capability of communities in developing countries is often less effective than in more affluent areas.

To put hurricane damage into perspective, Karl et al. (2008) showed the magnitude of total US damage costs from natural disasters over the period 1960-2005 (shown in Figure 3.8).

Damage due to hurricanes or tropical storms dominated the economic losses, topping $120B over this period.

Clearly, the combination of likely increases in the most intense tropical cyclones coupled with increasingly large populations in coastal areas where these storms make landfall is a recipe for trouble.

BIOLOGICAL CONSEQUENCES OF OCEAN WARMING

Seasonal cycles characterize the oceans with spring and summer maxima, and autumn and winter minima being typical for biological productivity. This annual cycle dominated by temperature-dependent respiration is most obvious in surface waters where temperatures fluctuate the most. With depth, these annual cycles generally decrease and become less obvious (Keeling et al., 2009).

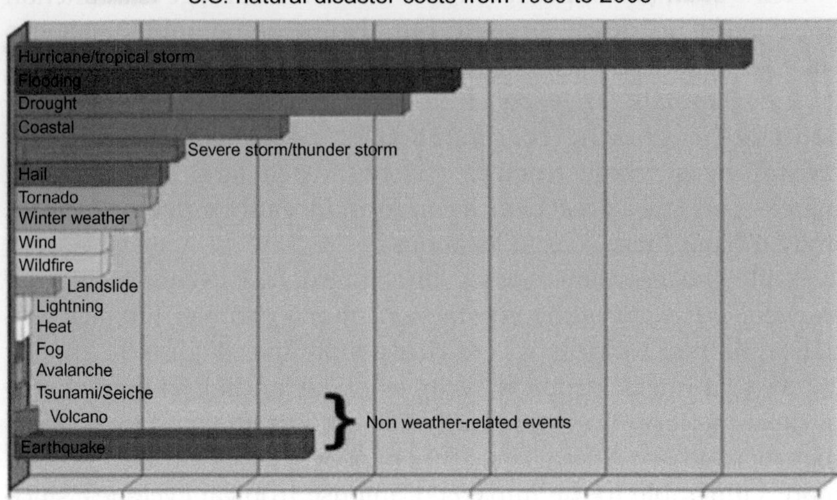

FIGURE 3.8 Damage costs (in billions of US dollars referenced to year 2005) for several different damage categories. *From Karl et al. (2008), figure box 1.1. © U.S. Global Change Research Program.*

Global warming related to our changing climate is poised to alter this seasonality that for thousands of years it has been the driving force in the life cycle of most organisms on earth. Much of the change will come from warming of surface and deep water as it affects:

- availability of dissolved oxygen,
- temperature-dependent respiration and metabolism, and
- thermal limits and distribution of organisms.

Several of the chapters in this book deal directly with biological consequences of ocean warming. In addition to temperature, interactions between changing physical processes like weather and currents (as described earlier) will also play a major role in determining the biological response. As part of US National Climate Assessment, Howard et al. (2013) succinctly summarized the impacts of climate change on marine organisms (Table 3.3). Realized and anticipated impacts will have substantial ecological, social, and economic effects around the globe.

TABLE 3.3 Key Finding for Impacts of Climate Change on Marine Organisms

Climate change impacts are being observed throughout ocean ecosystems, and there is high likelihood that these impacts will continue into the future
Observed impacts include shifts in species distributions and ranges, effects (primarily negative) on survival, growth, reproduction, health, and alterations in species interactions, among others
Impacts are occurring across a wide diversity of taxa and all ocean regions. However, high-latitude and tropical areas appear to be particularly vulnerable
There is high variability in the vulnerability and responses of marine organisms to climate change, leading to "winners" (i.e., species positively impacted) and "losers" (i.e., species negatively impacted)
Species with high tolerance for changes in temperature and other environmental conditions will likely experience fewer climate-related impacts and may therefore outcompete less tolerant species
Species that are highly vulnerable to climate change (e.g., corals, other calcifying organisms) will very likely experience negative impacts, resulting in potential declines
Climate change interacts with, and can exacerbate, the impacts of nonclimatic stressors (e.g., pollution, overharvesting, invasive species) on ocean ecosystems
Opportunities exist for ameliorating some of the impacts of climate change through reductions in nonclimatic stressors at local and regional scales
Effects of multiple stressors are difficult to predict given complex physiological effects and interactions between species
Past and current responses of ocean organisms to climate variability and climate change are informative, but extrapolations to future responses must be made with caution given that future environmental conditions are likely to be unprecedented
Observed responses often vary in magnitude across space and time
Potential threshold effects ("tipping points"), resulting in rapid ecosystem change, are an area of concern

Adapted from Howard et al. (2013).

Availability of Dissolved Oxygen

Availability of oxygen is a critical constraint on the functioning of marine ecosystems. As oceanic oxygen declines, habitability for aerobic organisms decreases rapidly (see Chapter 4). Predictions are that ocean warming and increased stratification of the upper ocean caused by global climate change will lead to declines in dissolved oxygen over broad areas of deeper ocean waters with implications for ocean productivity, nutrient cycling, carbon cycling, and suitability of marine habitat (Deutsch et al., 2011; Doney et al., 2011). Ocean models predict declines of 1-7% in the global ocean oxygen inventory over the next century, with declines continuing for a thousand years or more into the future (Keeling et al., 2009; Shaffer et al., 2009). An important consequence will be the expansion in the area and volume of oxygen minimum zones (OMZs), where oxygen levels are too low to support fisheries (Stramma et al., 2008). There could be a 50% increase in OMZs by 2100 (Arrigo, 2007; Riebesell et al., 2007).

Temperature-Dependent Respiration and Metabolism

Temperature has a profound effect on physiological processes. For cold-blooded organisms, temperature is the master factor that controls respiration and metabolic rates, along with many other cellular functions (Somero, 2011). Higher temperatures compress the thermal window with greater stress leading to increased metabolic oxygen demand and ultimately to oxygen deficiency at the cellular level (Figure 3.9, from Pörtner and Farrell, 2008). As global warming continues, this three-way interaction of temperature, oxygen, and metabolism will become increasing difficult for organisms to balance at both the cellular and population levels.

All organisms display tolerance limits that, when exceeded, have negative consequences (Pörtner and Farrell, 2008; Somero, 2011). Extreme or prolonged high- or low-temperature events can lead to sublethal effects, such as reduced growth and changes in the timing and magnitude of reproductive output (Figure 3.9).

Thermal Limits and Distribution of Organisms

Corals are among the most vulnerable organisms to increases or decreases in temperature. As little as 1-2 °C increase above summer maximums will severely stress some coral species leading to the ejection of symbiotic zooxanthellae and coral bleaching. If populations of zooxanthellae are unable to reestablish, corals decline and eventually die (Hoegh-Guldberg et al., 2007). Globally, as much as 75% of all coral reefs are threatened due to the interactive effects of

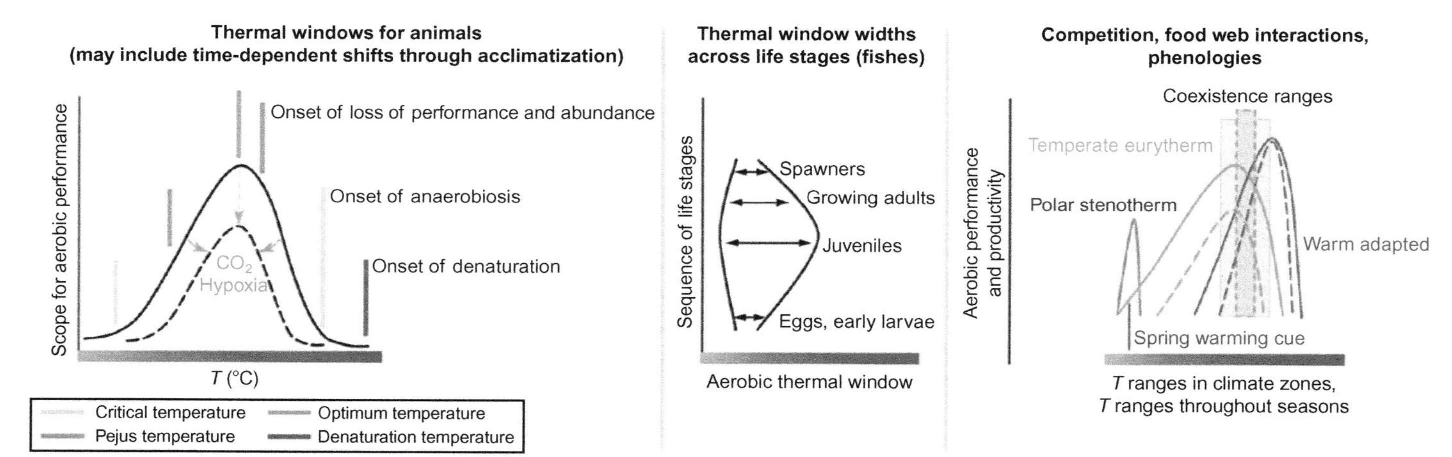

FIGURE 3.9 Temperature effects on aquatic animals. The thermal windows of aerobic performance (left) display optima and limitations by pejus (pejus means "turning worse"), critical, and denaturation temperatures, when tolerance becomes increasingly passive and time-limited. Seasonal acclimatization involves a limited shift or reshaping of the window by mechanisms that adjust functional capacity, endurance, or protection. Positions and widths of windows on the temperature scale shift with life stage (middle). Acclimatized windows are narrow in stenothermal species, or wide in eurytherms, reflecting adaptation to climate zones. Windows still differ for species whose biogeographies overlap in the same ecosystem (right, examples arbitrary). Warming cues start seasonal processes earlier (shifting phenology), causing potential mismatch with processes timed according to constant cues (light). Synergistic stressors like ocean acidification and hypoxia narrow thermal windows according to species-specific sensitivities (broken lines), modulating biogeographies, coexistence ranges, and other interactions. *From Karl et al. (2008), figure 1.3. © U.S. Global Change Research Program.*

multiple stressors from climate change including warming and ocean acidification (Carpenter et al., 2008; Chapter 2), and also from pollution associated with growing human population (see Chapters 6 and 8).

By 2030, half of all coral reefs will experience severe bleaching due to the thermal stress associated with increasing water temperatures and more than 95% by 2050, based on current trajectories of greenhouse gas emissions (Burke et al., 2011). Loss of coral reefs will have cascading effects on biodiversity and valuable ecosystem services (Carpenter et al., 2008). The converse to stress and mortality of sessile species would be movement of mobile species. As higher latitude waters warm, the range of many species will shift poleward, which will have major consequences for fisheries and the cost of fishing (Sumaila et al., 2011; Chapters 7 and 11).

SUMMARY AND TAKE-HOME MESSAGES

The ocean is warming. Exactly how much warmer the ocean will become and how rapidly this will happen are not known, and will be dependent on the choices we make about our collective resource use. Carbon dioxide emissions from human activities in particular are a key driver for ocean warming. Even though our knowledge of the impacts of ocean warming is nowhere near complete, it is sufficiently comprehensive to be able to draw a number of conclusions:

- A warming ocean will contribute to increases in extreme weather;
- Tropical cyclones are projected to become less frequent, but more intense—and thus more damaging—in the future;
- Climate related warming will increase stress on coastal and ocean ecosystems by pushing them nearer thermal limits, thus making them more susceptible to other stressors such as pollutants and invasive species;
- Changing temperatures will lead to spatial shifts in habitat suitability for fishes and will favor some species over others in a wide range of habitats.

Ocean warming is a very serious issue in its own right. However, since many of the effects of warming on marine ecosystems are exacerbated by other threats such as acidification, pollution, or overuse of resources, adaptation strategies need to be developed in a much more holistic framework than is currently being used. Chapter 12 provides suggestions for policies and strategies that put the effects of a warming ocean into this broader context.

References

Alexander, L. V., et al. (2006), Global observed changes in daily climate extremes of temperature and precipitation, *J. Geophys. Res.*, **111**(D5), D05109, http://dx.doi.org/10.1029/2005jd006290.

Allen, M. R., and W. J. Ingram (2002), Constraints on future changes in climate and the hydrologic cycle, *Nature*, **419**(12 September), 224–232.

Arrigo, K. R. (2007), Carbon cycle: marine manipulations, *Nature*, **450**(7169), 491–492.

Bender, M. A., T. R. Knutson, R. E. Tuleya, J. J. Sirutis, G. A. Vecchi, S. T. Garner, and I. M. Held (2010), Modeled Impact of Anthropogenic Warming on the Frequency of Intense Atlantic Hurricanes, *Science*, **327**(5964), 454–458, http://dx.doi.org/10.1126/science.1180568.

Bengtsson, L., K. I. Hodges, M. Esch, N. Keenlyside, L. Kornblueh, J. -J. Luo, and T. Yamagata (2007), How may tropical cyclones change in a warmer climate? *Tellus A*, **59**(4), 539–561.

Bindoff, N. L., et al. (2007): Observations: Oceanic Climate Change and Sea Level. In: *Climate Change 2007: The Physical Science Basis. Contribution of Working Group I to the Fourth Assessment Report of the Intergovernmental Panel on Climate Change* [Solomon S., D. Qin, M. Manning, Z. Chen, M. Marquis, K. B. Averyt, M. Tignor and H. L. Miller (eds.)]. Cambridge University Press, Cambridge, UK; New York NY, USA.

Burke, L., K. Reytar, M. Spalding, and A. Perry (2011), Reefs at Risk Revisited *Rep.*, 114 pp., World Resources Institute, Washington, DC.

Carpenter, K. E., et al. (2008), One-third of Reef-Building Corals Face Elevated Extinction Risk from Climate Change and Local Impacts, *Science*, **321**(5888), 560–563, http://dx.doi.org/10.1126/science.1159196.

Deutsch, C., H. Brix, T. Ito, H. Frenzel, and L. Thompson (2011), Climate-Forced Variability of Ocean Hypoxia, *Science*, http://dx.doi.org/10.1126/science. 1202422.

Doney, S. C., et al. (2011), Climate Change Impacts on Marine Ecosystems, *Annual Review of Marine Science*, **4**(1), 11–37, http://dx.doi.org/10.1146/annurev-marine-041911-111611.

Elsner, J. B., J. P. Kossin, and T. H. Jagger (2008), The increasing intensity of the strongest tropical cyclones, *Nature*, **455**(7209), 92–95, http://dx.doi.org/10.1038/nature07234.

Emanuel, K. (2005a), Increasing destructiveness of tropical cyclones over the past 30 years, *Nature*, **436**(7051), 686–688, http://dx.doi.org/10.1038/nature03906 http://www.nature.com/nature/journal/v436/n7051/suppinfo/nature03906_S1.html.

Emanuel, K. (2005b), Meteorology: Emanuel replies, *Nature*, **438**(7071), E13, http://dx.doi.org/10.1038/nature04427.

Field, C. B., et al. (eds.) (2012), *IPCC, 2012: Managing the Risks of Extreme Events and Disasters to Advance Climate Change Adaptation. A Special Report of Working Groups I and II of the Intergovernmental Panel on Climate Change*, 582 pp., Cambridge University Press, Cambridge, UK.

Gualdi, S., E. Scoccimarro, and A. Navarra (2008), Changes in Tropical Cyclone Activity due to Global Warming: Results from a High-Resolution Coupled *General Circulation Model, J. Clim.*, **21**(20), 5204–5228, http://dx.doi.org/10.1175/2008JCLI1921.1.

Hoegh-Guldberg, O., et al. (2007), Coral Reefs Under Rapid Climate Change and Ocean Acidification, *Science*, **318**(5857), 1737–1742, http://dx.doi.org/10.1126/science.1152509.

Howard, J., et al. (2013), Oceans and Marine Resources in a Changing Climate, *Oceanogr. Mar. Biol. An Annu. Rev.*, **51**, 71–192.

Karl, T.R., G. A. Meehl, C. D. Miller, S. J. Hassol, A. M. Waple, and W. L. Murray (2008): CCSP 2008: weather and climate extremes in a changing climate. Regions of focus: North America, Hawaii, Caribbean and U.S. Pacific Islands. In: *A Report by the U.S. Climate Change Science Program and the Subcommittee on Global Change Research Rep.*, Department of Commerce, NOAA's National Climatic Data Center, Washington, DC, 164.

Keeling, R. F., A. Körtzinger, and N. Gruber (2009), Ocean Deoxygenation in a Warming World, *Annual Review of Marine Science*, **2**(1), 199–229, http://dx.doi.org/10.1146/annurev.marine.010908.163855.

Knutson, T. R., J. L. McBride, J. Chan, K. Emanuel, G. Holland, C. Landsea, I. Held, J. P. Kossin, A. K. Srivastava, and M. Sugi (2010), Tropical cyclones and climate change, *Nature Geoscience*, **3**(3), 157–163, http://dx.doi.org/10.1038/ngeo779. http://www.nature.com/ngeo/journal/v3/n3/suppinfo/ngeo779_S1.html.

Kossin, J. P., K. R. Knapp, D. J. Vimont, R. J. Murnane, and B. A. Harper (2007), A globally consistent reanalysis of Hurricane variability and trends, *Geophys. Rese. Lette.*, **34**(4), L04815, http://dx.doi.org/10.1029/2006gl028836.

Landsea, C. W. (2005), Meteorology: hurricanes and global warming, *Nature*, **438** (7071), E11–E12.

Lloyd, I. D., and G. A. Vecchi (2011), Observational Evidence for Oceanic Controls on Hurricane Intensity, *J. Clim.*, **24**(4), 1138–1153, http://dx.doi.org/10.1175/2010jcli3763.1.

Min, S. -K., X. Zhang, F. W. Zwiers, and G. C. Hegerl (2011), Human contribution to more-intense precipitation extremes, *Nature*, **470** (7334), 378–381, http://www.nature.com/nature/journal/v470/n7334/abs/10.1038-nature09763-unlocked.html-supplementary-information.

Murakami, H., B. Wang, and A. Kitoh (2011), Future Change of Western North Pacific Typhoons: Projections by a 20-km-Mesh Global Atmospheric Model*, *J. Clim.*, **24**(4), 1154–1169, http://dx.doi.org/10.1175/2010JCLI3723.1.

O'Gorman, P. A., and T. Schneider (2009), The physical basis for increases in precipitation extremes in simulations of 21st-century climate change, *Proc. Natl Acad. Sci.*, **106**(35), 14773–14777, http://dx.doi.org/10.1073/pnas.0907610106.

Pall, P., T. Aina, D. A. Stone, P. A. Stott, T. Nozawa, A. G. J. Hilberts, D. Lohmann, and M. R. Allen (2011), Anthropogenic greenhouse gas contribution to flood risk in England and Wales in autumn 2000, *Nature*, **470**(7334), 382–385, http://www.nature.com/nature/journal/v470/n7334/abs/10.1038-nature09762-unlocked. html-supplementary-information.
Peixoto, J. P., and A. A. Oort (1996), The Climatology of Relative Humidity in the Atmosphere, *J. Clim.*, **9**(December), 3443–3463.
Pielke, Jr., R. A., J. Gratz, C. W. Landsea, D. Collins, M. A. Saunders, and R. Musulin (2008), Normalized hurricane damage in the United States: 1900–2005, *Natural Hazards Review*, **9**(1), 29–42, http://dx.doi.org/10.1061/(asce)1527-6988(2008) 9:1(29).
Pörtner, H. O., and A. P. Farrell (2008), Physiology and climate change, *Science*, **322**(5902), 690–692, http://dx.doi.org/10.1126/science.1163156.
Riebesell, U., et al. (2007), Enhanced biological carbon consumption in a high CO_2 ocean, *Nature*, **450**(7169), 545–548, http://www.nature.com/nature/journal/v450/n7169/suppinfo/nature06267_S1.html.
Roemmich, D., W. John Gould, and J. Gilson (2012), 135 years of global ocean warming between the Challenger expedition and the Argo Programme, *Nature Clim. Change*, **2**(6), 425–428, http://www.nature.com/nclimate/journal/v2/n6/abs/nclimate1461.html-supplementary-information.
Saunders, M. A., and A. S. Lea (2008), Large contribution of sea surface warming to recent increase in Atlantic hurricane activity, *Nature*, **451**(7178), 557–560, http://dx.doi.org/10.1038/nature06422.
Shaffer, G., S. M. Olsen, and J. O. P. Pedersen (2009), Long-term ocean oxygen depletion in response to carbon dioxide emissions from fossil fuels, *Nature Geoscience*, **2**(2), 105–109, http://www.nature.com/ngeo/journal/v2/n2/suppinfo/ngeo420_S1.html.
Somero, G. N. (2011), Comparative physiology: a "crystal ball" for predicting consequences of global change, *Am. J. Physiol. Regul. Integr. Comp. Physiol.*, **301**(1), R1–R14, http://dx.doi.org/10.1152/ajpregu.00719.2010.
Stramma, L., G. C. Johnson, J. Sprintall, and V. Mohrholz (2008), Expanding Oxygen-Minimum Zones in the Tropical Oceans, *Science*, **320**(5876), 655–658, http://dx.doi.org/10.1126/science.1153847.
Sumaila, U. R., W. W. L. Cheung, V. W. Y. Lam, D. Pauly, and S. Herrick (2011), Climate change impacts on the biophysics and economics of world fisheries, *Nature Clim. Change*, **1**(9), 449–456, http://dx.doi.org/10.1038/nclimate1301.
Webster, P. J., G. J. Holland, J. A. Curry, and H. -R. Chang (2005), Changes in Tropical Cyclone Number, Duration and Intensity in a Warming Environment, *Science*, **309**(16 September), 1844–1846.
Zhao, M., I. M. Held, S. -J. Lin, and G. A. Vecchi (2009), Simulations of Global Hurricane Climatology, Interannual Variability, and Response to Global Warming using a 50-km Resolution GCM, *J. Clim.*, **22**(24), 6653–6678, http://dx.doi.org/10.1175/2009JCLI3049.1.

CHAPTER

4

Hypoxia

Robert J. Diaz, Hanna Eriksson-Hägg*[†], *Rutger Rosenberg*[‡]

*Department of Biological Sciences, Virginia Institute of Marine Science, College of William and Mary, Gloucester Point, Virginia, USA

[†]Baltic Nest Institute, Stockholm Resilience Centre, Stockholm University, Kräftriket 2B, Stockholm, Sweden

[‡]Kristineberg Marine Research Station, University of Gothenburg, Gothenburg, Sweden

KEY MESSAGES

- Between the 1950s and 1960s, eutrophication has emerged as a problem that threatens and degrades coastal ecosystems, alters fisheries, and impacts human health in many areas of the world. Hypoxia is one of the most acute symptoms of eutrophication.
- The global extent of eutrophication and its threats to human health and ecosystem services are just beginning appreciated, but much remains unknown relative to social and economic consequences. Virtually all of our information on hypoxia and its effects is from North America and Europe. We know very little about conditions in the most populated parts of the planet (India, Asia, and Indonesia) or oceanic islands (mostly in the Pacific).
- Close to 1000 areas around the world have been identified as experiencing the effects of eutrophication. Of these, over 600 have problems with hypoxia, but through nutrient and organic

loading management, about 70 systems can be classified as recovering.
- Threat of declining oxygen is more than of academic interest. It is one of the many stressors that are a looming "perfect storm" that threatens to seriously degrade ecosystem services we have come to depend upon. Losses in water quality and nutrient cycling services from hypoxia are likely in the billions of US dollars.
- The importance of maintaining adequate levels of oxygen in our coastal and ocean systems is summarized by the motto of the American Lung Association: "if you can't breathe nothing else matters."
- Climate model predictions and observations reveal regional declines in oceanic dissolved oxygen, which are influenced by global warming.
- If we farm the land and sea, govern, and do business as usual, our oceans will suffer and so will we as a global society.

INTRODUCTION

Human population is expanding exponentially, recently passing 7 billion, and will likely exceed 8-10 billion by the year 2050. This expansion has led to extensive modification of landscapes at the expense of ecosystem function and services (Table 4.1) that we have come to rely upon, including pervasive effects on coastal primary production from excess nutrients to overfishing. Long-term records of nutrient discharges provide compelling evidence of a rapid increase in the fertility of many coastal ecosystems starting in the 1960s. On a global basis, by 2050, coastal marine systems are expected to experience, from today's levels, at least a doubling in both nitrogen and phosphorus loading, with serious consequences to ecosystem structure and function (Foley et al., 2005; Gruber and Galloway, 2008).

The question asked by Foley et al. is: Are land-use activities degrading the global environment in ways that undermine ecosystem services, which in turn undermine human welfare? When it comes to dissolved oxygen the answer is yes. In marine ecosystems,

TABLE 4.1 Ecosystem Services Affected by Excess Nutrients

Provisioning Services	Supporting Services
Food (N and P)	Soil fertility (N and P)
Fiber (N and P)	Photosynthesis (P)
Freshwater quality (N)	Primary production (P)
	Waste treatment (N)
	Nutrient cycling (N)
Cultural Services	**Regulations Services**
Aesthetic values (N)	
Cultural heritage values (N)	Climate regulation (N)
Recreation and ecotourism (N)	Biodiversity (N)

Negative (N) indicates the service would be reduced. Positive (P) indicates the service would be increased.

oxygen depletion has become a major structuring force for communities and energy flows at global scales.

Eutrophication can be defined as an increasing rate of primary production and organic carbon accumulation in excess of what an ecosystem is normally adapted to processing. It is one part of a complex of multiple stressors that interact to shape and direct ecosystem-level processes (Cloern, 2001). The visible ecosystem response to eutrophication is the excessive growth of algae and vegetation in coastal areas as primary production increases in direct response to nutrient enrichment. One of the most serious threat from this eutrophication is the reported decrease in oxygen in bottom waters created by the increased flux of organic matter to the sea bed, which can lead to "dead zones" (Diaz and Rosenberg, 2008). In the past, this eutrophication-induced low oxygen or hypoxia was mostly associated with rivers, estuaries, and bays. But dead zones have now developed in continental sea, such as the Baltic Sea, Kattegat, Black Sea, Gulf of Mexico, and East China Sea.

Much of the sensitivity of organisms to low oxygen is related to the fact that oxygen is not very soluble in water and that small changes in oxygen concentration lead to large percentage differences. For freshwater at 20 °C, 9.1 mg of oxygen (O_2) will dissolve in a liter of water, so a 1-mg O_2/l drop is about a 11% decline in saturation. In addition, oxygen solubility is strongly dependent on temperature and the amount of salt dissolved in the water. Saturation declines about 1 mg O_2/l from 20 to 26 °C and about 2 mg O_2/l from freshwater to seawater at similar temperatures (Benson and Krause 1984). So, depending on temperature and salinity, water contains 20-40 times less oxygen by volume and diffuses about 10,000 times more slowly through water than air (Graham 1990).

Thus what appear to be small changes in oxygen can have major consequences to animals living in an oxygen limited milieu. Physiologically, higher temperatures also increase metabolic requirements for oxygen and increase rates of microbial respirations and, therefore, oxygen consumption. For salmonid fishes, oxygen can become limiting at relatively high values and even air saturation can be limiting at higher temperatures (Fry, 1971). Concentrations of dissolved oxygen below 2-3 mg O_2/l are a general threshold value for hypoxia for marine and estuarine organisms and 5-6 mg O_2/l in freshwater. However, species and life stages differ greatly in their basic oxygen requirements and tolerances (Vaquer-Sunyer and Duarte, 2008).

The relatively low solubility of oxygen in water combined with two principal factors leads to the development of hypoxia and at times anoxia. These factors are water column stratification that isolates the bottom water from exchange with oxygen-rich surface water and decomposition of organic matter in the isolated bottom water that reduces oxygen levels. Both factors must be at work for hypoxia to develop and persist in deeper waters.

While unintended fertilization of marine systems, mainly from excess nitrogen, has been linked to many ecosystem-level changes associated with eutrophication, there are also natural processes that can lead to eutrophic-like conditions along continental margins and produce similar ecosystem responses. Coastal upwelling zones that occur along the western side of continents are highly productive but can also produce severe hypoxia from respiration of fluxed organic matter (<0.7 mg O_2/l). Oxygen minimum zones (OMZs) are another natural low oxygen phenomenon. They are persistent oxygen-depleted areas occurring primarily in the eastern Pacific Ocean, south Atlantic west of Africa, Arabian Sea, and Bay of

Bengal at intermediate depths about 200-1000 m (Helly and Levin, 2004). Where OMZs contact the bottom, the benthic fauna are adapted to oxygen concentrations as low as 0.1 mg O_2/l. This is in stark contrast to the faunal response to eutrophication-induced hypoxia in coastal and estuarine areas where oxygen concentrations of <1 mg O_2/l lead to mass mortality and major change in community structure (Levin et al., 2009).

THE HEART OF THE PROBLEM

Overenrichment of our water primarily with nutrients, such as nitrogen and phosphorus from fertilizers, has emerged as one of the leading causes of water quality impairment. These nutrients have led to excess production of organic matter that is called eutrophication. It is an indirect result of activities that support a rapidly expanding human population and directly related to agricultural practices, increased industrial activities, combustion of fossil fuels, and municipal sewage discharges, all of which contribute to the increased flow of nitrogen and phosphorus to terrestrial and aquatic environments. On a global basis, humans add more nitrogen to the land or ocean than is supplied by natural biological nitrogen fixation (Figure 4.1; Gruber and Galloway, 2008).

Most of these nutrients end up in the sea and fuel eutrophication, which increasingly threatens the health of coastal ecosystems and fisheries as well as human health. Excess nutrients lead to excessive growth of phytoplankton and algae, which have the potential to lead to severe secondary impacts. Harmful algal blooms and hypoxia are two of the most prominent impacts associated with eutrophication. Harmful algal blooms, often referred to as red tides, can cause fish kills and shellfish poisoning in humans. Hypoxic or dead zones stress aquatic ecosystems and can also cause fish kills and altered food webs.

Virtually all the ocean's food-provisioning ecosystem services for humans require oxygen to support organism growth and production. Oxygen is just absolutely necessary to sustain the life of all the fishes and invertebrates we have come to depend upon. By the early 1900s, dissolved oxygen was a topic of interest in research and management, and by the 1920s, it was recognized that a lack of

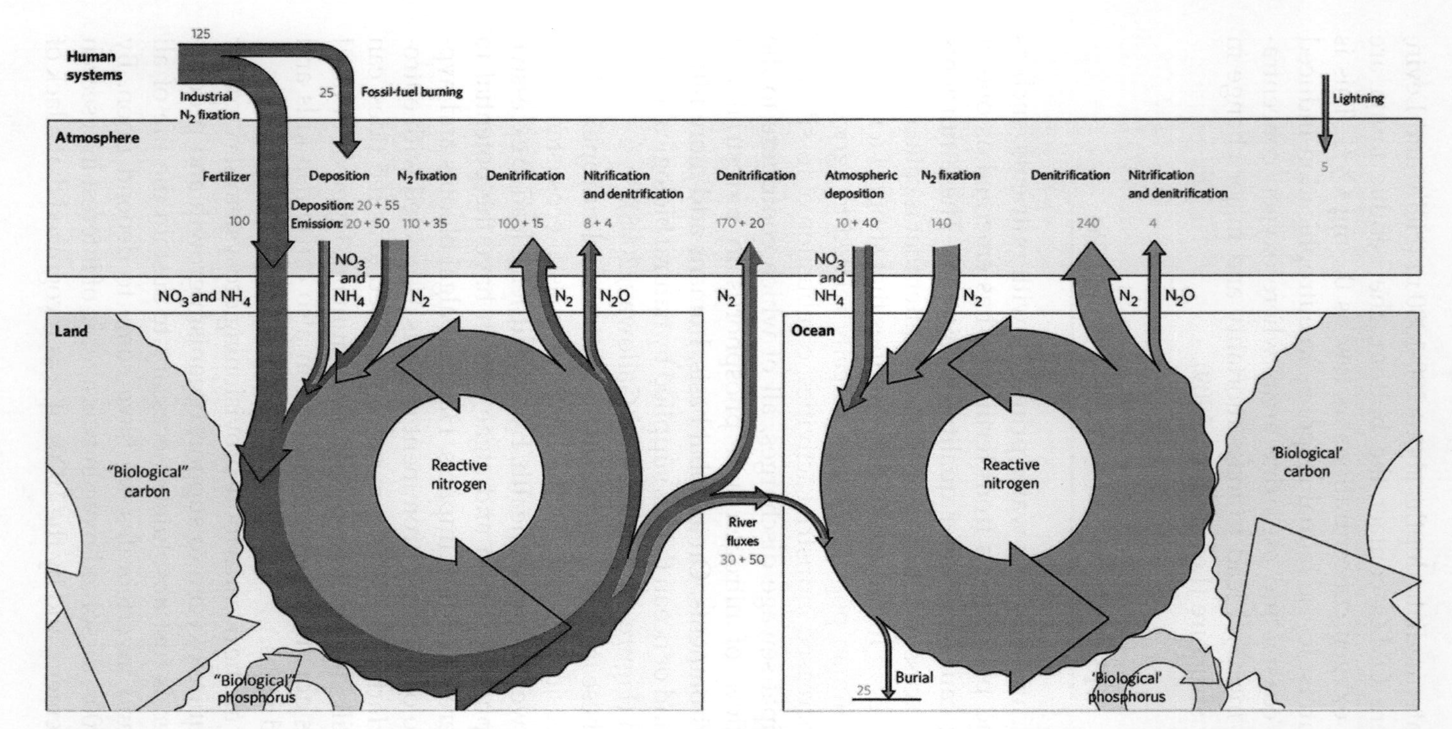

FIGURE 4.1 Depiction of the global nitrogen cycle on land and in the ocean from Gruber and Galloway (2008). Major processes that transform molecular nitrogen into reactive nitrogen are shown. There is also a tight coupling between the nitrogen cycles on land and in the ocean with those of carbon and phosphorus. Blue fluxes denote "natural" (unperturbed) fluxes; orange fluxes denote anthropogenic perturbation. Numbers are in Tg N per year and are for the 1990s.

oxygen was a major hazard to fishes. But it was not obvious that dissolved oxygen would become critical in shallow coastal systems until the 1970s and 1980s when large areas of low dissolved oxygen started to appear with associated mass mortalities of invertebrate and fishes. From the middle of the twentieth century to today, there have been drastic changes in dissolved oxygen concentrations and dynamics in marine coastal waters. Diaz and Rosenberg (1995) noted that no other environmental variable of such ecological importance to estuarine and coastal marine ecosystems as dissolved oxygen has changed so drastically, in such a short period of time.

Accounts of environmental problems related to low dissolved oxygen predate our ability to measure oxygen concentration in water. For example, the Drammensfjord in Norway appears to have been persistently hypoxic and anoxic since at least the 1700s based on foramaniferan proxies (Alve, 1995). Even in this small fjord with extended residence time of deepwater, historic anoxia has been made worse over the last two centuries by eutrophication. Improvements were observed only after reductions in organic loading. Another example would be the Mersey Estuary, England, which had poor water quality and hypoxia since at least the 1850s but is now recovered through concerted management efforts (Jones, 2006).

GLOBAL PATTERNS IN HYPOXIA

Since the 1960s, alarming trends of declining oxygen concentrations have immerged both in coastal areas and in the open oceans (see Diaz and Rosenberg, 2008; Keeling et al., 2010; Conley et al., 2011). Many of these trends have been linked to human activities. So, how did we get to where we are? In 2009, an interesting article proposed that there are planetary boundaries for a safe operating space for humanity (Rockström et al., 2009). While oxygen was not one of the nine boundaries discussed, it is influenced by many of the processes discussed (Table 4.2). The expanding size of our human population has led to three of the boundaries being crossed, which are climate change, rate of biodiversity loss, and the nitrogen cycle. Of these, alterations to the nitrogen cycle have the most direct consequences for dissolved oxygen followed by climate change. The more nutrients added to the sea, the more organic matter will

TABLE 4.2 Planetary Boundary Processes (Rockström et al., 2009) and How Exceeding Them will Affect Oxygen

Earth-System Process	Parameters	Proposed Boundary	Current Status	Preindustrial Value	Consequences for Oxygen
Climate change	(i) Atmospheric carbon dioxide concentration (parts per million by volume)	350	387	280	More carbon dioxide in water reduces oxygen concentration
	(ii) Change in radiative forcing (Watts per meter squared)	1	1.5	0	Warmer water holds less oxygen
Rate of biodiversity loss	Extinction rate (number of species per million species per year)	10	>100	0.1-1	Lower oxygen will stress more species
Nitrogen and phosphorus cycle	(i) Amount of N2 removed from the atmosphere for human use (millions of tons per year)	35	121	0	More N and P entering coastal systems will increase primary production, which will in turn decompose and lower oxygen increasing hypoxia
	(ii) Quantity of P flowing into the oceans (millions of tons per year)	11	8.5-9.5	~1	
Stratospheric ozone depletion	Concentration of ozone (Dobson unit)	276	283	290	Unknown
Ocean acidification	Global mean saturation state of aragonite in surface sea water	2.75	2.9	3.44	More acidic waters contain less oxygen

Global freshwater use	Consumption of freshwater by humans (km^3/year)	4000	2600	415	Reduced river flow would improve low oxygen conditions
Change in land use	Percentage of global land cover converted to cropland	15	11.7	Low	More cropland leads to more nutrient runoff and increased primary production which will in turn decompose and lower oxygen increasing hypoxia
Atmospheric aerosol loading	Overall particulate concentration in the atmosphere on a regional basis		To be determined		Unknown
Chemical pollution	Amount emitted or concentration of persistent organic pollutants, plastics, endocrine disrupters, heavy metals, and nuclear waste in the global environment, or the effects on ecosystem and functioning of Earth systems		To be determined		Unknown

be produced, which will create a greater oxygen demand when it is decomposed potentially leading to more hypoxia.

To bring this problem of low oxygen into focus, it has taken this unintended consequence of the green revolution less than 40 years to significantly alter the global nitrogen cycle on land and in our coastal systems. It was in the mid-1940s that the global use of industrial fertilizer surpassed all natural forms of fertilizer for growing corps. This laid the groundwork for the rapid rise in the areas affected by hypoxia in the 1970s and 1980s. This lag is the time it took for excess organic matter from primary production to build up and overwhelm an ecosystem's assimilative capacity, to the detriment of higher trophic levels but to the benefit of microbes. By comparison, it has taken over 100 years for the industrial revolution to significantly alter the global carbon cycle.

Since the 1960s, the number of hypoxic systems has about doubled every 10 years up to 2000 (Figure 4.2). Prior to 1960, there were about 40 systems with reports of eutrophication-related hypoxia. During the 1960s, another 25 systems were added. The 1970s saw estuarine and coastal ecosystems around the world becoming over-enriched with organic matter from expanding eutrophication and the number of oxygen-depleted ecosystems jumped from 65 to 135. In the 1980s, many more systems reported hypoxia for the first time bringing the total to 280. An additional 165 hypoxic areas were reported in the 1990s. By the end of the twentieth century, hypoxia had become a major, worldwide environmental problem with only a small fraction of systems (70) showing signs of improvement. At the end of the first decade of the twenty-first century, another 190 sites reported, bringing the total to about 640. An additional over 400 coastal sites globally were identified as areas of concern that currently exhibit signs of eutrophication and are at risk of developing hypoxia (Diaz et al., 2010; Conley et al., 2011).

There are signs of a slowing growth in hypoxic systems, mostly because North America and Europe are well studied and about completely reported. We predict that as more data become available on oxygen conditions in Asia, the Indo-Pacific, and Pacific Islands, the number of dead zone will more than double to over 1000. Our prediction is based on the very strong correlation between human population centers and the presence of hypoxia. The global distribution of coastal oxygen depletion is either centered on major

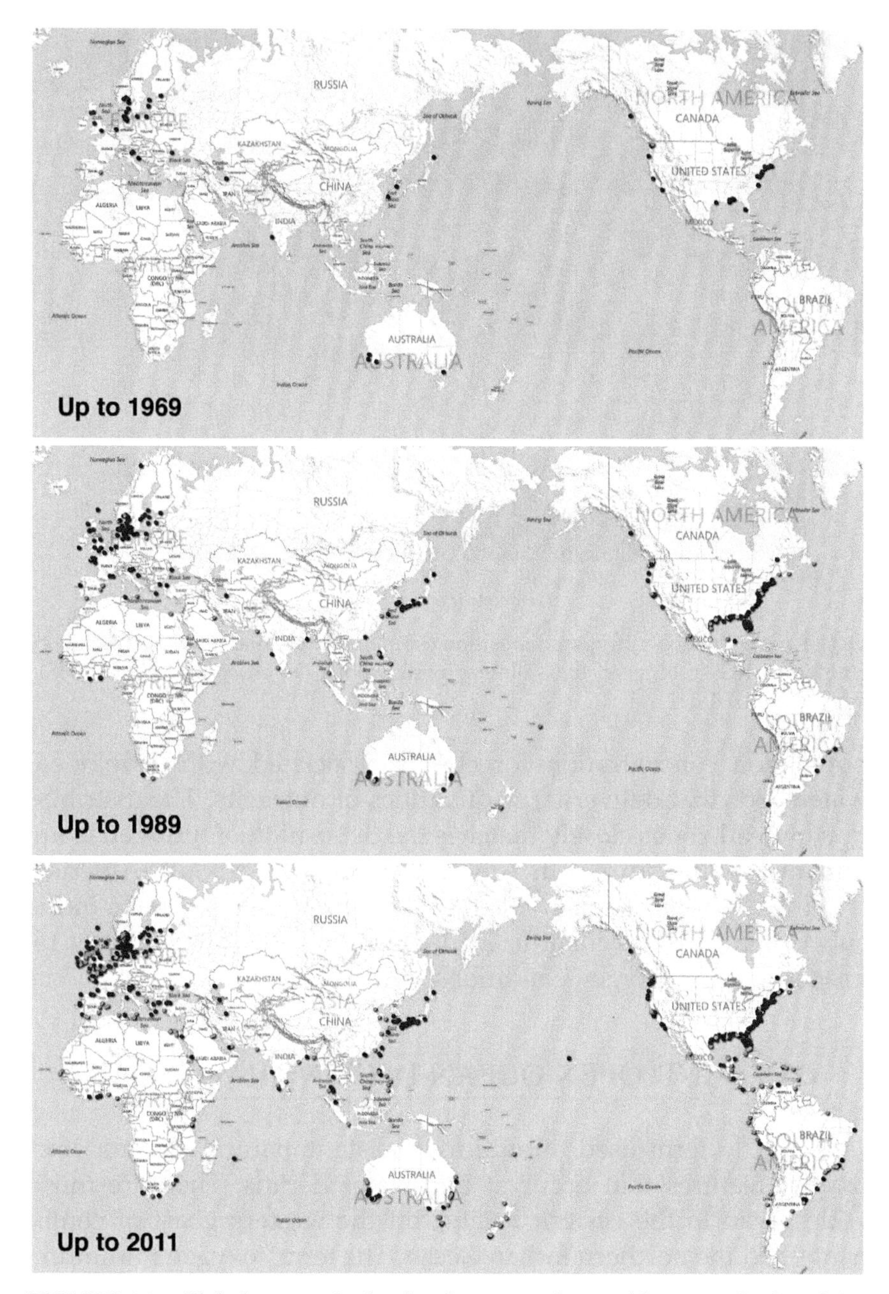

FIGURE 4.2 Global pattern in the development of coastal hypoxia. Each red dot represents a documented case related to human activities. Number of hypoxic sites is cumulative through time. *Based on Diaz et al. (2010) at http://www.wri.org/project/eutrophication. © World Resources Institute.*

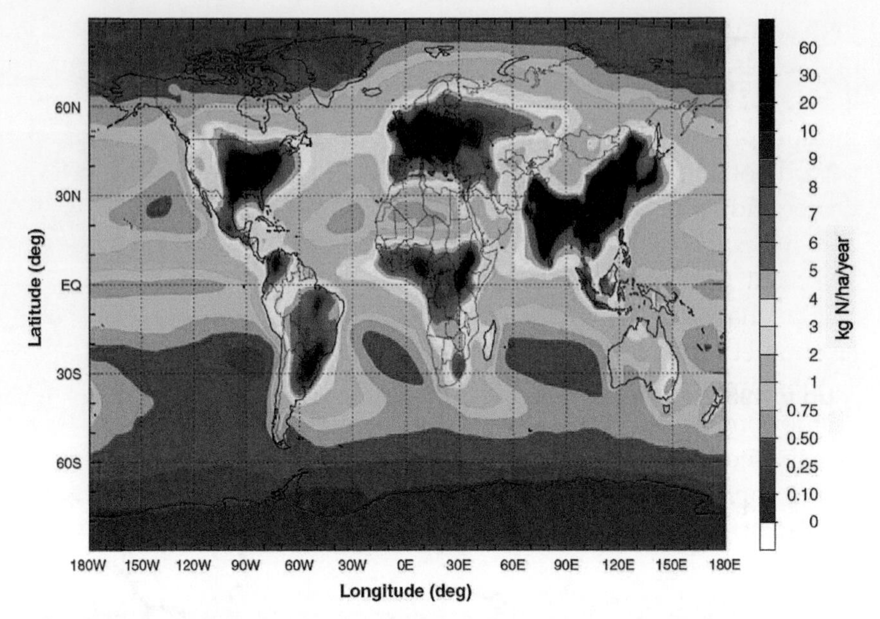

FIGURE 4.3 Global nitrogen deposition estimated from total N (NO*y* and NH*x*) emissions, 105 Tg N/year, by Galloway et al. (2008). *From Galloway et al. (2008), figure 2. © AAAS.*

population concentrations or closely associated with developed watersheds that deliver large quantities of nutrients. The distribution of dead zones closely matches the deposition of nitrogen from human activities in North America, Europe, and South America (Figure 4.3). While some of the highest deposition rates are in India and China, there is insufficient information on water quality to properly assess oxygen conditions.

OMZs AND OPEN OCEAN DECLINE IN OXYGEN

OMZ is a term used to refer to persistent, natural, oceanic low oxygen features that occur at mid-water depths. They are most widespread in the eastern Pacific, off the western coast of continents, and the northern Indian Ocean. The term "oxygen minimum layer" is sometimes used to refer to mid-water layers, exhibiting reduced oxygen relative to waters above and below. Such layers are ubiquitous in the global ocean due to isolation from sources of oxygenation, but often do not reach hypoxic levels. For example, in the central Gulf of Mexico and eastern North Atlantic Ocean, oxygen levels are depressed but not seriously low (Kamykowski and Zentara, 1990). OMZs and OMLs (Oxygen Minimum Layers) cover million of square kilometers of the oceans and persist for long periods of time (at greater than decadal scales), but have variable upper and lower depth boundaries controlled by natural processes and cycles (Helly and Levin, 2004). However, this is about to change as recent expanding trends have been noticed in several Pacific OMZs that are linked to climate change (Stramma et al., 2008; Keeling et al., 2010).

Most important for near shore fisheries is where OMZs contact the bottom (Figure 4.4). Globally, about 1,150,000 km^2 of sea floor is exposed to oxygen concentrations <0.7 mg O_2/l from OMZs (Helly and Levin, 2004). The specialized species that live in OMZs differ from shallower species exposed to seasonal or episodic hypoxia in having much lower oxygen tolerance thresholds, morphological adaptations to maximize respiratory surface, specialist rather than opportunistic lifestyles, and potential to utilize chemosynthesis-based nutritional pathways (Levin, 2003).

OMZs often occur in association with upwelling regions and supply oxygen depleted but nutrient-rich water that fuels upwelling-

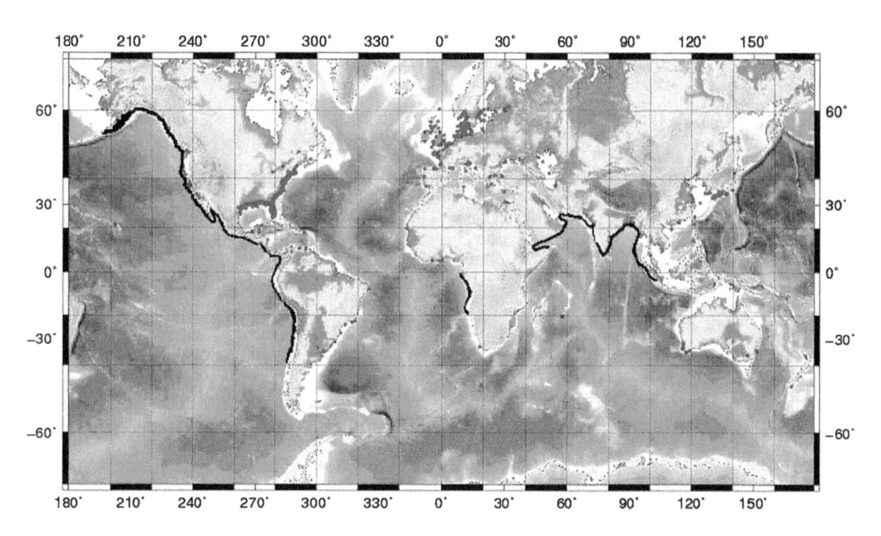

FIGURE 4.4 World distribution of outer shelf areas threatened by hypoxia from upwelling or OMZs (black lines) and eutrophication-induced coastal hypoxia (red dots). *From Levin et al. (2009), figure 7. © Biogeochemical Society.*

induced productivity. This immense primary production associated with upwellings supports about 80% of global fisheries catch (Pauly and Christensen, 1995), but it is also a primary contributor to persistent hypoxia under upwelling regions. However, hypoxia associated with coastal upwelling is not as long-lived and stable as that associated with OMZs. Upwelling can interact strongly with OMZs to produce intense continental shelf hypoxia and have long been known as a cause of mass mortality events. These are frequently observed off the coast of Chile (Fuenzalida et al., 2009), west Africa (Monteiro et al., 2008), and India (Banse, 1959). Since 2004, intense inner shelf hypoxia has been observed off Oregon and Washington, USA, likely linked to climate-driven changes in wind and currents and has caused mass mortality (Grantham et al., 2004; Chan et al., 2008).

Demersal and pelagic fisheries do benefit from the enhanced organic production associated with upwelling when oxygen concentrations remain high. For example, demersal fisheries (scallop, hake, and octopus) flourish under better oxygenated El Niño conditions. On the other hand, the resulting oxygen depletion, whether from mid-ocean OMZs or coastal upwelling or their interaction, affects mid-water plankton and pelagic organisms creating areas of low biodiversity and hostile environmental conditions for many commercially valued fisheries resources (Ekau et al., 2009). Monsoon-driven upwelling along the western coast of India migrates onshore and offshore depending on winds and currents. When pushed onshore, there are declining catches of fishes and prawns and diminished demersal fisheries (Banse, 1959). Recent intensification in the severity of low oxygen associated with this upwelling zone is related to a combination of climate change and land-derived nutrients and will most likely further impact inshore fisheries (Naqvi et al., 2006).

The upper depth limit of OMZs has major implications for fisheries. Expansion of OMZs toward the surface in the eastern tropical Pacific has limited the depth distribution of tropical pelagic marlins, sailfish, and tunas into a narrower surface layer of oxic water about 50-100 m thick (Prince and Goodyear, 2006). The high-performance physiology of these fishes leads to a relatively high hypoxic threshold (Brill, 1996), making any reduction in oxygen problematic. Declining oxygen and expansion of the OMZ in the tropical northeast Atlantic Ocean toward the surface is also restricting usable habitat of billfishes and tunas. From 1960 to 2010, Stramma et al. (2012) found hypoxia-based habitat compression to decreasing their suitable habitat by 15%. The combination of shallowing of OMZs encroaching onto outer continental shelfs and increased coastal eutrophication-induced hypoxia will eventually reduce suitable habitat space for both pelagic and demersal fishes.

ENVIRONMENTAL CONSEQUENCES OF HYPOXIA

Relative to oxygen, we do not know where the boundaries are on catastrophe for setting a planetary boundary (Table 4.2). The amount, severity, and duration of hypoxia and anoxia must all be factored in. While there are at times spectacular events such as mass mortalities of sessile organisms, there is little information on population-level effects of hypoxia and anoxia. We do know that when a dead zone forms, oxygen in the water is so low that fish, crabs, shrimp, and other marine life will swim away to areas of higher oxygen concentration. This is the origin of the term dead zone, a place where fishermen cannot find anything to catch. This escape must come at a cost to the individuals in terms of lost growth potential, increased predation risk, increase susceptibility to fishing gear, etc. (Table 4.3). A combination of other stressors from climate change, pollution, and habitat destruction also produces similar responses to dead zones that further complicate identifying population-level responses to hypoxia. We do know that hypoxia influences harvest, through effects both on processes underlying production such as growth and mortality (including effects on juveniles before they are subject to fishing mortality) and on processes influencing catchability (e.g., emigration, avoidance behavior).

Earliest account of hypoxia-stressed systems is from European fjords, such as the Drammensfjord mentioned above, and rivers with population and industrial centers, such as the Mersey Estuary where a combination of factors including hypoxia led to the elimination of salmon by the 1850s (Jones, 2006). In the United States, the earliest account of ecological stress associated with hypoxia comes from Mobile Bay, Alabama, where in the 1850s, hypoxic bottom water pushed by tides and wind into shallow water caused mobile organisms to migrate and concentrate at the water's edge. These

TABLE 4.3 Generalized Response of Populations to Hypoxia and Potential Economic Effect

Factor	Result	Economy
Mortality	Loss of stock, may take years to recover	Lower landings Increased time fishing
Reduced recruitment	Smaller populations, effect may be long lasting	Lower landings Increased time fishing
Reduced growth	Smaller individuals	Lower individual value
Poor body condition	Weaker individuals	Lower value
Increased Migration	Energy resources diverted to movement	Smaller individuals Increased time fishing
Aggregation	Exposure to increased risks of predation and exploitation	Less time fishing But easier to catch
Altered behavior	More/less susceptible to fishing gear	Increased or decreased catch ability

events became known as "Jubilees" as it was easy for people to pick up the hypoxia-stressed fish and crabs (May, 1973). In all cases, hypoxia caused the movement of mobile fauna, which has energetic consequences, and reduced or eliminated the trophic base for bottom-feeding species, which has adverse consequences for a system's higher level energy flows (Baird et al., 2004).

There is a similarity of faunal response across systems to varying types of hypoxia that range from beneficial to mortality (Diaz and Rosenberg, 1995; Vaquer-Sunyer and Duarte, 2008). Consequences of low oxygen are often sublethal and do affect growth, immune response, and reproduction. Mobile fauna have to contend with two simultaneous problems, loss of habitat as they are forced to migrate into higher oxygen waters, and reduced or changed prey resources for feeding. Fauna that cannot move initiate a graded series of behaviors to survive and will eventually die as oxygen declines or extends through time. The result is a hypoxia-based habitat compression when hypoxia overlaps with essential habitat such as nursery areas, feeding grounds, or deeper and cooler refuge waters during the summer (Coutant, 1990). Hypoxic habitats avoided are lost to the ecosystem for varying lengths of time and are not productive for fisheries.

If it is important for fish and shrimp to reach critical nursery or feeding areas at certain times in their life cycle, then hypoxia may affect population dynamics by delaying arrival to spawning or feeding grounds. In such cases, the cost of delayed migration in terms of population mortality and production is not known. For example, in 1976, continental shelf hypoxia in the New York-New Jersey Bight blocked the northward migration of bluefish (*Pomatomus saltatrix*). Fish that encountered the hypoxic zone did not pass through or around it, but stayed to the south waiting for hypoxia to dissipate and then continued their migration but delayed by weeks. On the continental shelf of the northern Gulf of Mexico, hypoxia interfered with the migration of brown shrimp (*Farfantepenaeus aztecus*) from inshore wetland nurseries to offshore feeding and spawning grounds. Juvenile brown shrimp leaving nursery areas migrates farther offshore when hypoxia was not present and was compressed inshore when hypoxia was present. In avoiding hypoxia, brown shrimp aggregated both inshore and offshore of low oxygen areas losing about a quarter of their shelf habitat for as much as 6 months (Rabalais et al., 2010).

Small and large systems exposed to long periods of hypoxia and anoxia have lower annual secondary production with productivity a function of how quickly benthos can recruit and grow during periods when oxygen is normal. The extreme case would be the perennial hypoxic/anoxic areas of the Baltic Sea that cover about 70,000 km^2 where benthic invertebrate production is near zero. Under normal oxygen condition, this area of the Baltic Sea should be producing about 1.3 million metric tons (mt) wet weight of potential benthic prey for bottom-feeding predators. In Chesapeake Bay, which has about 3500 km^2 of seasonal hypoxia that lasts about 3 months, about 75,000 mt wet weight of potential prey for fish and crab predators is lost or translocated. In the northern Gulf of Mexico, severe seasonal hypoxia covers about 20,000 km^2 and leads to approximately a 210,000 mt wet weight loss of prey from the fisheries forage base (Diaz and Rosenberg 2008). The question remains as to whether a system can recover the secondary production lost to hypoxia during periods of normal oxygen. The Chesapeake has about 9 months to recover and the northern Gulf of Mexico 6 months.

The elimination of benthic prey and hypoxia-based habitat compression can have profound effects on ecosystem functions as organisms die and are decomposed by microbes. Up food chain energy transfer is inhibited in areas where hypoxia is severe as benthic resources are killed directly and mobile predators avoid the area (Figure 4.5). As mortality of benthos occurs, microbial activities quickly dominate energy flows (Baird et al., 2004). This energy diversion tends to occur in ecologically important places and at the most inopportune time for predator energy demands, and causes an overall reduction in an ecosystem's functional ability to transfer energy to higher trophic levels and renders the ecosystem potentially less resilient to other stressors (Diaz and Rosenberg, 2008). Systems reporting mass mortality provide primary examples of degradation in trophic structure.

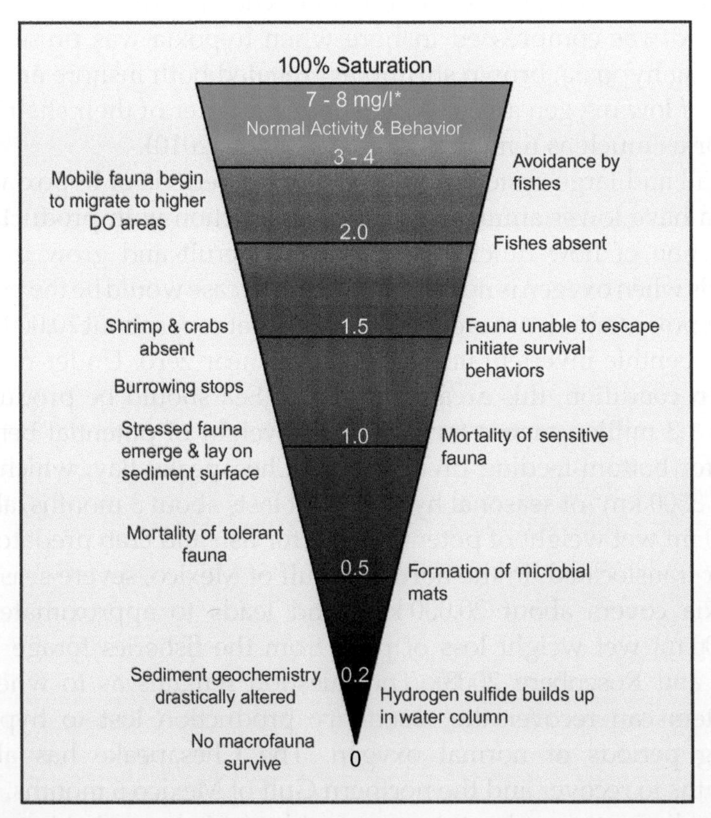

FIGURE 4.5 Range of behavior and ecological impacts as dissolved oxygen levels drop from saturation to anoxia.

Hypoxia has clear mortality effects on sessile, and at times mobile, organisms but its population-level effects in coastal environments remain uncertain. Much of the evidence supporting negative effects from hypoxia comes from laboratory experiments, localized effects in nature, fish kills, and our intuition that lack of oxygen can lead to dire consequences (Rose et al., 2009). Conclusive evidence of widespread population-level response to hypoxia is lacking. Quantifying the effects of hypoxia on fish populations, whether large or small, is critical for effective management of coastal ecosystems and for cost effective and efficient design of remediation actions. The potential for interaction of direct and indirect effects and subtle changes in vital rates (such as reproduction and recruitment) leading to population responses complicates field study and management, but does not excuse us from quantifying the population losses due to hypoxia. As coastal ecosystems continue to decline their capacity to deliver ecosystem services will also decline (Millennium Ecosystem Assessment, 2005). At some point, the consequences of hypoxia will become obvious.

ECONOMIC CONSEQUENCES OF HYPOXIA

Economic effects attributable to hypoxia are subtle and difficult to quantify even when mass mortality events occur. However, from assessing ecological effects of hypoxia, it is known that populations can experience a range of problems that at some point will negatively affect economic interests (Table 4.3). Much of the problem in assess economic consequences is related to the multiple stressors acting on targeted commercial populations (habitat degradation, overexploitation, pollution) and also factors that stress fisher's economics (aquaculture, imports, economic costs of fishing, fisheries regulations).

Because of their devastating effects on stocks and fishermen, it is possible to estimate effects of mass mortality events. Losses from hypoxia-related mortality of oysters in Mobile Bay, USA, in the early 1970s were US$500,000 (in 1970 dollars), but greater economic losses were associated with the declining stocks and poor recruitment of oysters (*Crassostrea virginica*) associated with recurring severe hypoxic (May, 1973). Estimated losses, both actual and to the resources, to marine-related industries from the New York

Bight hypoxic event in the summer of 1976 were over US$570 million (Figley et al., 1979). Much of this loss was in surf clams (*Spisula solidissima*) that accounted for >US$430 million. Factored by the area of hypoxia (987 km^2), the 1976 event cost about US $580,000 per km^2 for resources and fisheries-related activities and US$165,000 per km^2 for just the resources lost.

Lack of identifiable economic effects in fisheries landing data does not imply that declines would not occur should conditions worsen. In the northern Gulf of Mexico, brown shrimp landing appears to be inversely related to the area of hypoxia (O'Connor and Whitall, 2007). Whether this relationship will remain linear or transform into a catastrophe function at some critical point is not known (Kemp et al., 2009). Other large systems have suffered serious ecological and economic consequences from seasonal hypoxia; most notable are the Kattegat with localized loss of catch and recruitment failures of Norway Lobsters (*Nephrops norvegicus*) in the late 1980s and northwest continental shelf of the Black Sea which suffered regional loss of bottom fishery species also in the 1980s (Karlson et al., 2002; Mee, 1992; Mee, 2006). With the abatement of hypoxia in the early 2000s, Black Sea benthic communities and fish stocks are recovering but the trophic relationships and transfers are not.

Economic valuation of losses from hypoxia seems small relative to the total value of fisheries, but the key point is that losses from hypoxia are measureable in economic terms (Huang and Smith, 2011). For example, the valuation of recreational fishing relative to hypoxia in the Patuxtant River, a tributary in Chesapeake Bay, showed that as oxygen declined to mild hypoxic levels total losses for striped bass (*Morone saxatilis*) fishing was about US$10,000 with a net present value of about US$200,000. If the same water quality was allowed to occur in the entire Chesapeake Bay, the net present value of the losses due to hypoxia would be >US$145 million (Lipton and Hicks, 2003). A similar analysis of recreational fishing in northeast and middle Atlantic regions found that, overall, as oxygen declined, capture rate of fish declined (Bricker et al., 2006). The finding of effects of hypoxia on recreational fishing for one species would lead one to believe that similar effects are likely being experienced by commercial fishers. This has been documented for the brown shrimp fisheries in North Carolina. Huang et al. (2010) found that hypoxia may reduce annual harvest by about 13%, valued at about 1.2 million dollars annually.

Experience with other hypoxic zones around the globe shows that both ecological and fisheries effects become progressively more severe as hypoxia worsens (Caddy 1993; Diaz and Rosenberg 1995). This would lead one to believe that at some point economic loses will become more obvious and costly. However, currently, the direct connection of hypoxia to fisheries landings at large regional scales is weak, likely related to a number of factors that include confounding effects of eutrophication, overfishing, and compensatory mechanisms that alter or mask effects of hypoxia on landings. Breitburg et al. (2009) found a hint of a connection with a possible decline in landings of benthic species in systems where $\geq$40% of the bottom area becomes hypoxic.

The most pervasive and measurable effects of hypoxia are on biogeochemical processes, which support ecosystem services of water quality and nutrient cycling (Millennium Ecosystem Assessment, 2005). Oxygen is a key regulator of these two services with hypoxia disrupting most biogeochemical processes. For example, the cycling of nitrogen is complex and dependent on the concentration of oxygen (Middelburg and Levin, 2009). Following the Costanza et al. (1997) approach to valuing global ecosystem services, which is problematic for many economists (Ackerman, Chapter 10), at least an order of magnitude approximation of the impacts of hypoxia can be made. Costanza et al. (1997) estimated that marine ecosystems provide at least US$21 trillion worth of services annually, with the majority of this total outside market valuation in contributions from waste treatment and nutrient cycling. Hypoxia covers about 4% (240,000 km^2) of all coastal and estuarine habitats annually (Diaz and Rosenberg, 2008). If we assume a complete loss in water quality and nutrient cycling services to hypoxia, the amount of loss is approximately US$170 billion. This represents about 0.8% of all marine ecosystems services.

GLOBAL CHANGE AND HYPOXIA

Since the early 2000s, many assessments of global environmental and resource health identify hypoxia as one of the factors threatening coastal and ocean life, for example, the Millennium Ecosystem Assessment (2005). In addition, climate model predictions and observations reveal regional declines in oceanic dissolved oxygen linked to global warming (Matear and Hirst, 2003; Gilbert et al., 2009; Deutsch et al., 2011). Understanding hypoxia and its effects on ecosystems requires several perspectives that start at a local level, move to regional, and finally to a global perspective. The most important scale is local for stressors like coastal development, nutrients, pollution, and eutrophication. This is the scale (<1 to 1000 km^2) at which impacts are most pronounced and we have the most information. At regional scales (>1000 to 1,000,000 km^2), it is a mix of influences from land and sea. This involves local land-based impacts and processes bleeding into regional seas, and large-scale open ocean processes that are at the boundaries of regions, such as upwelling and thermocline depth. Much of the problem with hypoxia at local and regional levels can be directly tied to concentrations of human populations and agriculture, both of which have significantly altered the global nitrogen cycle (Figure 4.1). Global scale factors that influence oxygen and hypoxia are changes in circulation patterns, climate, temperature, and pH.

Climate change, whether from global warming or from microclimate variation, will have consequences for eutrophication-related oxygen depletion that will progressively led to an onset of hypoxia earlier in the season and possibly extending it through time. The influence of multiple climate drivers needs to be considered to understand what future change to expect (Table 4.4). Climate change may make systems more susceptible to the development of hypoxia through its direct effects on water column stratification, precipitation patterns, and temperature. These effects will likely occur primarily though warming, which will lead to increased water temperatures and subsequent decrease in oxygen solubility. Warmer surface waters will extend and enhance water column stratification, a key factor in the development of hypoxia. Warmer water will increase organism metabolism, which is the key process for lowering oxygen concentrations. In addition to warming, future climate predictions include large changes in precipitation patterns. If changes in precipitation lead to increased runoff to estuarine and coastal ecosystems, stratification and nutrient loads are likely to increase and worsen oxygen depletion (Justić et al., 2007; Najjar et al., 2010). Conversely, if stratification decreases due to lower runoff or is disrupted by increased storm activity or intensity, the chances for oxygen depletion should decrease (Table 4.4).

TABLE 4.4 Influence of Climate Drivers on the Extent and Severity of Hypoxia

Climate Driver	Direct Effect	Secondary Effect	Influence on Hypoxia
Increased temperature	More evaporation	Decreased stream flow	+
		Land use and cover changes	+/−
	Less snow cover	More nitrogen retention	−
	Warmer water	Stronger stratification	+
		Higher metabolic rates	+
More precipitation	More stream flow	Stronger stratification	+
		More nutrient loading	+
	More extreme rainfall	Greater erosion of soil P	+
Less precipitation	Less stream flow	Weaker stratification	−
		Less nutrient loading	−
Higher sea level	Greater depth	Stronger stratification	+
		Greater bottom water volume	−
		Less hydraulic mixing	+
	Less tidal marsh	Diminished nutrient trapping	+
Summer winds and storms	Weaker, less water column mixing	More persistent stratification	+
	Stronger, more water Column mixing	Less persistent stratification	−
	Shifting wind patterns	Weaker/stronger upwelling potential	+/−

Modified from Boesch et al. (2007).

Much of how climate change will affect hypoxia in the coastal zone will depend on coupled land-sea interactions with climate drivers (Table 4.4). But the future pervasiveness of hypoxia will also be linked between management practices and expansion of agriculture. Land management will affect the nutrient budgets and concentrations of nutrients applied to land through agriculture. If in the next 50 years humans continue to modify and degrade coastal systems as in previous years (Halpern et al., 2008), human population pressure will likely continue to be the main driving factor in the persistence and spreading of coastal dead zones. For example, the expansion of agriculture for the production of crops to be used for food and biofuels will result in increased nutrient loading and expand eutrophication effects (Rabalais et al., 2007). Overall, climate drivers will tend to magnify the effects of expanding human population.

Climate-related changes in wind patterns are of great concern for coastal system as wind direction and strength influence the strength of upwelling/downwelling, which in turn affect stratification strength and delivery of less-oxygenated deepwater and nutrients into shallow coastal areas. Even relatively small changes in wind and current circulations could lead to large changes in the area of coastal seabed exposed to hypoxia. Changes in the pattern of upwelling on Pacific coast off the Oregon and Washington coasts due to shifts in winds that affected the California current systems appeared to be responsible for the recent development of severe hypoxia over a large area of the inner continental shelf (Grantham et al., 2004; Chan et al., 2008).

Thus the future status of hypoxia and its consequences for the environment, society, and economies will depend on a combination of climate change (primarily from warming, and altered patterns for wind, currents, and precipitation) and land-use change (primarily from expanded human population, agriculture, and nutrient loadings). Our expanding population has led to intensification of agriculture/aquaculture and threatens local and regional coastal systems everywhere. Expanding energy demands associated with our population now drive climate change that threatens all ecosystems at all scales from local to global.

RESTORATION AND THE FUTURE OF HYPOXIA

Recognizing the negative consequences of hypoxia and related coastal eutrophication, many nations have recently made major socioeconomic commitments for reducing nutrient loads to the adjacent estuaries, bays, and seas, upon which they depend. Worldwide, of the over 500 coastal hypoxia systems, about 55 have responded positively to remediation. All but one improved from management of point discharges. The northwest continental shelf of the Black Sea is the only exception, responding positively to reduction on nutrient runoff. Once the second largest anthropogenic hypoxic area on earth, it is now remediated through concerted efforts to reduce point discharges and runoff from agricultural lands (Mee, 2006; Langmead et al., 2008).

Much of the scientific interest and management concern about hypoxia include a focus on the general problem of eutrophication caused by the overenrichment of coastal waters from a combination of sewage/industrial discharge and nutrient runoff. Nutrients generally increase biological production, while hypoxia acts in the opposite direction, reducing biomass and habitat quality. Overall, the combination of stressors associated with eutrophication has and continues to degrade our coastal systems. The future effects of hypoxia and nutrient enrichment on food webs and fisheries will be strongly influenced by the extent to which these two factors co-occur. Unless the leakage of nutrients from land-based sources to the sea can be reduced, the future for our estuarine and coastal resources looks bleak. Where applied, nutrient management has reversed the effects of hypoxia. But concerted effort in the future will be needed to allow more systems to recover, particularly for those systems affected primarily by land runoff.

In addition to the complexity of interactions among physical, chemical, and biological processes, two key elements hinder a global assessment of the effects of hypoxia:

(1) Lack of details on the timing of occurrence and area covered by hypoxia
(2) Accessibility of data on various ecosystem components.

For hypoxia in Europe and North America, much is known about its occurrence in coastal areas, including spatial and temporal patterns. Less is known from the other continents, which have most of our population, and Pacific islands. For all systems, less is known about long-term trends, factors controlling dissolved oxygen depletion and replenishment, impacts on ecological processes, and economic losses. The World Resources Institute eutrophication project has made a start by compiling data on over 500 hypoxic areas and another 300 areas that are eutrophic and in danger of becoming hypoxic (Diaz et al., 2010). A similar effort by Conley et al. (2011) for the Baltic Sea coastal regions identified another 100 hypoxic areas. The conclusion of Conley et al. (2011) for the Baltic could easily be applied to the rest of the globe: "The Baltic Sea coastal zone displays an alarming trend with hypoxia steadily increasing with time since the 1950s effecting nutrient biogeochemical processes, ecosystem services, and coastal habitat." To formulate effective strategies for remediating coastal hypoxia, it is essential to have an understanding of what the specific drivers are and what responses to expect with various remediation approaches. Some of these drivers are nationally controlled, while others lack defined ownership/responsibility. However, management of hypoxia including the control of drivers of eutrophication is often a transboundary issue.

Acknowledgments

This work was supported by a grant from the Okeanos Foundation (http://www.okeanos-foundation.org/) to the Stockholm Environment Institute. We are grateful to anonymous reviewers for insightful comments that improved the quality of the chapter.

References

Alve, E. (1995), Benthic foraminiferal distribution and recolonization of formerly anoxic environments in Dramrnensfjord, southern Norway, *Mar. Micropaleontol.*, **25**, 169–186.

Baird, D., R. R. Christian, C. H. Peterson, and G. A. Johnson (2004), Consequences of hypoxia on estuarine ecosystem function: Energy diversion from consumers to microbes, *Ecol. Appl.*, **14**, 805–822.

Banse, K. (1959), On upwelling and bottom trawling off the Southwest coast off India, *Journal of the Marine Biological Association of India*, **1**, 33–49.

Benson, B. B., and D. Krause (1984), The concentration and isotopic fractionation of gases dissolved in freshwater in equilibrium with the atmosphere: 1. Oxygen, *Limnol. Oceanogr.*, **25**, 662–671.

Boesch, D. F., V. J. Coles, D. G. Kimmel, and W. D. Miller (2007): Ramifications of climate change for Chesapeake Bay hypoxia. In: *Regional impacts of climate change, four case studies in the United States, Pew Center on Global Climate Change* [Center Pew (ed.)]. Arlington, Virginia, 57–70.

Breitburg, D. L., D. W. Hondorp, L. W. Davias, and R. J. Diaz (2009), Hypoxia, nitrogen and fisheries Integrating effects across local and global landscapes, *Annual Review Marine Science*, **1**, 329–350.

Bricker, S., D. Lipton, A. Mason, M. Dionne, D. Keeley, C. Krahforst, J. Latimer, and J. Pennock (2006), *Improving methods and indicators for evaluation coastal water eutrophication: a pilot study in the Gulf of Maine, NOAA Technical Memorandum NOS NCCOS 20,* 81 pp., National Ocean Survey, Silver Springs, Maryland.

Brill, R. W. (1996), Selective advantages conferred by the high performance physiology of tunas, billfishes, and dolphin fish, *Comp. Biochem. Physiol.*, **113**, 3–15.

Caddy, J. F. (1993), Towards a comparative evaluation of human impacts on fishery ecosystems of enclosed and semi-enclosed seas, *Review of Fisheries Science*, **1**, 57–95.

Chan, F., J. Barth, J. Lubchenco, J. Kirincich, A. Weeks, H. Peterson, W. T. Mengl, and B. A. Chan (2008), Emergence of anoxia in the California Current Large Marine Ecosystem, *Science*, **319**, 920.

Cloern, J. E. (2001), Our evolving conceptual model of the coastal eutrophication problem, *Mar. Ecol. Prog. Ser.*, **210**, 223–253.

Conley, D. J., J. Carstensen, J. Aigars, P. Axe, E. Bonsdorff, T. Eremina, B. -M. Haahti, C. Humborg, P. Jonsson, J. Kotta, C. Lännegren, U. Larsson, A. Maximov, M. Rodriguez Medina, E. Lysiak-Pastuszak, N. Remeikaité-Nikiené, J. Walve, S. Wilhelms, and L. Zillén (2011), Hypoxia Is Increasing in the Coastal Zone of the Baltic Sea, *Environ. Sci. Technol.*, **45**, 6777–6783.

Costanza, R., R. d'Arge, R. de Groot, S. Farber, M. Grasso, B. Hannon, S. Naeem, K. Limburg, J. Paruelo, R. V. O'Neill, R. Raskin, P. Sutton, and M. van den Belt (1997), The value of the world's ecosystem services and natural capital, *Nature*, **387**, 253–260.

Coutant, C. C. (1990), Striped bass, temperature, and dissolved oxygen: a speculative hypothesis for environmental risk, *Trans. Am. Fish. Soc.*, **114**, 31–61.

Deutsch, C., H. Brix, T. Ito, H. Frenzel, and L. Thompson (2011), Climate-Forced Variability of Ocean Hypoxia, *Science*, **333**, 336–339.

Diaz, R. J., and R. Rosenberg (1995), Marine benthic hypoxia: A review of its ecological effects and the behavioural responses of benthic macrofauna, *Oceanogr. Mar. Biol. Annu. Rev.*, **33**, 245–303.

Diaz, R. J., and R. Rosenberg (2008), Spreading dead zones and consequences for marine ecosystems, *Science*, **321**, 926–928.

Diaz, R., M. Selman, and C. Chique (2010), Global Eutrophic and Hypoxic Coastal Systems, World Resources Institute. Eutrophication and Hypoxia: Nutrient Pollution in Coastal Waters, http://www.wri.org/project/eutrophication

Ekau, W., H. Auel, H. -O. Pörtner, and D. Gilbert (2009), Impacts of hypoxia on the structure and processes in the pelagic community (zooplankton, macro- invertebrates and fish), *Biogeosciences Discussion*, **6**, 5073–5144.

Figley, W., B. Pyle, and B. Halgren (1979): Socioeconomic impacts. In: *Oxygen depletion and associated benthic moralities in New York Bight, 1976*, NOAA Professional Paper 11 [Swanson R. L. and C. J. Sindermann (eds.)]. U.S. Government Printing Office, Washington, D.C., 315–322.

Foley, J. A., R. DeFries, G. P. Asner, C. Barford, G. Bonan, S. R. Carpenter, F. S. Chapin, M. T. Coe, G. C. Daily, H. K. Gibbs, J. H. Helkowski, T. Holloway, E. A. Howard, C. J. Kucharik, C. Monfreda, J. A. Patz, I. C. Prentice, N. Ramankutty, and P. K. Snyder (2005), Global consequences of land use, *Science*, **309**, 570–574.

Fry, F. E. J. (1971), *The effect of environmental factors on the physiology of fish*, Academic Press, New York, NY.

Fuenzalida, R., W. Schneider, J. Graces-Vargas, L. Bravo, and C. Lange (2009), Vertical and horizontal extension of the oxygen minimum zone in the eastern South Pacific Ocean, *Deep-Sea Res. II*, **56**, 992–1003.

Graham, J. B. (1990), Ecological, evolutionary, and physical factors influencing aquatic animal respiration, *Am. Zool.*, **30**, 137–146.

Grantham, B. A., F. Chan, K. J. Nielsen, D. S. Fox, J. A. Barth, A. Huyer, J. Lubchenco, and B. A. Menge (2004), Upwelling-driven nearshore hypoxia signals ecosystem and oceanographic changes in the northeast Pacific, *Nature*, **429**, 749–754.

Galloway, J. N., A. R. Townsend, J. W. Erisman, M. Bekunda, Z. Cai, J. R. Freney, L. A. Martinelli, S. P. Seitzinger, and M. A. Sutton (2008), Transformation of the nitrogen cycle Recent trends, questions, and potential solutions, *Science*, **320**, 889–892.

Gilbert, D., N. N. Rabalais, R. J. Diaz, and J. Zhang (2009), Evidence for greater oxygen decline rates in the coastal ocean than in the open ocean, *Biogeosciences Discussion*, **6**, 9127–9160.

Gruber, N., and J. N. Galloway (2008), An earth-system perspective of the global nitrogen cycle, *Nature*, **451**, 293–296.

Halpern, B. S., S. Walbridge, K. A. Selkoe, C. V. Kappel, F. Micheli, C. D'Agrosa, J. F. Bruno, K. S. Casey, C. Ebert, H. E. Fox, R. Fujita, D. Heinemann, H. S. Lenihan, E. M. P. Madin, M. T. Perry, E. R. Selig, M. Spalding, R. Steneck, and R. Watson (2008), A global map of human impact on marine ecosystems, *Science*, **319**, 948–952.

Huang, L., and M. D. Smith (2011), Management of an annual fishery in the presence of ecological stress: The case of shrimp and hypoxia, *Ecological Economics*, **70**, 688–697.

Huang, L., M. D. Smith, and J. K. Craig (2010), Quantifying the Economic Effects of Hypoxia on a Fishery for Brown Shrimp Farfantepenaeus aztecus, *Marine and Coastal Fisheries: Dynamics, Management, and Ecosystem Science*, **2**, 232–248.

Helly, J. J., and L. A. Levin (2004), Global distribution of naturally occurring marine hypoxia on continental margins, *Deep-Sea Res. I*, **51**, 1159–1168.

Jones, P. D. (2006), Water quality and fisheries in the Mersey estuary, England: A historical perspective, *Mar. Pollut. Bull.*, **53**, 144–154.

Justic', D., V. J. Bierman, Jr, D. Scavia, and R. D. Hetland (2007), Forecasting gulf's hypoxia: the next 50 years? *Estuaries and Coasts*, **30**, 791–801.

Kamykowski, D., and S. J. Zentara (1990), Hypoxia in the world ocean as recorded in the historical data set, *Deep-Sea Res. I*, **37**, 1861–1874.

Karlson, K., R. Rosenberg, and E. Bonsdorff (2002), Temporal and spatial large-scale effects of eutrophication and oxygen deficiency on benthic fauna in Scandinavian and Baltic waters–a review, *Oceanogr. Mar. Biol. Annu. Rev.*, **40**, 427–489.

Keeling, R. F., A. Körtzinger, and N. Gruber (2010), Ocean deoxygenation in a warming world, *Annual Review of Marine Science*, **2**, 199–229.

Kemp, W. M., J. M. Testa, D. J. Conley, D. Gilbert, and J. D. Hagy (2009), Temporal responses of coastal hypoxia to nutrient loading and physical controls, *Biogeosciences*, **6**, 2985–3008.

Langmead, O., A. McQuatters-Gollop, L. D. Mee, J. Friedrich, A. J. Gilbert, M. -T. Gomoiu, E. L. Jackson, S. Knudsen, G. Minicheva, and V. Todorova (2008), Recovery or decline of the northwestern Black Sea: a societal choice revealed by socio-ecological modelling, *Ecological Modelling*, **220**, 2927–2939.

Levin, L. A. (2003), Oxygen minimum zone benthos adaptation and community response to hypoxia, *Oceanogr. Mar. Biol. Annu. Rev.*, **41**, 1–45.

Levin, L. A., W. Ekau, A. J. Gooday, F. Jorissen, J. J. Middelburg, S. W. A. Naqvi, C. Neira, N. N. Rabalais, and J. Zhang (2009), Effects of natural and human-induced hypoxia on coastal benthos, *Biogeosciences*, **6**, 2063–2098.

Lipton, D., and R. Hicks (2003), The cost of stress: low dissolved oxygen and economic benefits of recreational striped bass (*Morone saxatilis*) fishing in the Patuxent River, *Estuaries*, **26**, 310–315.

Matear, R. J., and A. C. Hirst (2003), Long-term changes in dissolved oxygen concentrations in the ocean caused by protracted global warming, *Global Biogeochem. Cycles*, **17**, 1125. http://dx.doi.org/10.1029/2002GB001997.

May, E. (1973), Extensive oxygen depletion in Mobile Bay, Alabama, *Limnol. Oceanogr.*, **18**, 353–366.

Mee, L. (2006), Reviving dead zones, *Sci. Am.*, **295**, 78–85.

Mee, L. D. (1992), The Black Sea in crisis: a need for concerted international action. *Ambio*, **21**, 278–286.

Middelburg, J., and L. A. Levin (2009), Coastal hypoxia and sediment biogeochemistry, *Biogeosciences*, **6**, 1273–1293.

Millennium Ecosystem Assessment (2005), *Ecosystems and Human Well-being: Synthesis*, 155 pp., Island Press, Washington, DC.

Monteiro, P., A. van der Plas, J. -L. Melice, and P. Florenchie (2008), Interannual hypoxia variability in a coastal upwelling system Ocean-shelf exchange, climate and ecosystem-state implications, *Deep-Sea Res. I*, **55**, 435–450.

Najjar, R. G., C. R. Pyke, M. B. Adams, D. Breitburg, M. Kemp, C. Hershner, R. Howarth, M. Mulholland, M. Paolisso, D. Secor, K. Sellner, D. Wardrop, and R. Wood (2010), Potential climate-change impacts on the Chesapeake Bay, *Estuar. Coast. Shelf Sci.*, **86**, 1–20.

Naqvi, S. W. A., H. Naik, A. Pratihary, W. D'Souza, P. V. Narvekar, D. A. Jayakumar, A. H. Devol, T. Yoshinari, and T. Saino (2006), Coastal versus open-ocean denitrification in the Arabian Sea, *Biogeosciences*, **3**, 621–633.

O'Connor, T., and D. Whitall (2007), Linking hypoxia to shrimp catch in the northern Gulf of Mexico, *Mar. Pollut. Bull.*, **54**, 460–463.

Pauly, D., and V. Christensen (1995), Primary production required to sustain global fisheries, *Nature*, **374**, 255–257.

Prince, E. D., and C. P. Goodyear (2006), Hypoxia-based habitat compression of tropical pelagic fishes, *Fish. Oceanogr.*, **15**, 451–464.

Rabalais, N. N., R. E. Turner, B. K. SenGupta, D. F. Boesch, P. Chapman, and M. C. Murrell (2007), Hypoxia in the Northern Gulf of Mexico: Does the Science

Support the Plan to Reduce, Mitigate, and Control Hypoxia? *Estuaries and Coasts*, **30**, 753–772.

Rabalais, N. N., R. E. Turner, D. Justic', and R. J. Diaz (2009), Global change and eutrophication of coastal waters, *ICES J. Mar. Sci.*, **66**, 1528–1537.

Rabalais, N. N., R. J. Diaz, L. A. Levin, R. E. Turner, D. Gilbert, and J. Zhang (2010), Dynamics and distribution of natural and human-caused hypoxia, *Biogeosciences*, **7**, 585–619.

Rockström, J., W. Steffen, K. Noone, Å. Persson, F. S. Chapin, E. F. Lambin, T. M. Lenton, M. Scheffer, C. Folke, H. J. Schellnhuber, B. Nykvist, C. A. de Wit, T. Hughes, S. van der Leeuw, H. Rodhe, S. Sörlin, P. K. Snyder, R. Costanza, U. Svedin, M. Falkenmark, L. Karlberg, R. W. Corell, V. J. Fabry, J. Hansen, B. Walker, D. Liverman, K. Richardson, K. Crutzen, and J. A. Foley (2009), A safe operating space for humanity, *Nature*, **461**, 472–476.

Rose, K. A., A. T. Adamack, C. A. Murphy, S. E. Sable, S. E. Kolesar, J. K. Craig, D. L. Breitburg, P. Thomas, M. H. Brouwer, C. F. Cerco, and S. Diamond (2009), Does hypoxia have population-level effects on coastal fish? Musings from the virtual world, *J. Exp. Mar. Biol. Ecol.*, **381**, S188–S203.

Stramma, L., G. C. Johnson, J. Sprintall, and V. Mohrholz (2008), Expanding oxygen-minimum zones in the tropical oceans, *Science*, **320**, 655–658.

Stramma, L., E. D. Prince, S. Schmidtko, J. Luo, J. P. Hoolihan, M. Visbeck, D. W. R. Wallace, P. Brandt, and A. Körtzinger (2012), Expansion of oxygen minimum zones may reduce available habitat for tropical pelagic fishes, *Nature Climate Change*, **2**, 33–37.

Vaquer-Sunyer, R., and C. M. Duarte (2008), Thresholds of hypoxia for marine biodiversity, *Proc. Natl. Acad. Sci. U S A.*, **105**, 15452–15457.

5

Sea-Level Rise

Kevin J. Noone

Department of Applied Environmental Science, Stockholm University, Stockholm, Sweden

INTRODUCTION

The main goal of this chapter is to briefly describe the current state of knowledge regarding the causes of sea-level rise, summarize recent estimates of how much sea level may rise by the end of the century, and present some of the impacts of sea-level rise on society.

Research on sea-level rise has made very substantial progress in the last few years. It is not my intention to try to provide a very detailed accounting of recent scientific advances. Such a detailed and authoritative summary can be found in Church et al. (2010), chapters of which are referenced extensively here. Another recent reference is Gehrels (2009), which concentrates on the rate of sea-level rise. Here, I will attempt to reflect the recent scientific advances as accurately as I can, while trying to attain a level of detail appropriate for nonspecialists.

CAUSES OF SEA-LEVEL RISE

Changes in mean sea level at any given location are a combination of a number of processes:

- Warming seawater causing a decrease in its density (thermal expansion)

- Adding water to the oceans from the continents through melting or transport of ice from glaciers and ice caps, and transfer of groundwater
- Changes in ocean and atmospheric circulations
- Uplift or subsidence due to tectonic movements or rebound from the loss of glacier ice cover from the last ice age (glacial isostatic adjustment, GIA)
- Natural and human-induced subsidence (often through changes in groundwater levels)

All these processes act together in differing degrees at different locations to alter the relative level between land and sea. For this reason, it is important to distinguish between *absolute* and *relative sea-level change (RSLC)*. *Absolute* sea-level change refers to change measured by satellites relative to the center of the earth. RSLC refers to change measured relative to the land by tide gauges. Both are negative (falling) in some parts of the world and positive (rising) in others. Both are influenced by changes in ocean volume, mass, and mass redistribution but differ locally because of differences in vertical land motion (e.g., land rebound from melted ice sheets, movement along faults, groundwater removal, dams, and reservoirs).

Absolute Versus Relative Sea-Level Change

Given the different processes that influence mean sea level, it is important to keep in mind that changes in sea level at any given location may be very different from the global average. The impacts of changes in sea level will also depend on local conditions, rather than the global mean (Boon, 2012; Sallenger et al., 2012). Figure 5.1 shows measurements of sea level from a number of coastal cities around the world.

Observations for all of the locations show different degrees of variability and different overall trends. Seattle, Honolulu, and Wellington all show a fair amount of interannual variability, but clear upward trends in sea level. In contrast, relative sea level in Stockholm has decreased by roughly 6% over the last century. The fact that the relative sea level in Stockholm (or some other specific locations) is decreasing is not inconsistent with an increase in

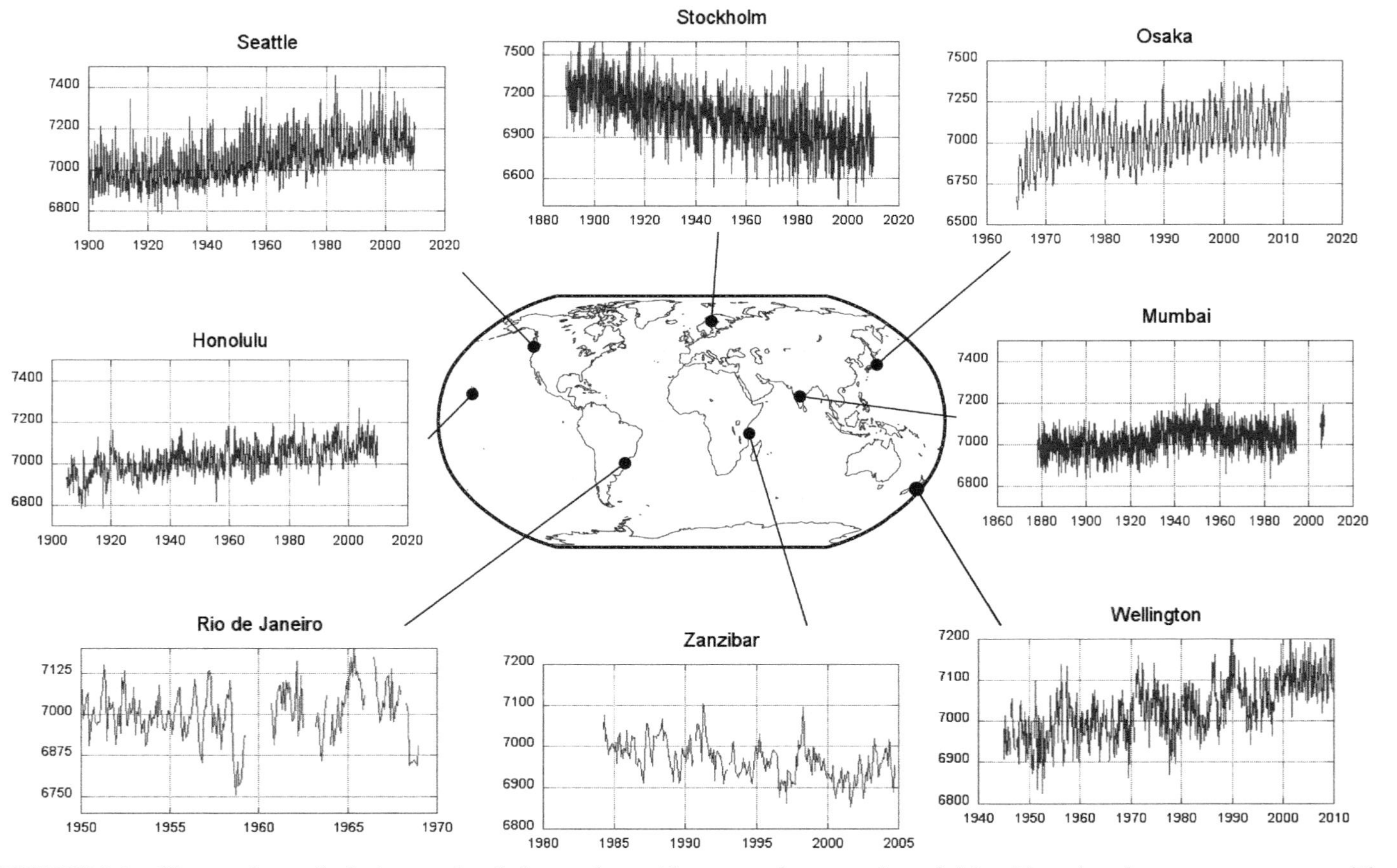

FIGURE 5.1 Observations of relative sea-level change from tide gauges for a number of cities. Note that the measurements at different locations started at different times. *Source: Permanent service for mean sea level (http://www.psmsl.org/) accessed March 2011.*

global sea level. The reason for the decrease in Stockholm is because the land itself has been rising faster than sea level in this period. The Scandinavian Peninsula is still rebounding after the disappearance of the ice sheet that covered it during the latest ice age.

Thermal Expansion/Density Changes

As the surface temperature of the Earth increases, some of the energy associated with the warming is transferred into the oceans. Recent advances in accounting for biases and multiple sources of uncertainty in measurements from *expendable bathythermographs* (XBTs—instruments that measure profiles of temperature in the ocean) have shown a significant warming trend of 0.64 W m^{-2} between 1993 and 2008 (Lyman et al., 2010). This amounts to a substantial increase in the heat content of the oceans, shown in Figure 5.2, taken from Trenberth (2010).

Given the huge volume of water in the oceans, even the small thermal expansion caused by heating water can result in substantial changes in sea level. The uppermost 4 m of the ocean can store as much heat as the entire atmosphere. Because of its large thermal inertia, it will take a very long time for the upper ocean to "lose" the energy now being stored there. Domingues et al. (2008)

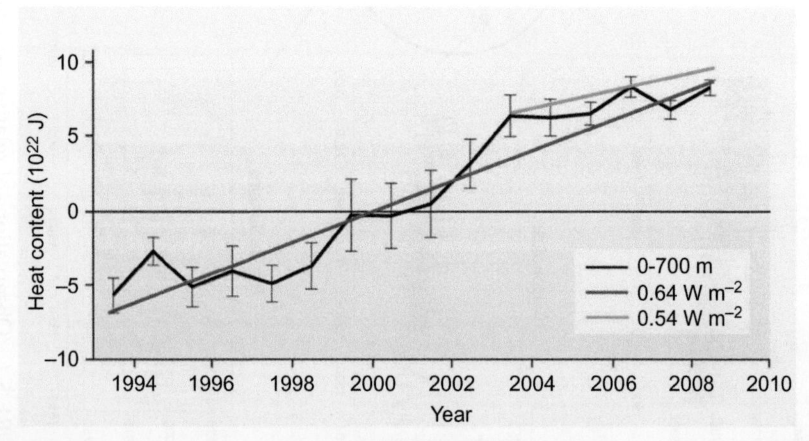

FIGURE 5.2 Time series of oceanic heat content from the surface to 700 m (Trenberth, 2010). *Reprinted with permission from* Nature.

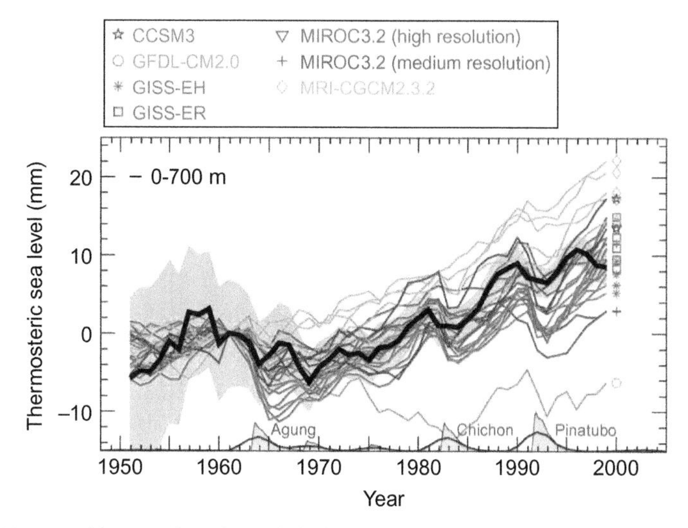

FIGURE 5.3 Observed and modeled global average thermal expansion of the upper 700 m of the ocean. *Adapted from Domingues et al. (2008).*

compared observations of the thermal expansion of ocean water with model calculations over the period 1950-2000 (shown in Figure 5.3).

Figure 5.3 shows changes in the thermosteric sea level (changes due to thermal expansion) over time from 1950 to 2000. In the figure, the thick black line shows the mean values derived from observations, and the gray-shaded area represents one standard deviation error estimates (Domingues et al., 2008). Results from several model calculations[1] are shown in the figure. All models include both natural variations (such as volcanic eruptions) and human-induced changes in the climate system. Stratospheric aerosol loadings from major volcanic eruptions (on an arbitrary scale) are shown at the bottom of the figure.

The observations show decadal-scale variability over the entire period and an upward trend since roughly 1970. Most of the models

[1] Models: National Center for Atmospheric Research, Boulder, CO (CCSM3); Geophysical Fluid Dynamics Laboratory, Princeton, NJ (GFDL-CM2.0); Goddard Institute for Space Studies in New York (GISS-EH, GISS-ER); Frontier Research Center for Global Change in Japan (MIROC-CGCM2.3.2 medium resolution, MIROC-CGCM2.3.2 high resolution); Meteorological Research Institute in Japan (MRI-CGCM2.3.2).

(CCSM3, GISS, MIROC3.2) capture both the trend and much of the interannual variability that is seen in the observations—though with more within- and between-model variation than is seen in the measurements. One model (the GFDL-CM2.0) does not show the same trend or interannual variability as the observations or the other models. The MRI-CGCM2.3.2 model does show an increasing trend in thermal expansion, but does not capture the same interannual variability that is seen in the observations.

While dipping perhaps a bit too deeply into details, this comparison illustrates a number of important points: (1) both observations and model calculations (with one exception) show that the oceans have been expanding due to heating over the 1950-2000 period; (2) there is clear interannual and decadal-scale variability in thermal expansion during this period; (3) the cooling effects of major volcanic eruptions can be seen in both measurements and model calculations; (4) while consistent with the observations, the differences between the various model calculations show that more work needs to be done to improve the descriptions of heat transport to and within the oceans.

Interestingly, both measurements and most of the models show decreases in thermal expansion following major volcanic eruptions—consistent with the observations of surface temperature and our understanding of how these volcanic eruptions influence the Earth's energy balance.

Glacier and Ice Sheet Melting

Freshwater is stored on the continents in the form of small glaciers and ice caps, as well as the very large ice sheets covering most of Greenland and Antarctica. While we take the major ice sheets for granted, they were not always present. The Antarctic ice sheet started to form around 35 million years before present, and the Greenland ice sheet "only" a few million years ago (Zachos et al., 2001) when atmospheric CO_2 concentrations decreased during the Late Pliocene era (Lunt et al., 2008). For the last few million years—during the period of human evolution—they have played a critical role in the Earth's climate. Currently, glaciers and ice sheets contain 29 million km^3 of ice (shown in Figure 5.4), mostly in Antarctica and Greenland.

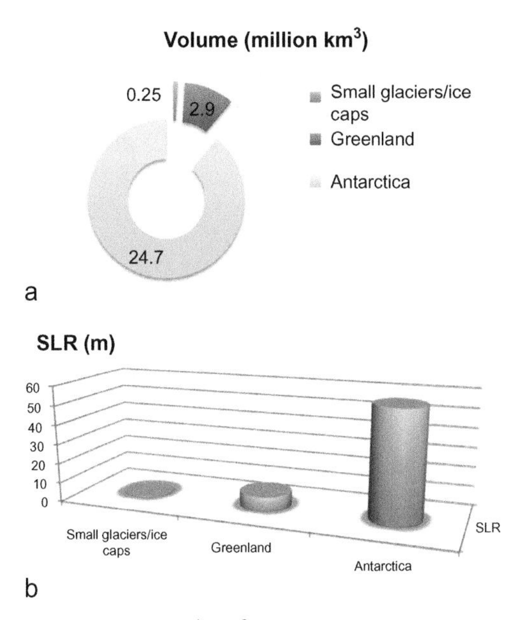

FIGURE 5.4 (a) Ice volume (10^6 km^3). (b) Sea-level rise if all ice were to melt.

Put into perspective, if all this ice were to melt, small glaciers and ice caps would cause sea level to rise by 0.6 m, the Greenland ice sheet would raise sea levels by 7.3 m, and melting all the ice on Antarctica would cause a 56.6-m increase in sea level (Steffen et al., 2010).

How rapidly the major ice sheets will respond to increases in global mean surface temperature remains unclear. Measurements of the gain or loss of ice mass in Greenland and Antarctica since the 1990s suggest that dynamical process other than simple melting has caused acceleration of the transport of water mass from the continents to the ocean. For example, meltwater can percolate down to the base of glaciers and act as a lubricant, causing the glaciers to flow more rapidly. Ice shelves (large expanses of ice partially floating on seawater but anchored to the land) act as a kind of cork, holding back the glaciers that feed into them. If they thin or break up (such as the Larsen B ice shelf in 2002—an area roughly the size of the US state of Rhode Island), then the glaciers flowing into them can flow faster, transferring ice into the oceans more rapidly. Were this trend to continue into the future, the response times of the major ice sheets would be more rapid than our previous estimates (Steffen et al., 2010).

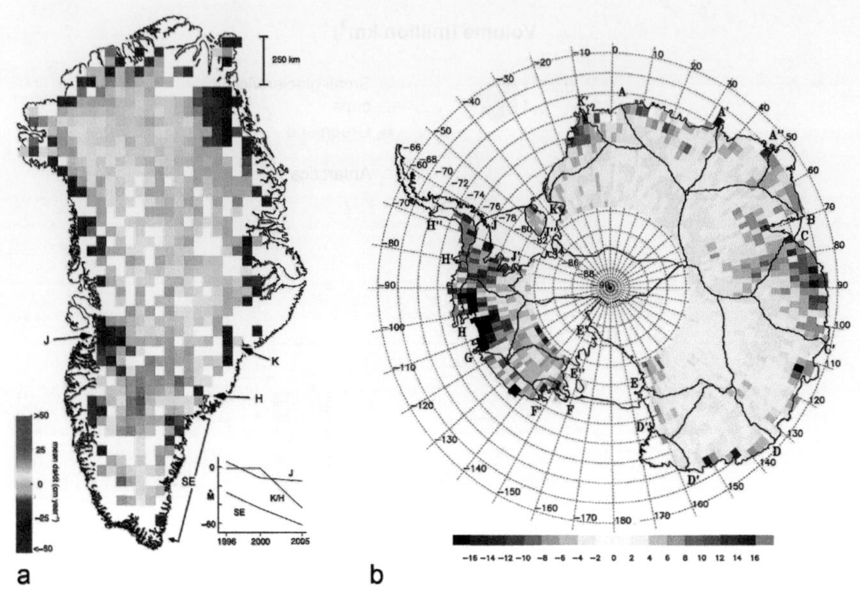

FIGURE 5.5 Change in elevation (cm $year^{-1}$) for (a) Greenland and (b) Antarctica. *Reprinted from Lemke et al. (2007) and Davis et al. (2005).*

Figure 5.5 shows satellite measurements of the change in elevation of the Greenland (panel a; Lemke, 2007) and the Antarctic ice sheets (panel b; Davis et al., 2005).

Both Greenland and Antarctica have areas that have gained mass and others that have lost. Greenland tends to increase in mass at the high-elevation center of the ice sheet and lose it at the edges, while Antarctica has gained mass in the eastern regions while losing it in the west. Gains in mass are caused by increases in precipitation, while losses are caused by melting or the physical flow of ice from the land into the ocean.

In terms of sea-level rise, it is the net gain or loss of mass that is important. Steffen et al. (2010) summarize the rate of mass loss for Greenland and Antarctica from a number of studies, examples of which are shown in Table 5.1.

For Greenland, the mass loss increased by a factor of 7 in the decade between the mid-1990s and the mid-2000s. For Antarctica, the ice loss nearly doubled in the same decade, almost entirely due to changes in West Antarctica and the Peninsula, with little change in East Antarctica.

TABLE 5.1 Mass Loss (Gt Year^{-1}) from Greenland and Antarctica for Different Time Periods

Time Period	Greenland	Time Period	East Antarctica	West Antarctica	Antarctic Peninsula
1994-1999	−27	1996-2004	+4(±61)	−106(±60)	−28(±45)
1998-2004	−80	1996		−86(±59)	
ca. 2000	−100	2006		−132(±60)	−60(±46)
ca. 2005	−200				

Melting glaciers also contribute to sea-level rise. Observations for over 300 glaciers around the world have been compiled and made available. An example of a global analysis from data in Dyurgerov and Meier (2005) is shown in Figure 5.6.

The left vertical axis in Figure 5.6 shows the yearly change in glacier volume. The right vertical axis is the cumulative contribution of glacier melting to relative sea level—effectively the sum over time of the amount of water in the oceans contributed by glacier melting, and how it influences relative sea level. There is clearly a lot of variability in glacier melt rates, and decreases in melt rate can be seen after major volcanic eruptions. Since 1960, melting glaciers have contributed more than 20 mm to sea-level rise. As can be seen in the "Summary" section, melting glaciers have been one of the largest contributors to sea-level rise.

Changes in Land Storage

The term *land storage* in this chapter refers to water stored on land in the form of snow, surface water (e.g., lakes, rivers, artificial reservoirs, marshes), or subsurface water (e.g., groundwater, liquid water trapped in soils, permafrost). Changes in land storage of water can be caused by changes in climate and by human modification of the land. Some of these human modifications are as follows:

- Damming of rivers
- Extraction of water from lakes

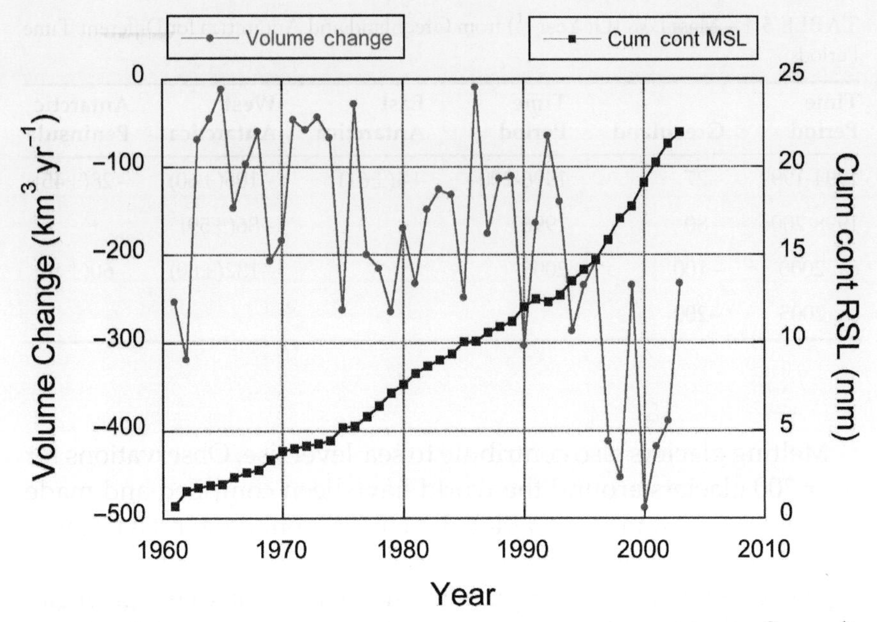

FIGURE 5.6 Time series of the change in volume (left vertical axis; km^3 year^{-1}) and the cumulative contribution to relative sea level (right vertical axis, mm) from small glaciers.

- Mining of groundwater
- Irrigation
- Wetland drainage
- Deforestation

Milly et al. (2010) provide a recent analysis of the contributions of terrestrial water storage to sea-level rise and variability. Their results are summarized in Figure 5.7.

Interestingly, the net effect of changes in land storage of water on sea-level rise is close to zero, since the major processes tend to cancel out each other (at least during the 1990s). The decrease in runoff to the oceans caused by damming rivers was balanced by the increase in flow from groundwater mining. Likewise, the decrease in flow from changes in snowpack, soil water, and shallow groundwater was offset by changes in storage in 15 of the world's largest lakes.

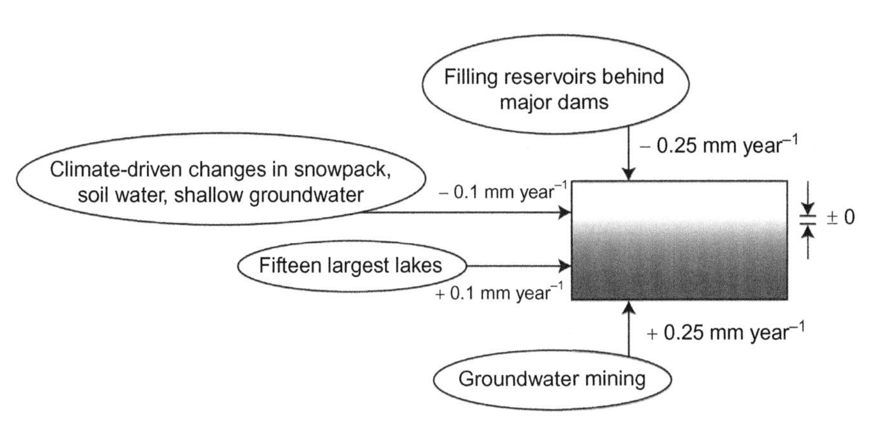

FIGURE 5.7 Estimated contributions from terrestrial water sources to sea-level change for the 1990s.

Summary

A considerable amount of research has been done over the last decades to measure and explain changes in global sea level. An overall picture was developed by Milly et al. (2010) from the IPCC 4th Assessment Report (Bindoff, 2007) and subsequent sources and is summarized in Table 5.2. Global mean sea level rose by approximately 1.8 mm $year^{-1}$ over the last five decades, doubled to 3.1 mm $year^{-1}$ in the 1990s, and was 2.5 mm $year^{-1}$ in the period 2003-2007.

TABLE 5.2 Summary of the Sources of Sea-Level Rise for Three Different Time Periods

Source	1961-2003 (mm $Year^{-1}$)	1993-2003 (mm $Year^{-1}$)	2003-2007 (mm $Year^{-1}$)
Observed change	1.8	3.1	2.5
Thermal expansion	0.4	1.6	0.35
Melting glaciers	0.5	0.8	1.1
Melting ice sheets	0.2	0.4	1
Land storage (liquid water)	–	0	–
Residual	0.7	0.3	0.1

The fractional contributions of the various sources to global sea-level rise are shown in Figure 5.8. The "residual" fraction is the difference between the observed sea-level rise rate and the sum of the four different sources. This can be interpreted as the "unexplained" amount of sea-level rise and is suspected to come primarily from melting of the large ice sheets in Greenland and Antarctica (Steffen et al., 2010). Panels (a)-(c) present data from Table 5.2 as fractions of the observed sea-level rise. Panel (d) is adapted from Domingues et al. (2008) and shows the different contributions to sea-level rise as a function of time for the period 1961-2003.

An intriguing result from these studies is the observation that the relative contributions of the various sources of sea-level rise change with time, and also that with time, the residual (unexplained) fraction has decreased—we have become better at quantifying the sources of sea-level rise. Thermal expansion of the upper 700 m of the oceans shows clear variability on 5- to 10-year timescales and can also be linked to volcanic eruptions. The contribution from liquid water from terrestrial sources (e.g., groundwater, lakes, marshes) is variable, but averages out to be roughly zero over this time interval. Glaciers and ice caps have been the dominant source of sea-level rise since about the late 1970s.

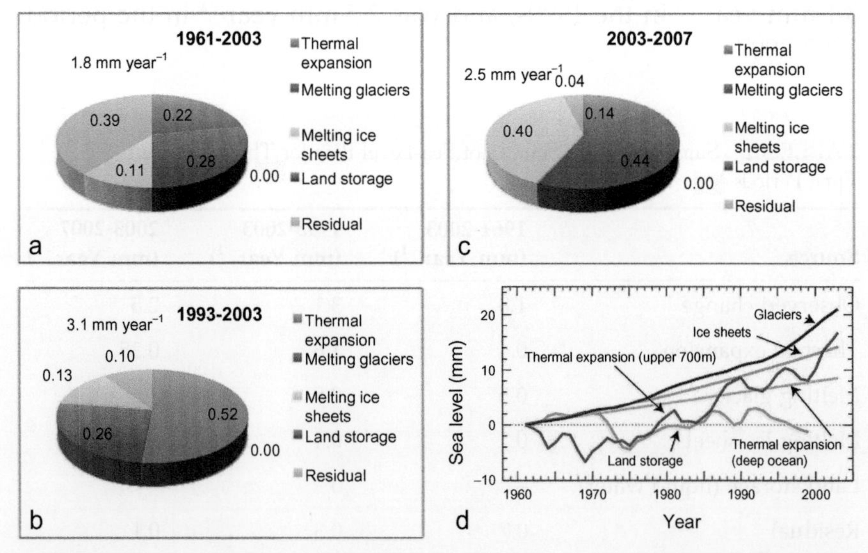

FIGURE 5.8 Sources of sea-level rise as a function of time. *Adapted from Steffen et al. (2010) and Domingues et al. (2008).*

OBSERVATIONS OF SEA-LEVEL RISE

Past Sea-Level Changes

When trying to understand how complex systems behave, it is a great advantage to have different independent sources of information on which to draw conclusions. In the following sections, two different, independent sets of observations of sea-level rise are described: direct measurements from tide gauges and remote-sensing measurements from satellites.

Direct Measurements: Tide Gauges

If observations from tide gauges around the world are put onto a consistent basis, a picture of the global mean sea-level rise can be obtained. Woodworth et al. (2009) compare five different time series of global sea-level rise derived from tide gauge measurements. Figure 5.9 shows two of these: the longer of the time series (back to 1700) comes from Jevrejeva et al. (2008), while the series from 1870 to 2007 is updated from Church and White (2006). Both investigations use the same underlying data (the Permanent Service for Mean Sea-Level compilation), but use different techniques for analysis.

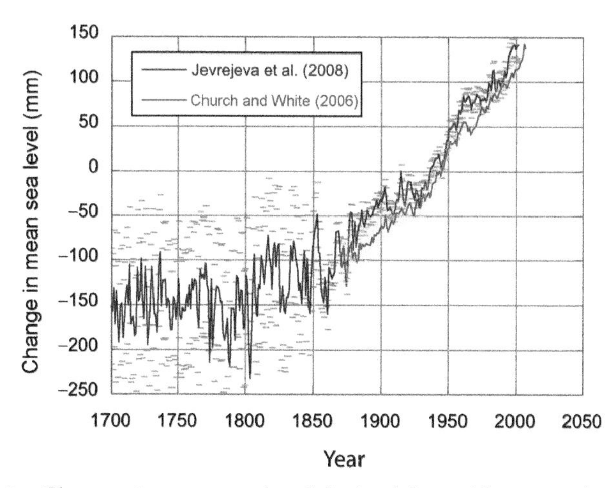

FIGURE 5.9 Changes in mean sea level derived from tide gauge data. Gray bars indicate uncertainties in the Jevrejeva et al. analysis; yellow bars for the Church and White analysis.

Both analyses show a consistent increase in sea level over the last two centuries, with an increase of about 250 mm between 1850 and 2000. For the entire twentieth century, the rate of sea-level rise was 1.7 ± 0.3 mm year^{-1} (Church and White, 2006). Sea level rose by 6 cm in the nineteenth century and by 19 cm in the twentieth century (Jevrejeva et al., 2008). The further back in time the data goes, the larger the uncertainties become. In their error analysis, Jevrejeva et al. (2008) account for both station representativity and between-region errors (Jevrejeva et al., 2006). Church and White (2006) account for serial correlation, uncertainties in corrections for glacial rebound, and uncertainties induced by the statistical method they used.

Satellite Measurements

Satellites provide another independent source of information on sea level, in addition to tide gauge measurements. However, reliable satellite measurements of global sea level did not become available until the early 1990s.

Figure 5.10 shows data from the TOPEX/Poseidon and Jason altimeters (Leuliette et al., 2004). Like the tide gauge data, the satellite observations show a clear upward trend in sea level, with year-to-year variability. The trend from the satellite observations over this period is 3.2 ± 0.8 mm year^{-1} (Leuliette and Willis, 2011).

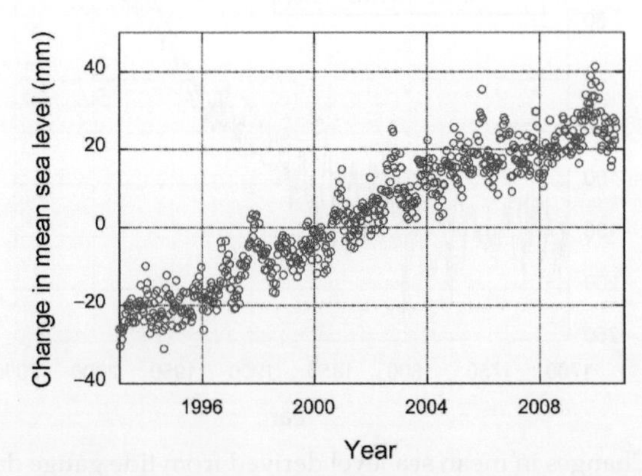

FIGURE 5.10 Changes in global mean sea level derived from satellite data available from the University of Colorado (http://sealevel.colorado.edu).

While global sea level is rising, the rate of sea-level rise depends on location. Figure 5.11 shows sea-level trends derived from a number of satellite instruments (Nicholls and Cazenave, 2010) for the period October 1992 to July 2009. There is a good deal of variability in the rate of sea-level rise in the Pacific, with maximum values in the western Pacific and minimum values in the eastern Pacific. The rate of sea-level rise at any given location will also vary with time. Despite the spatial and temporal variability, satellite measurements show a clear increase in globally averaged sea level, and together with the tide gauge measurements (corrected for GIA) tell us that sea-level rise has been accelerating.

Summary of Observations

A considerable amount of work has been done over the last decade to put observations of sea-level rise into a consistent framework. Direct measurements from tide gauges around the world have been analyzed; accounting for vertical land motion, the uneven spatial distribution of measurement sites and improvements in the statistical methods used to analyze the data have been made. New satellite instruments have provided an independent measure of sea level. A number of conclusions can be drawn from these observations:

- Independent measurements show that sea level is rising
- Sea levels have risen by about 250 mm since 1850
- Sea-level rise is accelerating

SEA LEVEL RISE IN THE FUTURE

Projections of Sea-Level Rise

Figure 5.12 shows four different projections for future sea-level rise. Panel (a) comes from the 3rd IPCC Assessment Report (AR3) (Church et al., 2001), panel (b) is from the 4th IPCC Assessment Report (AR4) (Meehl et al., 2007), panel (c) comes from Rahmstorf (2007), and panel (d) is taken from Church et al. (2008). The IPCC projections represent a consensus view at two different times (2001 and 2007) developed from a number of different

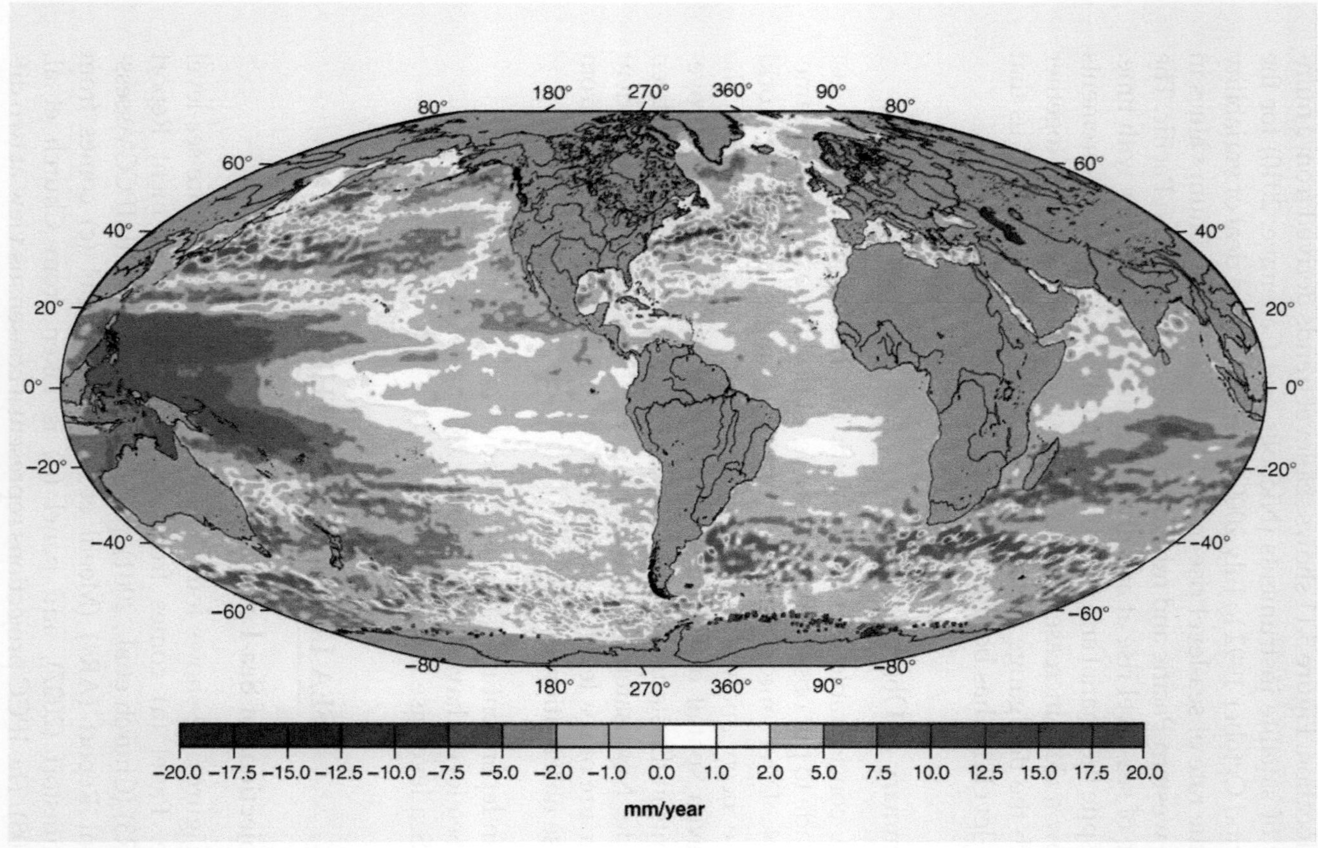

FIGURE 5.11 Rates of sea level derived from satellite measurements. *From Nicholls and Cazenave (2010), figure 2. © AAAS.*

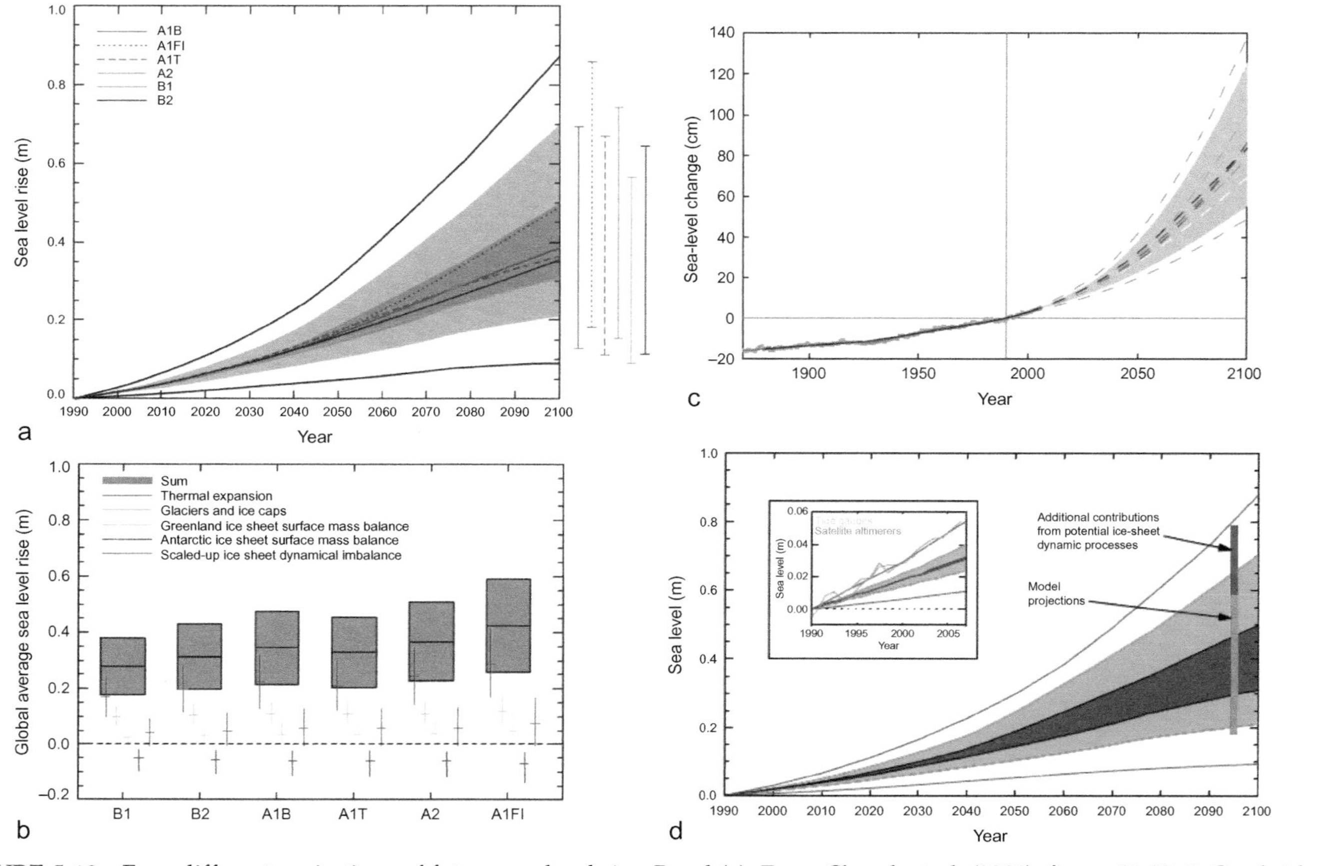

FIGURE 5.12 Four different projections of future sea-level rise: Panel (a): From Church et al. (2001), figure 11.12. © Cambridge University Press. Panel (b): From Meehl et al. (2007), figure 10.33. © Cambridge University Press. Panel (c): From Rahmstorf (2007), figure 4. © AAAS. Panel (d): From Church et al. (2008), figure 6. © Springer.

global climate models. The projections in Rahmstorf (2007) are made using a semiempirical approach using observed temperature and sea-level changes and projecting this observed relationship into the future. The Church et al. (2008) projections augment the IPCC work by adding an additional contribution from melting glaciers and ice caps in an attempt to account for observed melting that is currently not properly captured in climate models.

The intentions with presenting these figures together are to

1. show how better observations of and increased understanding about the causes of sea-level rise influence our ability to project changes into the future—learning more about how the system works allows for refinement of our predictions;
2. show that there is always a range of possibilities for future sea-level rise. This range depends on what assumptions are made about future population, demographics, technological and socioeconomic development, as well as different assumptions about how parts of the climate system work now and in the future.

Perhaps the greatest known source of uncertainty in projecting future sea-level rise is the extent to and rate at which melting glaciers, ice caps, and ice sheets will contribute in the future. The current rate of ice melt and flow of glacier ice into the ocean is not adequately captured in the climate models used in the IPCC assessments. If this current melting rate continues into the future, then the IPCC projections are likely to be underestimates. Another major uncertainty is estimating the rate at which ice flows from the continents (especially from Greenland and Antarctica) into the ocean. Like dropping an ice cube into a glass of water immediately raises the water level, transferring ice from land into the ocean will raise sea level even before the ice melts.

Three recent studies using semiempirical methods project intervals of sea-level rise by 2100 of 0.4-1 m (Horton et al., 2008), 0.9-1.3 m (Grinsted et al., 2010), and 0.6-1.6 m (Jevrejeva et al., 2010). Pfeffer et al. (2008) attempted to put an upper bound on how rapidly melting and ice flow into the ocean from Greenland could raise sea level. They concluded that sea-level rise of more than 2 m by 2100 was not tenable. A 2-m rise by 2100 was physically possible, but would require all of the variables influencing the

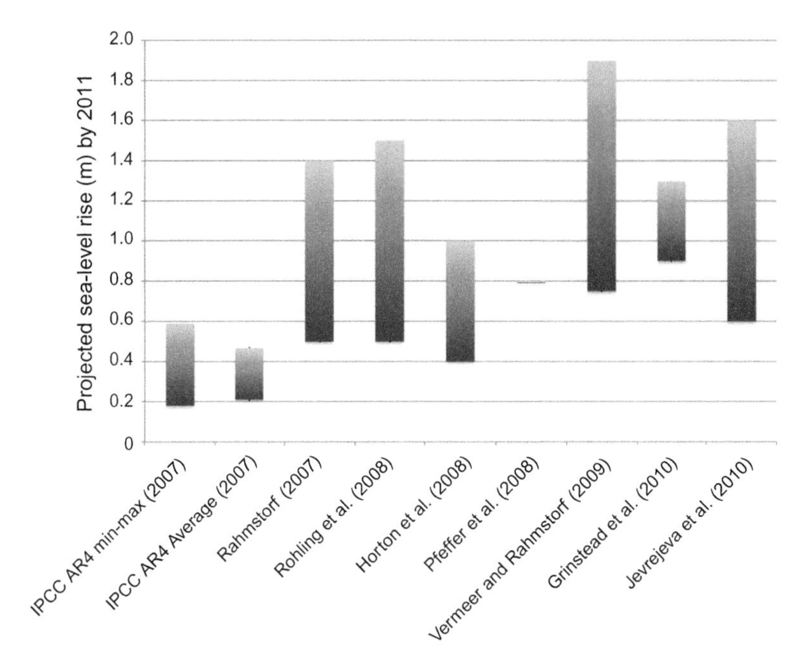

FIGURE 5.13 A summary of recent projections of sea-level rise.

melting or transport of ice into the ocean to be at their maximum values. They concluded that a more plausible value would be about 0.8 m. These results are summarized in Figure 5.13.

At this point, obtaining a very precise projection for sea level at the end of the century is beyond our abilities. The entire range of projections (from the IPCC AR3 and onwards) includes the interval from 0.2 to 2 m, with the central value for these projections increasing with time. Many of the post-IPCC AR4 projections rely on semi-empirical models, which have limitations when projecting past the interval in which data are available (Church et al., 2011; von Storch et al., 2008). Clearly, improved and more comprehensive observations of ice melt and flow are necessary, along with improvements of how these processes are described in climate models.

IMPACTS OF SEA-LEVEL RISE

Sea-level rise will impact all coastal areas, but to differing extents. It is an existential issue for some small island states, since some islands will disappear entirely with even modest increases in sea level. As of 1994, 1.88 billion people (33.5% of the world's population) lived within 100 vertical meters of sea level (Cohen and Small, 1998), and about 23% of the world's people live within 100 km of a coast (Nicholls et al., 2007). On a finer scale, Anthoff et al. (2006) estimate that about 145 million people live within 1 m of mean high water (more than 70% of whom are in Asia), and 268 and 397 million live within 5 and 10 m, respectively.

Types of Impacts

The physical impacts of sea-level rise include the following:

- The disappearance of some low-lying islands
- Submergence and increases flooding of coastal land
- Saltwater intrusion of surface and subsurface waters
- Increased erosion
- Habitat destruction in coastal areas

The effects, interacting factors, and possible adaptation approaches for these impacts have been summarized by Nicholls (2011) and are shown in Table 5.3.

The interacting factors include those related to climate change and others related more to land and resource management. In the table, the adaptation approaches are coded as protection (P), accommodation (A), and retreat (R). Examples of "hard" approaches are dikes, seawalls, and reinforced structures, while "soft" approaches include management, policy, and institutional considerations. These impacts will occur at different times, with impacts such as coastal erosion or habitat degradation often lagging inundation and flooding. These impacts will be exacerbated by the additional effects of extreme storms, discussed in Chapter 3. Rising sea levels will cause an increase in extreme water levels, which also will be increased if hurricane and severe storms become stronger.

Vulnerability to Sea-Level Rise

Nicholls and Cazenave (2010) examined the vulnerability of different regions to coastal flooding due to sea-level rise. The highest risk areas are coastal zones with high population density, low

TABLE 5.3 Impacts, Interacting Factors, and Possible Adaptation Approaches to Sea-Level Rise

Natural System Effect		Possible Interacting Factors		Possible Adaptation Approaches
		Climate	**Nonclimate**	
1. Inundation/ flooding	a. Surge (flooding from the sea)	Wave/storm climate, erosion, sediment supply	Sediment supply, flood management, erosion, land reclamation	Dikes/surge barriers/closure dams [P—hard], dune construction [P—soft], building codes/flood-proof buildings [A], land-use planning/hazard mapping/flood warnings [A/R]
	b. Backwater effect (flooding from rivers)	Runoff	Catchment management and land use	
2. Wetland loss (and change)		CO_2 fertilization, sediment supply, migration space	Sediment supply, migration space, land reclamation (i.e., direct destruction)	Nourishment/sediment management [P—soft], land-use planning [A/R], managed realignment/forbid hard defenses [R]
3. Erosion (of "soft" morphology)		Sediment supply, wave/storm climate	Sediment supply	Coast defenses/seawalls/land claim [P—hard], nourishment [P—soft], building setbacks [R]
4. Saltwater Intrusion	a. Surface waters	Runoff	Catchment management (overextraction), land use	Saltwater intrusion barriers [P], change water extraction [A/R]
	b. Groundwater	Rainfall	Land use, aquifer use (overpumping)	Freshwater injection [A], change water extraction [A/R]
5. Impeded drainage/higher water tables		Rainfall, runoff	Land use, aquifer use, catchment management	Drainage systems/polders [P—hard], change land use [A], land-use planning/hazard delineation [A/R]

From Nicholls (2011), Table 2.

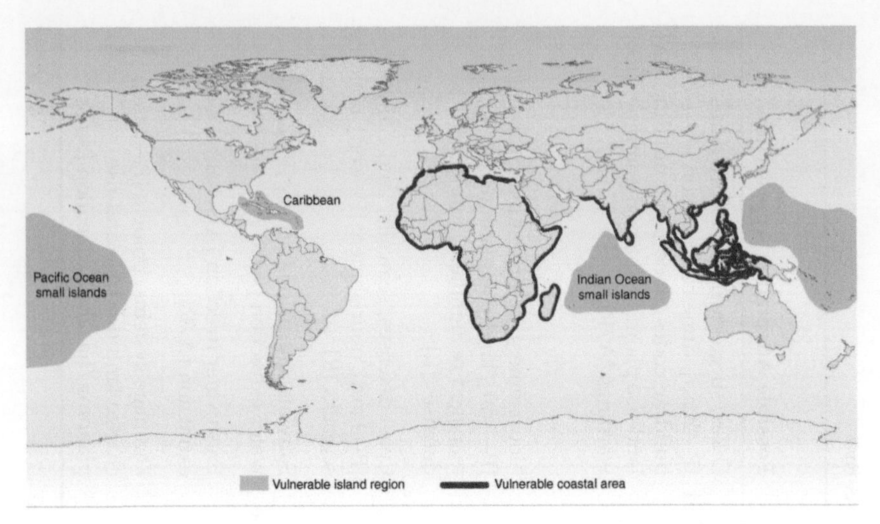

FIGURE 5.14 Regions vulnerable to flooding caused by sea-level rise. *From Nicholls and Cazenave (2010), figure 3. © AAAS.*

elevations, high rates of land subsidence, and limited adaptive capacity. Figure 5.14 is a map showing the locations of these vulnerable regions.

It is worthwhile to point out an important distinction in terms of trying to define vulnerability to sea-level rise; the impacts of sea-level rise are very asymmetrical in many important ways. For instance, there are fundamental differences in vulnerability between continental coastal locations and islands. In large, wealthy countries, coastal communities would at least in theory be able to relocate inland, given adequate preparation and support. In reality, however, relocating parts of any major coastal city would be an immense undertaking.

Such an option does not exist for some small island states, which may become uninhabitable or even entirely disappear. It is one thing to be forced to relocate within your own local region or country; it is another entirely to lose your country. In developing countries with very high populations in low-lying delta areas such as Bangladesh, sea-level rise may threaten the ability of the state to function (Byravan and Rajan, 2010).

Assessing the impacts of sea-level rise is a good example of the limitations of our current mainstream socioeconomic framework.

The monetary costs of paying to relocate a relatively small population of people from a disappearing island nation would be small compared to the costs of mitigating the cause of the relocation—increases in sea level due to global warming caused mostly by anthropogenic greenhouse gas emissions. However, the social costs of the loss of national identity and cultural history are not included in this monetary framework, but need to be accounted for.

How strongly these different impacts affect various socioeconomic sectors was examined by Nicholls (2011) and is shown in Table 5.4. Inundation and flooding affect all the socioeconomic sectors considered, while coastal erosion mainly influences infrastructure, tourism, and biodiversity, with weaker or not established impacts on other sectors.

Examples of health effects include the release of toxins from coastal landfills, increases in disease due to damaged coastal infrastructure such as sewers, or even mental health problems caused by

TABLE 5.4 The Strength of the Impact of Different Effects of Sea-Level Rise on Several Socioeconomic Sectors

Coastal Socioeconomic Sector	Sea Level Rise Natural System Effect (Table 5.1)				
	Inundation/ Flooding	**Wetland Loss**	**Erosion**	**Saltwater Intrusion**	**Impeded Drainage**
Freshwater resources	X	x	–	X	X
Agriculture and forestry	X	x	–	X	X
Fisheries and aquaculture	X	X	x	X	–
Health	X	X	–	X	x
Recreation and tourism	X	X	X	–	–
Biodiversity	X	X	X	X	X
Settlements/ infrastructure	X	X	X	X	X

X, strong; **x**, weak; –, negligible or not established.
From Nicholls (2011), table 2. © The Oceanography Society.

increased flooding. Aquaculture is an example of a socioeconomic sector adversely affected by several of the threats to the global oceans: sea-level rise, acidification, pollution, and hypoxia. It is very important to analyze the potential interactions between these effects and prepare strategies to deal with them (discussed in more detail in Chapter 12).

Economic Consequences of Sea-Level Rise

The economic impacts of sea-level rise will be discussed in detail in Chapter 10. The intention here is to present results from recent publications that illustrate the kinds of global-scale economic analyses that have been done and give an impression of the magnitude of the economic impacts.

People living in coastal areas close to sea level have essentially two choices about how to respond to increasing sea level: fight or flee. Fighting involves improving or building new coastal defenses—seawalls, dikes, and other built infrastructure. Fleeing involves relocating people from low-lying areas to higher ground. The choice is "simplified" for people living on some low-lying islands, since for them fighting is not an option.

Anthoff et al. (2010) examined the economic consequences of three different levels of sea-level rise: 0.5, 1, and 2 m above 2005 sea level. They use an integrated assessment model (FUND: The Climate Framework for Uncertainty, Negotiation, and Distribution), driven by four of the SRES[2] scenarios (A1, A2, B1, and B2) for socioeconomic and demographic development. They also use a control scenario in which population and GDP are kept constant at 1995 levels (C1995). In their simulations, they assume that sea level increases linearly with time until 2100.

They examine four damage components in their economic analysis: (1) the value of dryland lost; (2) the value of wetland lost; (3) the cost of protection (e.g., dikes) against rising sea levels; and (4) the cost of relocating people displaced from lost dryland. In their calculations, they assume that the number of people forced to migrate is a function of the average population density in each country and the area of dryland lost in that country. The economic

[2] Special Report on Emissions Scenarios; see Chapter 1.

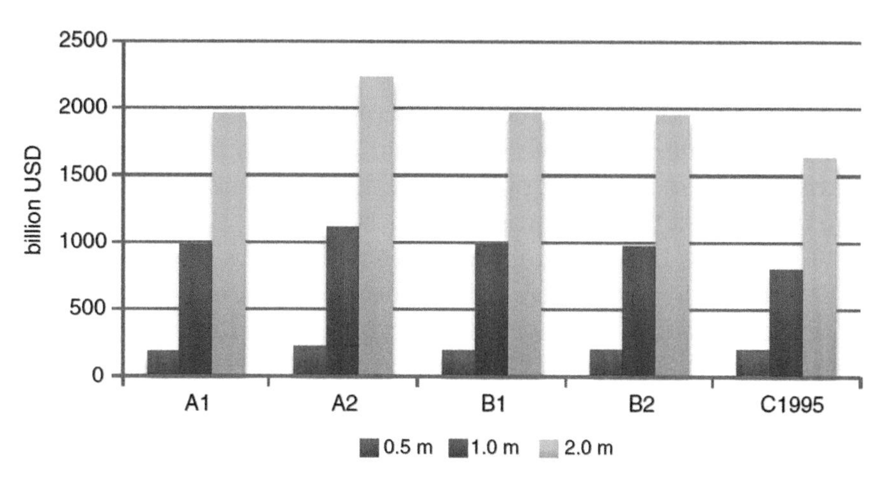

FIGURE 5.15 Total damage costs for 0.5, 1, and 2 m of sea-level rise for five different socioeconomic scenarios. *From Anthoff et al. (2010), figure 2. © Springer.*

cost of the people displaced is calculated as three times the average per capita income for the country in question. Figure 5.15 shows an example of their results.

With their assumptions, protection costs against a 0.5-m increase in sea level would be between about 170 and 200 billion USD. If sea level were to rise by 1 m, costs would become about five times as large, roughly 1 trillion USD. Costs would roughly double again to about 2 trillion USD if sea-level rise were 2 m. These costs can be compared with estimates of approximately 3 trillion USD in stimulus investments announced as a result of the economic downturn in 2008 (Robins et al., 2009a,b).

Looking "under the hood" in their calculations, protection costs are the most important component of total costs, independent of which socioeconomic scenario is chosen. Other costs vary more widely between scenarios. Furthermore, costs vary greatly between regions, with South Asia, South America, East Asia, North America, Europe, and Central America being the regions dominating total costs. As in most cost-benefit analyses for long-term issues, the results are sensitive to what discount rate is assumed. The central assumption for the pure time rate of preference in the Anthoff et al, (2010) calculations is 1%. They performed a sensitivity analysis assuming values of 0.1% and 3%, which showed that the results could be changed by roughly a factor of 2 up or down, depending on the discount rate assumed.

They conclude that the benefits of protection increase substantially with time. Compared to a scenario with no protection, substantial costs in terms of population displacement and land loss can be avoided through protection measures.

A weakness in cost-benefit analyses is that they assume perfect knowledge and a proactive approach to protection, while experience would tell us that protection is more often done in reaction to a disaster the near avoidance of a disaster. Recognizing this historical limitation is critical from the point of view of low-lying island states, since the protection option may well disappear along with the islands.

Summary and Take-Home Messages

- Globally averaged sea level has risen by about 25 cm since the 1800s
- Global sea-level rise is accelerating
- Sea-level rise is caused by melting glaciers and ice caps, loss of ice from major ice sheets on Greenland and Antarctica, thermal expansion of the oceans, and changes in terrestrial storage
- Projections of the amount of sea-level rise by the end of the twenty-first century vary from a minimum of roughly 0.2 m to a maximum of about 2 m
- The impacts of sea-level rise include the disappearance of some low-lying islands, submergence and increases flooding of coastal land, saltwater intrusion of surface and subsurface waters, increased erosion, and habitat destruction in coastal areas
- On a global scale, the costs of coastal protection and relocation of people from land areas lost to sea-level rise range from about 200 billion USD for an increase of sea level of 0.5 m and about 2 trillion USD for an increase of 2 m.
- Adaptation strategies for sea-level rise will need to be developed for local and regional conditions; how these are influenced by changes in global sea level must be taken into account.

Acknowledgments

This work was supported by a grant from the Okeanos Foundation (http://www.okeanos-foundation.org/). I am grateful to three anonymous reviewers for insightful comments that improved the quality of the chapter.

References

Anthoff, D., R. Nicholls, and R. Tol (2010), The economic impact of substantial sea-level rise, *Mitig. Adapt. Strat. Glob. Change*, **15**(4), 321–335, http://dx.doi.org/10.1007/s11027-010-9220-7.

Anthoff, D., R. J. Nicholls, R. S. J. Tol, and A. Vafeidis (2006), *Global and regional exposure to large rises in sea-level: a sensitivity analysis Rep.* 31 pp., Tyndall Centre for Climate Change Research, Norwich, UK.

Bindoff, N. L., et al. (2007): Observations: Oceanic Climate Change and Sea Level. In: *Climate Change 2007: The Physical Science Basis. Contribution of Working Group I to the Fourth Assessment Report of the Intergovernmental Panel on Climate Change* [Solomon S., D. Qin, M. Manning, Z. Chen, M. Marquis, K. B. Averyt, M. Tignor and H. L. Miller (eds.)]. Cambridge University Press, Cambridge, UK; New York NY, USA.

Boon, J. D. (2012), Evidence of Sea Level Acceleration at U.S. and Canadian Tide Stations, Atlantic Coast, North America, *J. Coastal Res.*, **28**(6), 1437–1445, http://dx.doi.org/10.2112/JCOASTRES-D-12-00102.1.

Byravan, S., and S. C. Rajan (2010), The Ethical Implications of Sea-Level Rise Due to Climate Change, *Ethics & International Affairs*, **24**(3), 239–260, http://dx.doi.org/10.1111/j.1747-7093.2010.00266.x.

Church, J., N. White, T. Aarup, W. Wilson, P. Woodworth, C. Domingues, J. Hunter, and K. Lambeck (2008), Understanding global sea levels: past, present and future, *Sustainability Science*, **3**(1), 9–22, http://dx.doi.org/10.1007/s11625-008-0042-4.

Church, J. A., J. M. Gregory, P. Huybrechts, M. Kuhn, K. Lambeck, M. T. Nhuan, D. Qin, and P. L. Woodworth (2001): *Changes in Sea Level. In: Climate Change 2001: The Scientific Basis. Contribution of Working Group 1 to the Third Assessment Report of the Intergovernmental Panel on Climate Change* [Houghton J. T., Y. Ding, D. J. Griggs, M. Noguer, P. J. van der Linden, X. Dai, K. Maskell and C. A. Johnson (eds.)]. Cambridge University Press, Cambridge, U.K, pp. 639–693.

Church, J. A., J. M. Gregory, N. J. White, S. M. Platten, and J. X. Mitrovica (2011), Understanding and projecting Level Change, *Oceanography*, **24**(2), 130–143.

Church, J. A., and N. J. White (2006), A 20th century acceleration in global sea-level rise, *Geophys. Res. Lett.*, **33**(1), L01602, http://dx.doi.org/10.1029/2005gl024826.

Church, J. A., P. L. Woodworth, T. Aarup, and W. S. Wilson (Eds.) (2010), Understanding Sea-Level Rise and Variability, 428 pp., Wiley-Blackwell, Chichester, UK.

Cohen, J. E., and C. Small (1998), Hypsographic demography: The distribution of human population by altitude, *Proc. Natl. Acad. Sci. U S A.*, **95**(24), 14009–14014.

Davis, C. H., Y. Li, J. R. McConnell, M. M. Frey, and E. Hanna (2005), Snowfall-Driven Growth in East Antarctic Ice Sheet Mitigates Recent Sea-Level Rise, *Science*, **308**(5730), 1898–1901, http://dx.doi.org/10.1126/science.1110662.

Domingues, C. M., J. A. Church, N. J. White, P. J. Gleckler, S. E. Wijffels, P. M. Barker, and J. R. Dunn (2008), Improved estimates of upper-ocean warming and multi-decadal sea-level rise, *Nature*, **453**(7198), 1090–1093, http://dx.doi.org/10.1038/nature07080.

Dyurgerov, M. B., and M. F. Meier (2005), *Glaciers and the Changing Earth System: A 2004 Snapshot Rep.* 117 pp., INSTAAR, University of Colorado, Boulder, CO, USA.

Gehrels, R. (2009), Rising Sea Levels as an Indicator of Global Change. In: *Climate Change*, Elsevier, Amsterdam, 325–336, http://dx.doi.org/10.1016/B978-0-444-53301-2.00018-X.

Grinsted, A., J. Moore, and S. Jevrejeva (2010), Reconstructing sea level from paleo and projected temperatures 200 to 2100AD, *Clim. Dyn.*, **34**(4), 461–472, http://dx.doi.org/10.1007/s00382-008-0507-2.

Horton, R., C. Herweijer, C. Rosenzweig, J. Liu, V. Gornitz, and A. C. Ruane (2008), Sea level rise projections for current generation CGCMs based on the semi-empirical method, *Geophys. Res. Lett.*, **35**(2), L02715, http://dx.doi.org/10.1029/2007gl032486.

Jevrejeva, S., A. Grinsted, J. C. Moore, and S. Holgate (2006), Nonlinear trends and multiyear cycles in sea level records, *J. Geophys. Res.*, **111**(C9), C09012, http://dx.doi.org/10.1029/2005jc003229.

Jevrejeva, S., J. C. Moore, and A. Grinsted (2010), How will sea level respond to changes in natural and anthropogenic forcings by 2100? *Geophys. Res. Lett.*, **37**(7), L07703, http://dx.doi.org/10.1029/2010gl042947.

Jevrejeva, S., J. C. Moore, A. Grinsted, and P. L. Woodworth (2008), Recent global sea level acceleration started over 200 years ago? *Geophys. Res. Lett.*, **35**(8), L08715, http://dx.doi.org/10.1029/2008gl033611.

Lemke, P., et al. (2007): Observations: Changes in Snow, Ice and Frozen Ground. In: Climate Change 2007: *The Physical Science Basis. Contribution of Working Group I to the Fourth Assessment Report of the Intergovernmental Panel on Climate Change* [Solomon S., D. Qin, M. Manning, Z. Chen, M. Marquis, K. B. Averyt, M. Tignor and H. L. Miller (eds.)]. Cambridge University Press, Cambridge, UK and New York, NY, USA, 337–383.

Leuliette, E. W., R. S. Nerem, and G. T. Mitchum (2004), Calibration of TOPEX/Poseidon and Jason Altimeter Data to Construct a Continuous Record of Mean Sea Level Change, *Mar. Geodesy*, **27**(1), 79–94.

Leuliette, E. W., and J. K. Willis (2011), Balancing the sea level budget, *Oceanography*, **24**(2), 122–129.

Lunt, D. J., G. L. Foster, A. M. Haywood, and E. J. Stone (2008), Late Pliocene Greenland glaciation controlled by a decline in atmospheric CO2 levels, *Nature*, **454**(7208), 1102–1105, http://www.nature.com/nature/journal/v454/n7208/suppinfo/nature07223_S1.htmlhttp://dx.doi.org/10.1038/nature07223.

Lyman, J. M., S. A. Good, V. V. Gouretski, M. Ishii, G. C. Johnson, M. D. Palmer, D. M. Smith, and J. K. Willis (2010), Robust warming of the global upper ocean, *Nature*, **465**(7296), 334–337, http://dx.doi.org/10.1038/nature09043.

Meehl, G. A., et al. (2007): Global Climate Projections. In: *Climate Change 2007: The Physical Science Basis. Contribution of Working Group I to the Fourth Assessment Report of the Intergovernmental Panel on Climate Change* [Solomon S., D. Qin, M. Manning, Z. Chen, M. Marquis, K. B. Averyt, M. Tignor and H. L. Miller (eds.)]. Cambridge University Press, Cambridge, UK and New York, NY, USA, 747–845.

Milly, P. C. D., A. Cazenave, J. S. Famiglietti, V. Gornitz, K. Laval, D. P. Lettenmaier, D. Sahagian, J. M. Wahr, and C. R. Wilson (2010): Terrestrial Water-Storage Contributions to Sea-Level Rise and Variability. In: *Understanding Sea-Level Rise and Variability* [Church J. A., P. L. Woodworth, T. Aarup and W. S. Wilson (eds.)]. Wiley-Blackwell, Chichester, UK, 226–255.

Nicholls, R. J. (2011), Planning for the impacts of sea level rise, *Oceanography*, **24**(2), 144–157.

Nicholls, R. J., and A. Cazenave (2010), Sea-Level Rise and Its Impact on Coastal Zones, *Science*, **328**(5985), 1517–1520, http://dx.doi.org/10.1126/science.1185782.

Nicholls, R. J., P. P. Wong, V. Burkett, J. Codignotto, J. Hay, R. McLean, S. Ragoonaden, and C. D. Woodroffe (2007): Coastal systems and low-lying areas. In: *Climate Change 2007: Impacts, Adaptation and Vulnerability. Contribution of Working Group II to the Fourth Assessment Report of the Intergovernmental Panel on Climate Change* [Parry M. L., O. F. Canziani, J. P. Palutikof, P. J. van der Linden and C. E. Hanson (eds.)]. Cambridge University Press, Cambridge, UK, 315–356.

Pfeffer, W. T., J. T. Harper, and S. O'Neel (2008), Kinematic Constraints on Glacier Contributions to 21st-Century Sea-Level Rise, *Science*, **321**(5894), 1340–1343, http://dx.doi.org/10.1126/science.1159099.

Rahmstorf, S. (2007), A Semi-Empirical Approach to Projecting Future Sea- Level Rise. *Science*, **315**(5810), 368–370, http://dx.doi.org/10.1126/science.1135456.

Robins, N., R. Clover, and C. Singh (2009a), *Building a green recovery Rep.*, 59 pp., HSBC Bank plc, London, UK.

Robins, N., R. Clover, and C. Singh (2009b), *A Climate for Recovery: The colour of stimulus goes green Rep.*, 47 pp., HSBC Bank plc, London, UK.

Rohling, E. J., K. Grant, C. Hemleben, M. Siddall, B. A. A. Hoogakker, M. Bolshaw, and M. Kucera (2008), High rates of sea-level rise during the last interglacial period, *Nature Geoscience*, **1**, 38–42.

Sallenger, A. H., K. S. Doran, and P. A. Howd (2012), Hotspot of accelerated sea-level rise on the Atlantic coast of North America, *Nature Clim. Change*, **2**(12), 884–888, http://www.nature.com/nclimate/journal/v2/n12/abs/nclimate1597.html-supplementary-information.

Steffen, K., R. H. Thomas, E. Rignot, J. G. Cogley, M. B. Dyurgerov, S. C. B. Raper, P. Huybrechts, and E. Hana (2010): Cryospheric Contributions to Sea-Level Rise and Variability. In: *Understanding Sea-Level Rise and Variability* [Church J. A., P. L. Woodworth, T. Aarup and W. S. Wilson (eds.)]. Wiley-Blackwell, Chichester, UK, 177–225.

Trenberth, K. E. (2010), Global change: The ocean is warming, isn't it? *Nature*, **465** (7296), 304, http://dx.doi.org/10.1038/465304a.

Vermeer, M., and S. Rahmstorf (2009), Global sea level linked to global temperature, *Proc. Natl. Acad. Sci.*, **106**, 21527–21532.

von Storch, H., E. Zorita, and J. González-Rouco (2008), Relationship between global mean sea-level and global mean temperature in a climate simulation of the past millennium, *Ocean Dynamics*, **58**(3), 227–236, http://dx.doi.org/10.1007/s10236-008-0142-9.

Woodworth, P. L., N. J. White, S. Jevrejeva, S. J. Holgate, J. A. Church, and W. R. Gehrels (2009), Evidence for the accelerations of sea level on multi-decade and century timescales, *Int. J. Climatol.*, **29**(6), 777–789.

Zachos, J., M. Pagani, L. Sloan, E. Thomas, and K. Billups (2001), Trends, rythms, and aberrations in global climate 65 Ma to present, *Science*, **292**, 686.

CHAPTER

6

Marine Pollution

Dan Wilhelmsson, Richard C. Thompson*[†], *Katrin Holmström*[‡], *Olof Lindén*[§], *Hanna Eriksson-Hägg*[¶]

***Swedish Secretariat for Environmental Earth System Science, Stockholm, Sweden**

[†]Marine Biology and Ecology Research Centre, School of Marine Science and Engineering, University of Plymouth Drake Circus, Plymouth, United Kingdom

[‡]Environment Department, Stockholm City, Box 8136, Sweden

[§]World Maritime University, Box 500, Malmö, Sweden

[¶]Baltic Nest Institute, Stockholm Resilience Centre, Stockholm University, Kräftriket 2B, Stockholm, Sweden

INTRODUCTION

Marine environment is submitted to contamination[1] that comes in many different forms, such as toxic chemicals (e.g., organic compounds, DDT, PCB, metals, pharmaceuticals, gas), solid waste (e.g., plastics), increased nutrient (e.g., nitrates and phosphates) and sediment inputs due to human activities (e.g., industry, agriculture, deforestation, sewage discharge, aquaculture), radioactivity, oil spills, and discarded fishing nets. Marine contamination changes the physical, chemical, and biological characteristics of the oceans and coastal zones, and potentially threatens marine organism, ecosystems, and biodiversity and affects thus the quality and

[1] Contamination is defined by the introduction of compounds or energy leading to concentration or level higher than the natural ones (Goldberg, 1995).

productivity of marine ecosystems. In this context, the contamination causing damage or negative impact on marine ecosystem is called pollution (see below). The ultimate effect of pollution on marine resources depends on the form, intensity (acute or chronic), and location of the contamination, with some marine environments, ecosystems, and species being more sensitive than others to pollution.

The 1982 United Nations Convention on the Law of the Sea defined marine pollution as "the introduction by man, directly or indirectly, of substances or energy into the marine environment ... which results or is likely to result in such deleterious effects as harm to living resources and marine life" (article 1.1.4). The definition has subsequently been included in, for example, the OSPAR and Helsinki Conventions and the UNEP Regional Seas Program. It is worth noting that, in addition to the influx of different kinds of substances into the oceans, the definition of marine pollution includes the input of energy, which here refers to thermal (e.g., discharge of cooling water from nuclear plants) and acoustic (or noise) pollution (Dotinga and Oude Elferink, 2000).

The three main sources of marine pollution are direct discharge as effluents and solid wastes from land or human activities at sea (e.g., shipping), runoff mainly via rivers, and atmospheric fallout. The relative contribution of each of these pathways to marine pollution varies greatly with substance and situation. The lack of data and the significant complexity of the natural processes, especially at the sea-land and sea-atmosphere boundaries, make quantitative estimates of these processes difficult and uncertain. To indicate the scientific status within these research areas, the authors of this chapter made a preliminary assessment of the scientific basis in relation to the pathways and impacts of pollutants in the marine environment and the results are presented in Figure 6.1.

This chapter first describes the different sources, pathways, and threats to marine ecosystems posed by some of the pollutants that are most frequently discussed within research and policy, notably excluding the discharge of nutrients and organic material into marine environments as this is treated in Chapter 4. The authors also attempt to assess the risks associated with marine pollution (also here excluding nutrients and organic material) at a global scale.

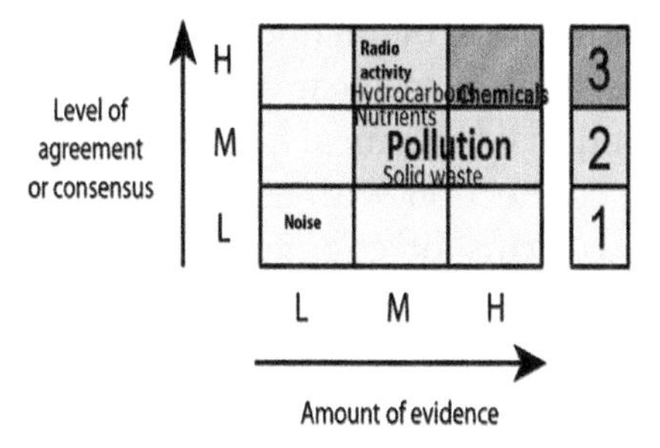

FIGURE 6.1 The authors' estimation of scientific basis in relation to the pathways and impacts of pollutants on global marine ecosystems, including a breakdown of the same estimation into different types of pollution (note that the assessment of the research status is limited to the relatively few investigated substances). Key: Amount of evidence: L = low, research is still in its infancy and comprehensive data are only available for a few species and areas; M = medium, available literature provides a number of solid, although often too site and species specific, studies; H = high, available literature provides data representing a broad range of organisms and ecosystems. Level of agreement or consensus: L = low, research data are inconclusive; M = medium, research data and conclusions are in general agreement; H = research data are largely unchallenged.

CHEMICAL POLLUTION—POPS AND METALS/TOXIC CHEMICALS

Sources and Pathways

This section describes the release of metals and persistent organic substances caused by human activities and the associated risks for chemical pollution of the ocean. Chemical pollution started early in human history. Evidence of European-wide pollution by lead, for example, can be traced back 2500 years to activities within the Roman Empire (Hong et al., 1994). For a long time, the fate of the chemicals and metals used and released was not considered, but it was commonly thought that the solution to pollution was dilution and to build higher chimneys and longer pipes. Today, however, it is recognized that chemical pollution of the environment can adversely affect human and ecosystem health.

The production and use of chemicals create emissions at all stages of their life cycle (production, use, and discarding), and the sources of chemical pollution of the ocean are thus large scale and diverse. The emission sources can be categorized as follows: intentional dissemination of chemical products (e.g., pesticides), unintentional dissemination of chemical products (sewage effluents, e.g., pharmaceuticals, leakage, wear and tear, e.g., PCB), and unintentional dissemination of chemical byproducts (impurities, formation and release in connection with combustion, e.g., dioxins).

The sources and forms of chemical pollution are of different importance in different regions. In the more industrialized regions, sources of chemical pollution include chemical industry, sewage treatment, landfills, and diffuse emissions from use of products within the society. In regions with high regulation on chemical pollution, the industrial emissions may have ceased, or production may have moved elsewhere, but emissions also occur through leaching and discarding of products. These coastal zones bordered by developed countries face pollutants such as pharmaceuticals and personal care products, flame retardants, plastic additives, etc. In less regulated parts of the world, industrial activities can lead to large-scale pollution. As example, the Manila Bay in Philippines is well known to be one of the pollution hot spots of East Asia Seas (Maria et al., 2009). After release, the pollutants can be transported away from the source regions with moving air masses through ocean currents or through biotic vectors, i.e., massive movements of organisms such as migrating birds (Wania, 1998). In that way, also remote and nonindustrialized areas become affected by chemical pollution. The importance of this long range, large-scale transport of pollutants can be illustrated by the high levels of certain pollutants found in remote areas, such as the Polar Regions (de Wit et al., 2004). In fact, the Arctic polar bears have some of the highest levels of toxic organic pollutants of any animal on the planet (Sonne, 2010).

Not all chemicals that are used and released in the society are of concern, but there are certain features of the chemical that are important in the understanding of its effects; it is the persistence and bioavailability, bioaccumulation potential, and toxicity of the chemical. The persistence of a substance determines its removal rate from the environment. Metals are inert, highly persistent, and continuously released in the environment from human activities as well as natural sources. Metal adsorption to particles and sedimentation is the major pathway of removal from the water column. Organic substances can be degraded biologically or via physicochemical processes. The degradation proceeds with different rates for different chemicals and the half-life can vary from hours to decades. High persistence enables concentrations of a compound to build up in the environment, and to some extent, it can also enable long-range transport.

The bioaccumulation potential of a substance describes its ability to accumulate in organisms in higher concentrations than in the surrounding environment. Organic compounds are usually characterized by their hydrophilic and hydrophobic properties expressed by the octanol water partition coefficient (K_{ow}). These properties drive the potential of molecules for bioaccumulation through the permeable surface of an organism (e.g., gills, digestive tracts, skins), when the substance is attracted toward a more lipid-rich environment (i.e., biological membrane) than the surrounding water. Regarding inorganic compounds such as trace element, the seawater physicochemical parameters determine the bioavailable fraction of element (according the chemical speciation of the element) that could be taken up by marine organisms through active or passive processes. Then, for all compounds that cannot be metabolized, organisms can accumulate contaminants with concentrations several orders of magnitude higher than the surrounding waters (Philips and Rainbow, 1993). The bioaccumulation efficiency results from the contributions of different uptake pathways (e.g., uptake from water, food, sediment) and the detoxification processes and elimination rate with respect to the species and the chemical. Beyond the individual scale, some toxics pass along the food chain and can be additively accumulated in the higher trophic level. This biomagnification of contaminants leads to very high concentrations in tissues of top predators as, for example, the Hg concentrations in tuna or the PCBs in polar bear. Hence, as a consequence, the bioaccumulation potential, discharging chemicals into the ocean, does not necessarily lead to dilution, but can result in concentration.

Effects on Marine Environment and Ecosystem Services

The main concern over chemical pollution is the potential for toxic effects in humans and wildlife. The toxicity of a substance can be investigated through biological tests. About 80,000-100,000 chemicals are registered for commercial use today (US EPA, 1998;

EU, 2001). Toxicity data exist for a few thousand of these chemicals, but there is virtually no knowledge of their combined effects. The toxicity of a chemical can be described as the ability of a chemical to enter, interfere with, or disrupt biological systems, like protein synthesis, hormonal regulation, or reproduction systems. For example, chemical substances may be allergenic and carcinogenic, toxic to reproduction, or genetically disrupting.

Toxic effects can be caused by direct exposure of an organism to chemicals in air, water, or soil. Effects can also occur through internal exposure to contaminants by bioaccumulation.

In the late 1960s and early 1970s, reproductive success of marine mammals and predatory birds started to fail in certain areas. The deleterious effects were later linked to the environmental contaminants like PCB and DDT (Ratcliffe, 1970; Jensen, 1972). Since wildlife concurrently is exposed to many different substances, it is, however, generally difficult to link observed effects to a single substance. Investigators have demonstrated immune dysfunction as a plausible cause for increased mortality among marine mammals. For example, it has been demonstrated that consumption by seals of prey that are contaminated with persistent organic pollutants may lead to vitamin and thyroid deficiencies and subsequent susceptibility to microbial infection and reproductive disorders (Ritter et al., 1995). There is evidence that chemical pollution may influence biodiversity, and increase organisms' vulnerability to other stresses, like diseases (e.g., Fisher et al., 1999, and see Khan, 2007 for references), and thus leading to a decline in the abundance of individuals in a population. For example, exposure to contaminants has been correlated with population declines in a number of marine mammals including the common seal, the harbor porpoise, bottlenosed dolphins, and beluga whales (Ritter et al., 1995).

Chemical pollution of the oceans can affect human health by direct exposure to pollutants associated with industrial and accidental point sources. But the greatest potential for large-scale effects on human health from chemical pollution is connected to consumption of contaminated seafood. For pollutants like dioxins and PCBs, and Hg, the major human exposure pathway is through intake of food, in general, and fish, in particular (Swedish Environmental Protection Agency, 2009). Environmental contaminant exposure in humans has been linked to negative health effects including sensitization, allergy, neurological development deficits, and disturbance of the thyroid hormone system (Grandjean and Landrigan, 2006; Grandjean et al., 2010). For populations that rely heavily on fish consumption in their diet this is of concern, and dietary intake of PCBs, for example, exceed national guidelines in a number of Arctic communities (AMAP, 2009).

Fish is a nutrient-rich food source with various health benefits for developing fetus and infants, and also for adults, for example, from fatty acids. However, the concerns raised regarding exposure to environmental contaminants via fish have led to restrictions and guidelines for fish consumption for certain consumer groups (Nesheim and Yaktine, 2007). The strictest guidelines apply to children and pregnant or breastfeeding women. As an example, EU has introduced rules that determine the highest concentrations of dioxins and dioxin-like PCBs that can be allowed in foodstuffs and in animal feeds (the commission's recommendation 2002/201/EU). Baltic herring and salmon often contain contaminants in higher concentrations than these set limits and, as a result, cannot be sold within EU (with the exception of human consumption in Finland and Sweden). Thus, as a consequence of chemical pollution, large parts of the population are now deprived of the health and economic benefits from fish consumption.

Chemical pollution can also impair recreational activities, such as angling if the fish is considered not safe to eat. Negative effects of chemicals on biodiversity, and, in particular, on higher trophic animals like marine mammals and birds may also have an impact on ecotourism and general quality of life.

Some of the most important physical and chemical properties of contaminants that determine how they are dispersed and partitioned in the environment are temperature dependent. Therefore, environmental levels of chemical pollutants can be influenced by climate change and climate variability. A comprehensive report on the climate change and persistent organic pollutants has been prepared by the Secretariat of the Stockholm Convention and the Arctic Council's Monitoring and Assessment Programme (UNEP/AMAP, 2011). The report reviews recent scientific findings on climate-change effects on persistent organic pollutants with a global perspective. The overall conclusion is that exposure and environmental fate of chemicals will be affected by a changing climate in various ways and with large uncertainties. Effects will mainly be visible on a regional scale, and on the global scale, a general conclusion was not possible to draw. Possible impacts of climate change on dissemination and effects of chemical pollution are further described below (UNEP/AMAP, 2011). An increased temperature will increase the volatility of organic chemicals and thus make them more mobile which could have an impact on the long-range transport of the chemical. A higher temperature can also increase the degradation rate of organic chemicals and contribute to a faster removal. However, degradation products of certain chemicals can sometimes be more toxic than the mother substance.

A changing climate will also lead to changes in wind fields and wind speed, and to less or more precipitation in certain areas. This will affect the deposition rate of pollutants, and the deposition may increase in areas with increased precipitation, or in areas in the wind direction of pollutant sources. Extreme events like storms and floods could have an impact of remobilization and subsequent bioavailability of pollutants in polluted soils or sediments. Climate change may also influence use patterns of chemicals and, for example, increase the demand for pesticides in certain areas. Changes of pollutant transport and fate as a result of climate change will have direct consequences on human and wildlife exposure.

Climate change will have an impact on biology of organisms (see also other chapters). A warmer climate will, for instance, affect cold-blooded organisms by enhancing the internal uptake in gills and the intestines. As pollutants can disrupt the immune and reproduction systems of organisms, they can put an extra stress on vulnerable animal species coping with a changing climate.

A changing climate can lead to changes in food web structures, like increased or decreased primary production, invasive species, or loss of key species. This may have consequences for the biomagnification of pollutants in food webs. Changes in food web structures may also alter the availability and quality of local food which may have consequences for bioaccumulation and human exposure of certain pollutants. Also climate-induced dietary shifts (e.g., between terrestrial and aquatic food webs) could largely alter both human and wildlife exposure to pollutants.

A consequence of the expected increased levels of carbon dioxide in the marine environment is ocean acidification, which is further described in Chapter 2. It is well known that pH interacts with

the toxicity of metals in aquatic organisms. The relationship between ocean acidification and toxicity of other pollutants is, however, less clear, but ocean acidification could affect the bioaccumulation potential of certain acidic organic compounds.

Regulations regarding chemical use and emissions exist on national, regional, as well as global levels in the form of complete bans, restricted use, or requirement for effective treatment of effluent. Two of the important global treaties are the Stockholm Convention on Persistent Organic Pollutants (2001) and the Geneva Convention on Long-Range Transboundary Air Pollution (LRTAP). LRTAP recognizes the existence of possible adverse effects, in the short and long term, from air pollution including transboundary disposal. The Stockholm Convention entered into force in 2004. Cosignatories agreed to outlaw nine of the so-called dirty dozen chemicals (Aldrin, Chlordane, Dieldrin, Endrin, Heptachlor, Hexachlorobenzene, Mirex, Toxaphene, and PCBs) and limit the use and unintentional production of three others. A set of new chemicals (brominated flame retardants and PFOS) were added to the Convention in 2009. There are also internationally recognized standards, guidelines, and other recommendations relating to food safety, with regard to contaminant and pesticide residue levels in food (see, e.g., Codex Alimentarius).

Bans and restrictions introduced against chemical pollutants have had considerable effect on concentrations in the environment. In the Baltic Sea, for example, the concentration of DDT has decreased at a rate of approximately 5-11% per year (in herring), since the end of the 1970s. Similarly, the concentration of PCB has decreased approximately with 5-10% per year, in herring and cod as well as in guillemot eggs and perch from the Baltic Sea, since the end of the 1970s (Bignert et al., 2011). However, other less-studied chemicals have been introduced and are currently used without restrictions.

OIL POLLUTION

Sources and Pathways

Petroleum, or crude oil, is a naturally occurring liquid that originates from the Carboniferous period about 200 to 400 million years ago. Plants, ferns, trees, and algae formed peat which over millions of years turned into petroleum (crude) oil, natural gas, and coal. Crude oil consists of a complex mixture of hydrocarbons of various molecular weights. In addition, nitrogen, oxygen, and sulfur occur in small quantities. The hydrocarbons consist of alkanes (paraffins) and cycloalkanes (naphthalenes), which are saturated hydrocarbons with straight or branched chains of hydrocarbon molecules. Alkanes and cycloalkanes normally constitute the dominant part of the oil, about 80%. These hydrocarbons have similar properties, but cycloalkanes have higher boiling points. The remaining hydrocarbons are aromatic, meaning that the molecules are unsaturated and made up of benzene rings. To this, group of molecules belongs the polysaturated aromatic hydrocarbons, some of which are known for their carcinogenic properties. One additional group of hydrocarbons that occur in varying amounts up to 10% in crude oil is the asphaltenes, which are molecules with relatively high weight. Oils consisting of a relatively high proportion of asphaltenes tend to be thick, almost like asphalt.

When the environmental aspects of oil spills in the marine environment are considered, it is important to recognize that the petroleum oil is a naturally occurring substance which for millions of years has seeped into the sea from seabed reservoirs.

In addition to natural seepage of petroleum hydrocarbons into the sea, anthropogenic activities contribute to oil contamination resulting directly from drilling operations and transportation and indirectly, for example, through fallout from the atmosphere. No recent estimates of these different sources can be found. However, based on the assessments made by the NRC (2003) and GESAMP (2001), updated with the information from ITOPF (2011), the following proportions are suggested (Table 6.1).

As a result of the combined efforts by the tanker industry and governments largely collaborating under the auspices of International Maritime Organization (IMO), the total amount of oil entering the oceans from tankers and tanker accidents has decreased significantly during the past decades. In the 1970s, on average, 25 major spills occurred with a total loss of between 200,000 and 700,000 tons per year. Gradually, these figures have dropped to an average 3-4 major spills per year with a total loss of 20,000-60,000 tons per year during the past 10+ years. This is the long-term

TABLE 6.1 Input of Petroleum Hydrocarbons into the Marine Environment (tons/year; GESAMP, 2001; NRC, 2003; ITOPF, 2011)

Source	Probable Range	Best Estimate	Percent
Natural seeps	20,000-2,000,000	200,000	10
Offshore production	200,000- 2,000,000	200,000	10
Ships: operational and deliberate release	100,000-1,000,000	500,000	26
Nontanker accidents	50,000-100,000	75,000	4
Tanker accidents	20,000-100,000	50,000	3
Atmosphere	50,000-500,000	300,000	16
Municipal and industrial wastes and runoff from land	200,000-1,500,000	600,000	31
Total	640,000-7,000,000	1,925,000	100

trend, but obviously in single years, this trend is broken due to events like the oil spills in the Gulf during the Iraq-Kuwait war or the Deepwater Horizon.

Small oil spills from shipping (defined as involving quantities less than 700 tons) usually occur as a result of human errors during routine operations especially during loading and discharging. However, deliberate (illegal) release of oil into the sea is also a relatively common problem. Accidents such as groundings and collisions are the most common causes of larger spills (larger than 700 tons) (Table 6.2).

An oil spill in the sea will undergo a series of chemical, biological, and physical processes which will lead to the degradation of the oil. The speed of these processes depends to a large extent on the character of the oil, the ambient temperature, and other environmental conditions. Heavy oils, i.e., oils with a high portion of hydrocarbons with high molecular weight, will degrade more slowly than lighter oils. In addition, at high temperatures, the degradation processes will go faster than at low temperatures. Important processes leading to the breakdown of an oil spill are evaporation, dispersion, dissolution, and sedimentation. Evaporation is a process where lighter compounds (molecules with four or fewer carbon atoms) enter a gaseous state. This process often removes significant portions of

TABLE 6.2 Causes of Oil Spill from Shipping (GESAMP, 2001; NRC, 2003; ITOPF, 2011)

Causes of Spills of Different Sizes	Small and Intermediate Spills (Less than 700 tons)	Larger Spills (More than 700 tons)
Groundings	19	34.2
Collisions	25	28.4
Loading/discharging	28	8.6
Hull failures	7	12.4
Fires and explosions	1	8.6
Bunkering	2	0
Other operations	5	0.3
Unknown	13	7.5

the oil in a few days after a spill, typically in the range of 40-75% of a medium to light crude oil (see, for example, Fingas, 1995). An example is the Ixtoc 1 blowout in the Gulf of Mexico in 1979-1980, when it was estimated that about 50% of the oil evaporated to the atmosphere and never reached the coastline (Jernelov and Linden, 1981). Dispersion is highly dependent on the state of the sea and in rough seas most of an oil spill may disperse into the water column more or less instantly. An example of this is the wreck of the tanker *Braer* in Shetland in 1993, when heavy weather conditions resulted in almost all of the oil being dispersed into the surrounding water resulting in little impact in littoral and near-shore benthic communities (Kingstone, 1999). Some oils are more easily dispersed than others, and the dispersion may be speeded up by the addition of chemical dispersants as a method to clean up the spill. Dispersed oil is more easily degraded by microorganisms.

Solvable components of the oil dissolve as the oil undergoes chemical stabilization in the water. Normally, only small fractions of the oil dissolve as most of the components that are dissolvable evaporate during the early phases of the spill. Portions of the dispersed oil will sediment to the seabed. The process of sedimentation depends on the density differences between the oil and the water. Another important factor is the incorporation of suspended sediment and detritus into the oil. Zooplankton may ingest small oil

droplets which will sink to the seabed after being excreted in fecal pellets. This process was documented after the *Arrow* spill in Chedabucto Bay (Conover, 1971) and the *Tsesis* in the Baltic Sea (Linden et al., 1979). In the process of sedimenting through the water column, the oil may accumulate on the top of denser water strata and spread with subsurface currents (NRC, 2003). This may explain the sudden appearance of oil in areas where no surface-drifting oil was observed, such as in the case of the Ixtoc 1 oil spill in Mexico (Jernelöv and Linden, 1981; Jernelöv, 2011).

Several reviews of the biodegradation of petroleum hydrocarbons have been published, for example, Atlas and Bartha (1992) and Heider et al. (1999), or the very recent report by the American Academy of Microbiology (2012). They show that dispersed oil is degraded by a number of different microorganisms. Heterotrophic bacteria and algae, fungus, and yeasts have been shown to be able to utilize the hydrocarbons as a carbon source to produce energy (so-called oxidative phosphorylation) while subsequently degrading the long-chain hydrocarbon molecules. A second degradation process is the metabolic detoxification of ingested oil droplets when organisms metabolize hydrocarbons to water-soluble products. At higher temperatures, these processes may degrade most of the oil in a relatively short time. Hence, the half-life of average medium crude oil may be less than 5 days, while at low temperatures, a larger portion of the dispersed oil can be expected to take significantly longer time and may sediment to the seabed. It should be pointed out that the local conditions determine the speed of the degradation. Factors of importance in addition to temperature and the composition of the oil are concentrations of nutrients in the water (P and N, in particular), the volume of water available for dilution, oxygen, light, pH, and salinity.

As mentioned above, the speed of different degradation processes are very much dependent on the ambient temperature. At higher temperatures, these processes are much more effective in eliminating the oil than at lower temperatures. This is an important aspect to keep in mind when assessing the risks in connection with Arctic oil exploration and transport. However, it should be pointed out that the investigations following the Deepwater Horizon blowout at 1500 m depth in the Gulf of Mexico where the temperature is only around 4 °C clearly show that low temperature is not as significant as previously thought in retarding the degradation of oil in the marine environment (American Academy of Microbiology, 2012).

Impacts on the Marine Environment and Ecosystem Services

Oil spills may under certain circumstances have serious effects on the marine ecosystem. In particular, the impacts on seabirds have been well documented (see, for example, NRC, 2003). Even relatively small single spills may cause significant mortality among seabirds such as long-tailed duck, eiders, and penguins. It is estimated that between 100,000 and 500,000 seabirds are killed due to oil spills every winter in the North and Baltic Seas. The oil is not only a threat to seabirds while it is floating at the surface or has contaminated shorelines. Cleaning of contaminated seabirds is sometimes attempted. However, the survival of birds that have undergone cleaning is often relatively low. However, there are cases when the cleaning of oiled birds have been successful, such as in connection with the rescue operations of the *Treasure* oil spill in South Africa in 2000 (Wolfaardt et al., 2008). The greatest threat to seabirds is when large numbers congregate during breeding or winter feeding. Under such conditions, single oil spills may cause long-term harm to entire populations. An illustrative case of this may be the Baltic population of guillemots (*Uria aalge*) where more than 90% of all the birds nest on the Stora and Lilla Karlsö west of Gotland in Sweden. Even a very small oil spill in or near this area during the period March to August could result in the elimination of most of this population.

Oil that is dispersed into the sea will come in contact with plankton. Plankton organisms contaminated with oil droplets on their surface and with ingested oil droplets have been observed in connection with a number of oil spills (Grose et al., 1977; Linden et al., 1979; Kingston et al., 1999). However, the presence of oil in the upper water column after oil spills is often patchy and transient. Field studies of the contamination of plankton after oil spills are therefore particularly difficult. The opinion among most experts is, however, that the impacts of single oil spills on plankton are of short-term nature (NRC, 2003). In most cases, the short-generation time of these organisms and the dilution of the oil in the water

column are likely to make such effects relatively short term. In the case of the *Tsesis* oil spill in the Stockholm archipelago, oil-contaminated plankton was observed during up to 4 weeks following the spill (Linden et al., 1979).

The littoral zone is often affected by oil spills. Impacts are often relatively severe, caused primarily by the physical properties of the oil; although in some cases, particularly with lighter oils, chemical toxicity may be a significant problem. The long-term impacts of oil spills in the littoral (intertidal) zone could be observed, for example, after the *Torrey Canyon* (Southward and Southward, 1978), Buzzards Bay (Sanders et al., 1980), and Exxon Valdez (see review in NRC, 2003) oil spills. Heavy oil contamination on shorelines may cause dramatic acute effects, such as extensive mortality of gastropod grazers and subsequent extensive proliferation of green algae that may prevent settlement of Fucoid algae, barnacles, and bivalve mussels for several years. However, the extent of damage can be difficult to predict against the background of natural fluctuations in species composition, abundance, and distribution in these habitats. The fact that many marine organisms reproduce through planktonic stages may speed up the recovery on locally impacted sites. However, species with limited dispersal will take longer to recolonize. There is often a debate about what constitutes recovery following an oil spill. With sophisticated chemical analysis, remaining fractions of the most persistent asphaltenes may be found many years—even decades—after the event (Boehm et al., 2004; Payne et al., 2008). However, there is a widespread acceptance among experts that biological recovery of shoreline ecosystems following oil spills under most circumstances is a matter of years rather than decades (NRC, 2003).

Oil from spills at the surface may under certain circumstances sink to the seabed in the form of oil droplets. Such droplets are subject to various degradation processes but temporary accumulation in the seabed sediment has been reported following several oil spills (Elmgren et al., 1983; Neff and Stubblefield, 1995; American Academy of Microbiology, 2012). Even relatively low levels of contamination of petroleum hydrocarbons, in the range of a few ppm of petroleum hydrocarbons, may cause impacts on sensitive organisms, for example, amphipods. Degradation of oil in seabed sediments may require several years and will vary according to, for example, prevailing temperature and oxygen conditions.

Marine mammals may be affected by oil spills although few systematic observations of such effects have been made. Whales and dolphins in the open sea are not likely to be particularly at risk. However, in shallow waters otters and seals can be (Hartung, 1995; Johnson and Garshelis, 1995; Wiens, 1995).

Oil spills may cause impacts on fish stocks under certain circumstances. For example, if light oil rich in toxic aromatic fractions become dispersed into confined waters in harbors or bays, the concentrations may become high enough to cause acute effects. This was observed in connection with the Amoco Cadiz oil spill in Brittany (see, for example, Spooner, 1978). However, the most common impact of oil spills on fisheries is the economic loss as a result of the disruption of the fishing activities while oil can be observed drifting in the water. In a number of cases, such disruption is the result of a precautionary ban on catching (and selling) fish and shellfish from the affected area to maintain market confidence (for a review, see, for example, ITOPF, 2004). Hence, in connection with the Gulf war in 1990-1991, there was a closure of the fishery in the upper Gulf for almost 2 years (Linden and Husain, 2002). Similarly, following the recent Deepwater Horizon event, authorities closed the fishery in an almost 10,000 km^2 large area around the Macondo well site. Aquaculture installations are also likely to suffer from oil spills and are particularly vulnerable since they cannot be readily relocated. In addition, cultivation equipment may be contaminated, providing a source for prolonged exposure to hydrocarbons. In connection with the Hebei Spirit oil spill in Korea, the contamination of shellfish cultures was extensive (ITOPF, 2008; Suh, 2011).

Coastal areas are often affected in connection with oil spills. Contaminated beaches as well as infrastructures, such as harbors and boats, require cleanup. The cleanup operations can be time consuming and labor intensive and therefore costly. In addition, shorelines contaminated with oil affect a number of recreational activities such as boating, bathing, and angling. The impacts on tourism including hotels and restaurants may be significant although short term. To provide figures for the economic impacts of oil spills based on quantities of spilt oil is problematic as the costs for cleanup, restoration, and compensation to those affected are highly dependent on a number of factors including the local conditions, the time of the year, and weather conditions at the time of the spill.

SOLID SUBSTANCES

A range of solid substances are both deliberately and unintentionally introduced into our oceans. Deliberate additions include legitimate actions undertaken with consent, such as redistribution of dredged sediment or mine tailings together with illegal actions such as dumping. The size of the item introduced into the oceans and the spatial extent of contamination are highly variable as are the potential effects on the environment. However, we can broadly consider introduction of solid substances in three categories: marine debris, sedimentation, and mine tailings, and the introduction/abandonment of hard structures. In some cases, the introduction of solid substances constitutes pollution, while in others, it may more appropriately be considered as a level of contaminations.

Marine Debris

In the context of this chapter, one of the most obvious and ironically largely unnecessary forms of pollution is the introduction of debris into the marine environment (Coe and Rogers, 1997). From a legislative and policy perspective, marine debris has recently been defined within the European Union as any form of manufactured or processed material discarded, disposed of, or abandoned in the marine environment. It consists of items made or used by humans and then deliberately or unintentionally lost to the sea including transport of these materials to the sea by rivers, drainage, sewage systems, or wind (Galgani et al., 2010).

Some of the most commonly reported items of marine litter are plastics, wood, metals, glass, rubber, clothing, and paper (Figure 6.2). Natural catastrophic events such as hurricanes, landslides, floods, and tidal waves can cause substantial, large scale, and indiscriminate introduction of all manner of items to the marine environment (Thompson et al., 2005). Putting such largely unavoidable introductions aside, there are some apparent trends in the types and quantities of debris entering our oceans. As a consequence of improved legislation, raised awareness, and port reception facilities, indiscriminate dumping of waste at sea is declining in many regions. There are, however, substantial accidental inputs

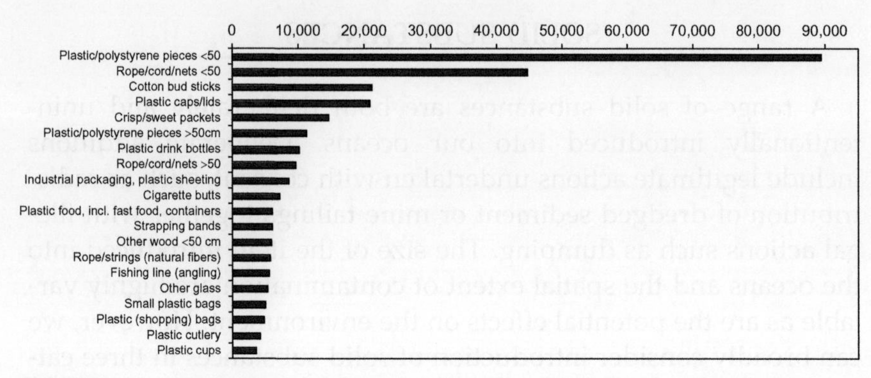

FIGURE 6.2 Total number of items of marine debris from selected reference beaches in Europe (OSPAR, 2009).

from loss of fishing equipment, including nets and accidental loss of cargo from vessels. Loss of fishing nets is of particular concern since they can continue to capture fish for long periods after they have been discarded, a process described as "ghost fishing" (Barnes et al., 2009; Ryan et al., 2009).

The proportion and the global extent of contamination of marine habitats by plastic debris have increased considerably in recent decades, partly reflecting the exponential increase in our use of plastic items and, in particular, their increasing usage for disposable and convenience items such as packaging. Plastic debris is now present in marine habitats from the poles to the equator and from the sea surface and intertidal to the deep sea (Barnes et al., 2009; Thompson et al., 2009; Browne et al., 2011). Marine animals are primarily affected by entanglement and ingestion of this debris, which can cause physical harm and even death (Derraik, 2002; Gregory, 2009). There is also concern that ingestion of plastics may present a toxicological hazard (Mato et al., 2001; Teuten et al., 2009; Hirai et al., 2011).

While plastics are a persistent and increasing proportion of marine debris, long-term trends in the quantities of plastic are highly variable and the temporal data that we have predominantly indicate increasing abundance or no apparent trend in abundance (Ryan et al., 2009; Law et al., 2010). Plastics are known to be incredibly durable materials that are likely to persist in the environment for hundreds if not thousands of years. Hence, gaps in our understanding of the ultimate sinks and impacts of plastic debris may account for the variability in abundance as monitored in the intertidal and sea surface. Three potential sinks are identified in the literature. For instance, there are reports of plastics becoming buried intertidal sediments (Williams and Tudor, 2001). In addition, while most plastics are buoyant in seawater, they can sink to the seabed as a consequence of fouling by marine organisms and the accumulation of sediment (Galgani et al., 2000). Hence, there are reports of quantities of plastic debris on the seabed that parallel those present at the sea surface (Barnes et al., 2009). There has also been considerable recent interest in the accumulation of small fragments of plastic described as microplastics (Thompson et al., 2004). Originally considered as truly microscopic, with fragments down to 1 μm being reported, this category now includes small fragments of plastic litter <5 mm (Arthur et al., 2009). These microplastics have formed either by the fragmentation of larger items or by direct introduction of plastic pellets and powders that are used in the manufacture of plastic items. Substantial quantities of these have entered the marine environment as a consequence of spillage and careless handling; for instance, there are reports of 10,000 pellets per m^2 on beaches in New Zealand (Gregory, 1978) and 100,000 small fragments per m^3 in Swedish coastal waters (Noren, 2008). An additional direct input of small fragments of plastic debris to the marine environment is from the use of plastic scrubbers and abrasives in cleaning products (typically, 200-300 μm in diameter) (Gregory, 1996). The majority of these particles are likely to pass through sewage treatment operations to aquatic habitats. The environmental impacts of microplastics are not yet fully understood. It has been suggested that these particles may cause physical damage to wildlife as a consequence of ingestion, and there is concern that ingestion may increase the transfer of chemicals to the food chain. This includes the transfer of potentially toxic chemicals that are incorporated into plastics during manufacture (phthalates, flame retardants, and antimicrobials) and the release of chemicals that have sorbed from the water column. More work is needed to establish the full environmental consequences of microplastics (GESAMP, 2010; Kershaw et al., 2011).

Looking to the future, while it seems certain that the potential demand for various types of plastic products will vary according to the climate scenarios, it is difficult to anticipate how this will influence the accumulation of marine debris. Since there are considerable carbon emissions associated with the production of plastic and potential savings in carbon emissions through recycling (STAP, 2011), we might anticipate a reduction in production of new plastic and an increase in recycling both of which could result in a reduction in the quantity of new items of debris entering the marine environment. What seems more certain is that unless we change our current patterns of production, use, and disposal, quantities of persistent plastic debris will increase.

Sedimentation and Mine Tailings

Increased sediment discharge and increased turbidity through runoff along many coastlines due to poor land-use practices and deforestation pose a significant threat to marine habitats, in particular, to coral and rocky reefs in many parts of the world (Airoldi, 2004; Halpern et al., 2007). Sedimentation can lead to smothering of the seabed, and extreme turbidity levels can be a stress factor for many marine organisms, for example, by reducing light (Orpin et al., 2004). Substantial quantities of sediment are also dredged from the seabed to maintain shipping channels or are disturbed and resuspended as a consequence of construction activities. Dredged material is typically deposited in areas of seabed nearby, while suspended material will settle in areas of reduced water flow. In addition, dredged material from harbors is frequently anoxic, due to the sheltered nature of harbor environments, and may also be contaminated with metals and persistent organic pollutants. The extent to which such contaminants are released increases as they become dispersed and oxygenated, for example, as a consequence of redistribution by tidal flow.

Mining operations on land also produce large quantities of residual material. For mines along the coast, this material is sometimes discharged to the sea, e.g., coal mines in Northumberland, UK and Copper mines in South Africa. In order to minimize damage inshore, material may be piped into deeper water as slurry. Wherever the material is introduced, it can result in smothering of benthic organisms, with potentially long-lasting effects from deposition. This is especially so for waste-containing substance like copper which is harmful to marine life (Figure 6.3). Additional

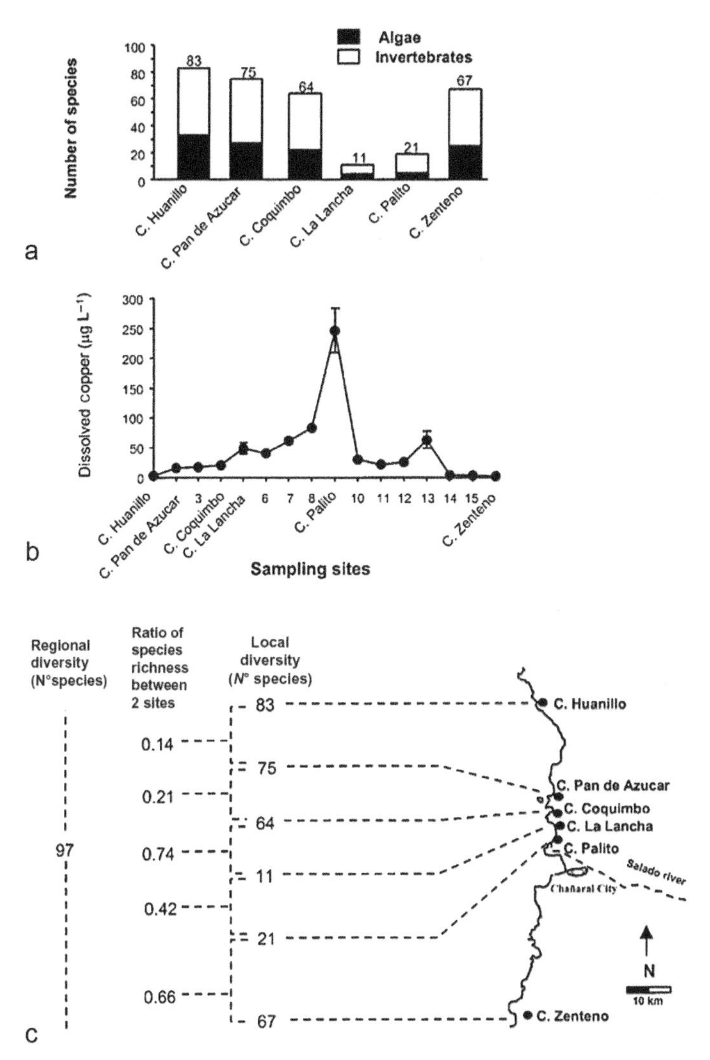

FIGURE 6.3 Patterns of biological diversity associated with discharge of minetailings form copper mines in Northern Chile. (a) Diversity profile from the northern Caleta Huanillo to the southern Caleta Zenteno. (b) Profile of dissolved copper along the same coastline. The copper value in Caleta Palito was obtained at the discharge, in the mixture zone. Values for copper at sites 10, 11, and 12 correspond to water samples taken at 200, 500, and 1000 m south of the discharge, respectively. (c) Detailed local diversity, values of species richness calculated for pairs of neighboring sites. *From OSPAR (2009), figure 2, page 109. © OSPAR Commission.*

sources of particulate industrial waste that have been discarded to the sea include fly ash produced as a relatively coarse residue from coal and oil-fired power stations and finer waste associated with extraction of clay.

While there appear to be reductions in chronic inputs of material from mining in many countries and increased considerations relating to dumping of dredged material, acute impacts from increasing construction of ports, harbors, sea walls, and renewable energy devices may lead to locally increased sedimentation. In addition, there is considerable wide-scale potential for changes in coastal dynamics and sediment redistribution as a consequence of increases in storm frequency and sea-level rise associated with climate change. While these latter effects are not strictly "pollution," they are likely to have substantial widespread effects on marine habitats and should be considered alongside changes in more directly anthropogenically driven sediment redistribution. There may be permanent changes in habitat type and associated biota where hard surfaces such as rocky reefs become smothered by sediment.

Looking to the future, some causes of sedimentation such as dumping of mine waste may become better regulated and decrease as a consequence, while others such as those associated with construction, land-use changes, and redistribution as a consequence of storms and flooding are likely to increase.

Introduction of Hard Structures

In terms of solid substances, we have considered fine particulates and items of marine litter ranging from less than a millimeter to fishing nets which can be hundreds of meters in length. A further category is the direct introduction and/or subsequent abandonment of hard structures. In many areas around the world, there are now derelict remains of oil and gas rigs, a classic example being the Gulf of Mexico where there are around 5000 platforms ranging in depth from about 3 m to over 1830 m. The process of decommissioning in many cases removes the upper part of the rigs leaving subsurface infrastructure in place, as part of the rigs to reefs program to create artificial reefs of potential benefit to fisheries (Macreadie et al., 2011). Although extraction of oil and gas is declining in many shallow-water locations, there are increases in deeper-water offshore operations.

In the context of climate change, increasing portions of our coastlines are being modified by introduction of hard structures to protect low-lying land from flooding. For example, 80% of the shoreline in some areas of the Adriatic has been modified from naturally sedimentary to being defended by hard structures (Airoldi et al., 2005). In addition, it is anticipated that over the next few decades, 10 thousands of renewable energy devices will be installed in the marine environment (Wilhelmsson et al., 2010). Historically, such marine engineering developments did not consider the potential for long-term environmental impacts beyond the operational lifetime of the structure and only recently has decommissioning started to be considered as part of the consenting process. The potential long-term impacts are a matter of debate; while it is clear that marine habitats will have been modified by the introduction of these structures, with altered patterns of sedimentation, tidal flows, and addition of hard epibenthic habitats, it is also important to consider the potential benefits that structures could offer through creation of artificial reefs (see Figure 6.4; Wilhelmsson et al., 2006; Langhamer and Wilhelmsson, 2009). These may provide a refuge, for example, from trawling and a source of food for fish stocks. While this should not be regarded as justification for the abandonment of derelict items of infrastructure on the seabed, there will undoubtedly be locations where the net environmental consequences of leaving a structure in place at the end of its serviceable life are lower than the consequences of its removal and disposal on land (Wilhelmsson et al., 2010; Macreadie et al., 2011).

FIGURE 6.4 An offshore wind turbine functioning as an artificial reef. *Photo. Dan Wilhelmsson.*

Looking to the future, it seems likely that there will be a progressive increase in the quantity of structures in our seas and likely an associated increase in the amount of derelict material remaining beyond the operational phase. While unattractive, the potential risks to the environment are probably not substantial. There is also the potential to consider potential ecological outcomes at the design and commissioning stage so as to influence assemblage composition of species colonizing structures and to introduce habitat for commercially exploited species (Martins et al., 2010).

RADIOACTIVE WASTE

Sources and Pathways

Since the mid-1940s, nuclear energy, plutonium production reactors, nuclear device testing, and weapon development have introduced man-made radioactivity into the environment in unprecedented quantities. As for most pollutants, the marine environment is a major recipient of this contamination, composed of a number of different radionuclides (e.g., ^{3}H, ^{14}C, ^{90}Sr, ^{99}Tc, ^{137}Cs, ^{210}Po, $^{238, 239, 240, 241}Pu$, ^{241}Am), through runoff, fallout from the atmosphere (e.g., from nuclear test explosions), or as liquid (e.g., reactor cooling water or outlets) and solid radioactive waste (e.g., nuclear submarines and national and international sea dumping in the 1940s to 1960s). For example, 60% of the global radioactive fallout from the first decades of thermonuclear device testing was deposited in the ocean (Aarkrog, 1998). In 1982, UNSCEAR reported that deposition of global fallout is about three times higher in the Northern hemisphere compared to the southern hemisphere (UNSCEAR, 1982). Contamination includes so-called anthropogenic radionuclides, which do not occur naturally, and so-called natural radionuclides that are also part of the natural background radioactivity.

Some radionuclides remain in the water column in soluble form, while insoluble radionuclides adhere to particles and are with time transferred to the seabed (e.g., Livingston and Anderson, 1983; Aarkrog, 1998). It has been shown that radionuclides are spread relatively widely by ocean currents (Aarkrog et al., 1987;

Aarkrog, 1998; Nakano and Povinec, 2003). Radioactive contamination in the ocean can also be transferred from the sea to terrestrial environments through spume and the food chain.

Effects on the Marine Environment and Ecosystem Services

Marine radioactive contamination can harm marine life and be concentrated in organisms consumed by humans, such as shell fish and algae (Preston and Jefferies, 1969; Beasley and Lorz, 1986; Aarkrog et al., 1987, Aarkrog, 1998, Nakano and Povinec, 2003; Børretzen and Salbu, 2009), although impacts on human health may be of most concern. A review by Glover and Smith (2003), for example, suggested that the past impacts of radioactive waste disposal and lost nuclear reactors have only had low and local impacts on the deep-sea-floor ecosystem. From a human health perspective, a major international study assessed the collective doses of naturally occurring and anthropogenic radionuclides by consumption of sea food (i.e., fish and shellfish ingestion that is believed to be the most important pathway) for each FAO fishing area, using radioactivity data for water and biota (Aarkrog et al., 1997). The main contributors to the collective dose from natural and anthropogenic radionuclides through marine foods are ^{210}Po (~70%) and ^{137}Cs (~80%), respectively (Aarkrog et al., 1997). It was suggested that the global mean doses from anthropogenic radionuclides are well below the accepted values for the public and do not present a significant health hazard, in concurrence with a subsequent study by Livingston and Povinec (2000). It was evident, however, that there is a considerable regional variation, with some "hot spots," such as areas contaminated by well-known point sources, which require special attention.

Data on naturally occurring radionuclides (represented by ^{210}Po in the studies) are less reliable, and although the collective doses from sea food seem lower than doses from the terrestrial environment and are below the accepted values (Aarkrog et al., 1997; Livingston and Povinec, 2000), the results need to be treated with special caution (Aarkrog et al., 1997).

In addition to a number of case studies, several global estimates have been made and these are, in general, agreement in terms of the global scale of the problem. Radioactive contamination of the oceans may well have serious consequences for marine life and human health through seafood consumption in particular localities and regions (Sazykina and Kryshev, 1997; Sazykina et al., 1998) as well as in relation to nuclear accidents, but the risk of significant impacts on ecosystems and humans at a global scale is low (Aarkrog et al., 1997; Livingston and Povinec, 2000).

It is suggested, however, that future research efforts should, in particular, focus on gaining a better understanding of the behavior of radionuclides in the marine environment and on improving the reliability of the dose factors for these radionuclides, as well as on the long-term behavior of radionuclides in different types of nuclear devices resting on the seabed in the world's oceans (e.g., sunken submarines, satellites, isotope batteries; e.g., Aarkrog, 1998; Nakamo and Povinec, 2003).

NOISE

Sources and Pathways

Sound travels very efficiently underwater, with about five times the speed of sound in air. Many parts of the oceans are naturally quite noisy, when considering all frequencies of sounds. In addition to biotic sounds such as those generated by shrimps, fish, and marine mammals, natural sound sources including breaking waves, earthquakes, and volcanic activity, as well as precipitation and lightning strikes on the surface (Wilson et al., 1985; Nystuen and Farmer, 1987; Weilgart, 2007).

These natural sounds are, however, increasingly overwhelmed by anthropogenic underwater noise, which has increased dramatically, with a doubling every decade in some areas, during the past 60 years (IWC, 2004; Samuel et al., 2005; Weilgart, 2007; Tougaard et al., 2009). Noise pollution in the oceans results from a variety of activities, with underwater explosions, seismic exploration, shipping (supertankers, merchant vessels, fishing vessels), the operation of naval sonar being the main noise source (see Weilgart, 2007 for references). Other sources include, for example, marine construction; seabed drilling; ice-breaking vessels; navigational-, fishery-, and research sonars; and the use of acoustic harassment devices to keep marine mammals away from fishing nets and fish farms (see Weilgart, 2007 for references). Underwater noise pollution of the oceans is likely to increase significantly in the future. For example, ship noise pollution in the Pacific Ocean is predicted to more than double by 2025 (McDonald et al., 2006). Further, if demand for offshore fossil fuels continues to increase, seismic exploration activities of the seafloor by the oil and gas industries can be expected to grow.

Effects on the Marine Environment and Ecosystem Services

Most marine animals use sound to a greater extent than vision for feeding, communicating (e.g., for maintaining group cohesion and often linked to reproduction), avoiding predators, navigating, and orientation (e.g., Hawkins, 1993; Croll et al., 2001; Noad et al., 2002; Amoser and Ladich, 2005; Simpson et al., 2005; Weilgart, 2007). For example, many species, including fish, marine mammals, and invertebrates, make use of calls during spawning (ICES, 2005), and fish larvae are believed to follow the sound of breaking waves to find reefs to settle (Simpson et al., 2005). Cetaceans (whales, dolphins, and porpoises) rely almost totally on sound and have highly specialized and sensitive hearing (Ketten, 2004). Toothed whales (Odontocetes), for example, use echolocation to orient themselves and to find food.

Noise pollution can, thus, adversely affect marine animals in various ways, with impacts on individuals ranging from disturbance of their ability to cope with the environment to instant mortality. In 2004, the UN Secretary-General stated that "there is a growing concern that noise proliferation poses a significant threat to the survival of marine mammals, fish and other marine species . . . flooding their world with intense sound interferes with (their) activities with potential serious consequences" (UN, Report on Oceans and the Law of the Seas, 2004).

Still, the effects of underwater noise on fish and mammals are relatively poorly studied (e.g., Popper and Hastings, 2009), although experiments and theoretical estimates are available for a number of species (Nedwell and Howell, 2003; Wahlberg and Westerberg, 2005; Thomsen et al., 2006; Kastelein et al., 2007). There are significant differences in the hearing and processing of sound among species, and the nature and detection level of different kinds of noise is largely unexplored (Thomsen et al., 2006; Kastelein et al., 2007). The response also depends on the life cycle stage, species (highly variable), and body size. Moreover, in terms of receptivity of noise, some species of fish are more sensitive to particle motion than to pressure (sound waves), and their responses to subsea noise and vibrations are poorly known (Thomsen et al., 2006). The impacts of produced noise on a particular species obviously also depend on the frequency, intensity, duration of the noise. Further, the propagation of underwater noise of a given nature depends on a variety of factors, such as depth, salinity, temperature, bottom topography, and seabed configuration (Nedwell and Howell, 2003; Wahlberg and Westerberg, 2005). Nevertheless, the hearing and avoidance threshold for a number of fish and mammal species have been investigated and modeled, and responses and impacts have been recorded (see, e.g., Figure 6.5).

For example, naval sonars used to detect submarines could affect fish and whales over an area of about 4 million km^2 (see Weilgart, 2007 for references). A seismic survey (using loud air guns), often including sounds generated every 10 s for weeks at a time, can raise noise levels 100 times for days within 300,000 km^2 and has been recorded as the loudest background noise in the ocean 3000 km from the source (Nieukirk et al., 2004, see Weilgart, 2007 for references). Worldwide at any given day, over 20 seismic survey ships are conducting surveys within the oil and gas sector only (MMC,

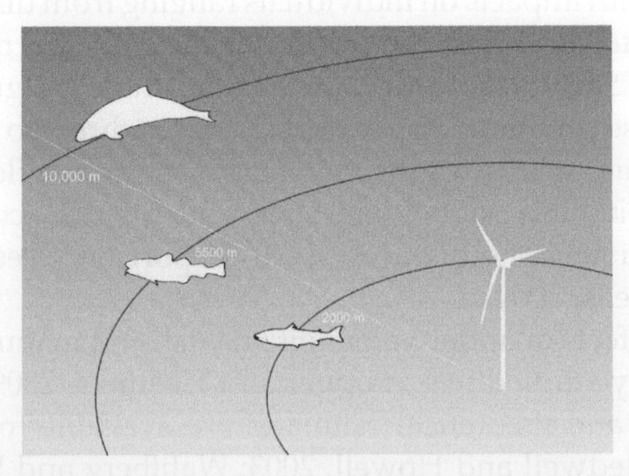

FIGURE 6.5 Example of potential noise disturbance of marine organisms: Suggested radii within which avoidance could be initiated by different species due to noise emitted during construction of offshore wind turbines. *From Wilhelmsson et al. (2010).*

2007). Fish, marine mammals, and sea turtles can be killed directly by the pressure wave close to the noise source (Hardyniec and Skeen, 2005; Nowacek et al., 2007; Snyder and Kaiser, 2009). Further, the evidence of that fatal whale mass strandings, after the whales have been injured in organs and hearing-related structures, can be caused by intense military sonars, in particular, when beaked whales are involved (Frantzis, 1998; Jepson et al., 2003; IWC, 2004, and see Weilgart, 2007 for references). Seismic air guns have also been blamed for causing internal injuries and subsequent mass strandings of both whales and giant squid.

The sublethal effects of anthropogenic noise may be as serious for populations as the lethal effects, by hampering the survival, growth, and reproduction of affected animals. Anthropogenic noise can, for example, mask natural sounds and thereby affect communication between animals (e.g., the acoustic range of blue whales may have decreased by 90% due to noise pollution; McDonald et al., 2006) and their ability to find prey and avoid predators or navigate (Miller et al., 2000; Erbe, 2002). Noise can cause avoidance by the animals, which means either habitat loss (which can include important feeding and spawning grounds) or extended migration routes (Engås et al., 1996; Tougaard et al., 2003; Thompsen et al., 2006). Animals can also indirectly be affected if their prey abandons the areas due to the noise disturbance (Weilgart, 2007). Other recorded sublethal effects range from behavioral disturbance, stress, and weakened immune systems, to damage of acoustic apparatus of fish and mammals, which are all effects that may increase mortality and hamper growth and reproduction (e.g., Morton and Symonds, 2002; Olesiuk et al., 2002; McCauley et al., 2003; IWC, 2004; Slotte et al., 2004; Weilgart, 2007; Popper and Hastings, 2009, and see Weilgart, 2007 for references).

Less research has been done on the effects of underwater noise on invertebrates (Moriyasu et al., 2004). Invertebrates constitute a diverse array of animal groups with an even wider range of morphologies represented compared to fish and marine mammals. Potential responses may vary greatly, and little is known about the effects on differing life cycle stages. For example, within the class of Malacostracans (e.g., crabs, lobsters, shrimps, krill), the observed responses to noise vary from no measurable reactions to increased mortality and reductions in growth and reproduction rates (Lagardère, 1982; Moriyasu et al., 2004). Furthermore, some mollusks seem sensitive to acute noises, while others seem very tolerant (Moriyasu et al., 2004).

As indicated, many local and incidental directly lethal impacts on marine mammal and fish assemblages or populations of noise have been and can be well documented and quantified through dedicated surveillance and follow-up efforts. In general, however, in particular, at a global scale, it is currently not possible to quantify these effects in isolation from other impacts, such as fishing, eutrophication, and inappropriate coastal development. Even local effects along specified coastal areas are difficult to assess due to the lack of background data on both fish and marine mammal populations. Weilgart (2007), for example, points out that, with the current monitoring effort and baseline data, major declines (even up to 50%) in marine mammal populations would only be detected in 30% of the cases. Furthermore, little scientific data are available on the severity of sublethal noise impacts, for example, masking of ambient sounds, stress, and hearing impairment, on survival and reproduction of individual fish and mammals, let alone on the impacts at population scale (e.g., Tyack, 2008).

It is noted, though, that direct mortality of individuals due to underwater noise may be serious for endangered species or populations of fish and whales. For marine mammals, for example, International Whaling Commission´s Scientific Committee (IWC, 2004) stated that "repeated and persistent acoustic insults (over) a larger area. . .should be considered enough to cause population level impacts." Noise is believed to contribute to the decline or lack of recovery of several whale species, and it is believed that a single military sonar event has killed or relocated an entire local whale population in Bahamas in 2000 (see Weilgart, 2007 for references). Even sublethal effects could put species survival at risk.

As for other threats and pressures on marine resources, the loss, weakening, or relocation of species populations, stocks, and ecosystem services due to noise pollution in the oceans have adverse effects on human societies, including, for example, impacts on fisheries resources and recreational, esthetic, and scientific values. No solid estimates of the risk of significant impacts on ecosystems and humans of noise pollution at a global scale are available, and as mentioned, large research gaps exist also in case of specific and local scales. Seen in isolation, though, and in comparison with other threats to the ocean, the risk at global scale should be low.

In spite of research gaps in terms of the magnitude of impacts of underwater noise pollution on individuals and populations, current scientific evidence and well-recorded incidents have formed the basis for a multitude of national and international requests for regulatory, awareness creating, and precautionary measures. Since 2004, the 1982 UN Convention of the Law of the Sea emphasizes the threat of underwater noise pollution, primarily to whales and other cetaceans, and the needs for regulations, research, and mitigation efforts in their annual reports to the General Assembly. The EU adopted Marine Strategy Framework Directive (2008), moreover, includes underwater noise in the definition of pollution and requests member states to address and control the impacts on marine populations. The IMO acknowledges that the noise from ships may have negative impacts on marine organisms (IMO Resolution A.720(17), 1991, as replaced by IMO Resolution A.982 (24), 2005).

PRESENT AND FORECASTED IMPACTS OF POLLUTION ON GLOBAL MARINE ECOSYSTEMS

As described in this chapter, pollutants are major stressors in many locations and impacts on marine species, habitats, and ecosystem services are evident in numerous regions. At a global scale, considering the ocean ecosystems as a whole and compared to other global threats, such as ocean acidification, overfishing, and increase of sea-surface temperatures, the significance of the impacts of marine pollution is less clear.

As part of the overall scope of this book, attempts to assess the risk of "large-scale changes in global marine ecosystems" due to the different threats described in the chapters were made. Two different climate-change scenarios developed by the International Panel on Climate Change were selected to set the scene for the assessments (RCP3 and RCP6; see Chapter 1 for further description). The climate-change scenarios have underlying storylines representing different socioeconomic, technological, environmental, and demographic developments, to add context for the scenario quantification related to the different emission-driving forces. While this was written the storylines were still under development by IPCC, and therefore, values developed for the most similar climate-change scenarios described in the previous IPCC report (IPCC, 2007) were used for the present assessments. In terms of pollution, however, the resolution and focus of the storylines, which specifically target greenhouse gas emission drivers, do not allow for solid predictions of the concurrent magnitude of marine pollution. At a finer resolution, each storyline can, namely, harbor different assumed pathways, in terms of, for example, production, transport, and consumption patterns of goods, services, and renewable energy.

One would only be able to speculate that, for example, population growth and growth of the global GNP per capita will lead to increased marine pollution, although the future use of fossil fuels will influence the extent of oil spills and chemical pollution (e.g., mercury from coal burning, Rockström et al., 2009). Shifting climate zones could introduce new types of pests to different regions and thereby increase the use of pesticides. Increased temperatures and changed precipitation patterns are likely to influence the behavior, pathways, and effects of pollutants, with increased or decreased impacts depending on pollutant and region (e.g., Lamon et al., 2009; Noyes et al., 2009; and described above). Other categorical conclusions could include that an expansion of renewable energy sources would add noise pollution in the ocean through offshore wind- and wave-power development (e.g., Wilhelmsson et al., 2010). On the other hand, if the decrease in the use of fossil fuels results in fewer offshore oil rigs and reduced shipping and seismic exploration of the sea floor, the noise pollution from these sectors would decrease.

The production, use, and dissemination of specific pollutants are, moreover, influenced by the future development of national and international agreements, policies, and priorities. Several possible development pathways to this regard can be encompassed within the IPCC scenarios, and which may also be decoupled from combating climate change. The present coarse assessment is therefore limited to the current estimated impacts of pollution on global marine ecosystems. Moreover, available scientific data, the level of coordination of data, and current monitoring efforts do not allow for a rigorous and generic quantification of the risks of substantial impacts on marine ecosystems at global scale. As, for example, Rockström and colleagues conclude with regard to chemical pollution, "Thresholds can be identified for only a few single chemicals or chemical groups and for only a few species, such as top predators."

Marine pollution poses significant threats to food quality and access and thereby to human health in many areas. In terms of consequences for marine species, it is, furthermore, clear that pollution of the oceans can have serious impacts in many localities and regions, and there is certainly a risk of widespread impacts of pollution at ecosystems scale (as described above and in, e.g., Islam and Tanaka, 2004; Halpern et al., 2007).

On a global average on the other hand, the overall pollution input in an area may be rare to occasional and have moderate effects on ecosystems, and affected global marine ecosystems may typically (but depending on the kind of pollution and affected ecosystem components) be reasonably resilient to pollution and have a relatively short recovery time (e.g., Halpern et al., 2007; Costello et al., 2010). For example, the individual impacts of the dumping of wastes, radioactive waste disposal, underwater noise, dumping of oil/gas structures, shipping (oil spills and other pollution), and sewage discharge at the deep sea floor, in pelagic waters, and on continental slopes have been suggested to be low and local to regional, and impacts from waste dumping, shipping, and sewage discharge were predicted to be moderate by 2025 (MA, 2005).

The distribution of pollution impacts is heterogeneous at global scale. For example, the reports of the Global International Waters Assessment (UNEP, 2006) conclude that pollution is a major concern in more than half of the regions investigated, although the study also included large lakes and riverine systems. Of the 66 regions included in the study, chemical pollution was considered severe in 9 and moderate in a further 35 regions, while solid waste was particularly severe in many regions in Africa, the Pacific Islands, and the Indonesian Seas and was considered the most severe pollution issue for the Indian Ocean Islands. Taking into account the spatial extent and distribution of both ecosystem types and pollution, it suggests that pollution has, in general, relatively small spatial extents when considering the oceans as a whole, while, naturally, posing a higher overall threat in many coastal areas and habitats (MA, 2005; Halpern et al., 2008).

Importantly, pollution impacts undermine the resilience of ecosystems to other stressors, such as those described in other chapters of this book (Jenssen, 2006; Khan, 2007; Piola and Johnston, 2008; Noyes et al., 2009; Russel et al., 2009; McLeod et al., 2010; Negri and Hoogenboom, 2011). Treating marine pollution in isolation though, the average risk of "large-scale changes in ocean ecosystems" due to this threat alone is moderate in comparison with several other single threats to the ocean (e.g., ocean acidification, industrial fishing), with a higher risk for coastal habitats than for the high/open and deep seas.

It should be noted that this is with reservation for the associated uncertainties underlying these assessments, which are, for example, delimited by what is known about the relatively small percentage of the chemicals that has been systematically investigated. In addition, the combined effects of chemicals (i.e., "cocktail effects") and synergies between different types of pollutants are virtually unknown.

References

Aarkrog, A., S. Boelskifte, H. Dahlgaard, S. Duniec, L. Hallstadius, E. Holm, and J. N. Smith (1987), Technetium-99 and Caesium-134 as long distance tracers in Arctic Waters, *Estuar. Coast. Shelf Sci.*, **24**, 637–647.

Aarkrog, A., M. S. Baxter, A. O. Bettencourt, R. Bojanowski, et al. (1997), A comparison of doses from 137Cs and 210Po in marine food: a major international study, *J. Environ. Radioact.*, **34**, 69–90.

Aarkrog, A. (1998), A retrospect of anthropogenic radioactivity in the global marine environment, *Radiat. Prot. Dosimetry*, **75**, 23–31.

Airoldi, L. (2004), Forecasting the effects of enhanced sediment loads to coastal areas: a plea for long-term monitoring and experiments, *Aquatic Conservation-Marine and Freshwater Ecosystems*, **14**, 115–117.

Airoldi, L., M. Abbiati, M. W. Beck, S. J. Hawkins, P. R. Jonsson, D. Martin, P. S. Moschella, A. Sundelof, R. C. Thompson, and P. Aberg (2005), An ecological perspective on the deployment and design of low-crested and other hard coastal defence structures, *Coastal Engineering*, **52**, 1073–1087.

AMAP (2009): Arctic monitoring of human media for persistent organic pollutants in the regions of the Central and Eastern Europe and the Western European and other groups. In: *First regional monitoring report, Western Europe and other states group (WEOG) region.* Global monitoring plan for persistent organic pollutants under the Stockholm convention, article 16 on effectiveness evaluation.

American Academy of Microbiology (2012), Microbes and Oil Spills, Report, http://www.dfo-mpo.gc.ca/science/publications/article/2012/01-25-12-eng.html.

Amoser, S., and F. L. Ladich (2005), Are hearing sensitivities of freshwater fish adapted to the ambient noise in their habitats? *J. Exp. Biol.*, **208**, 3533–3542.

Arthur, C., J. Baker, and H. Bamford (2009), Proceedings of the international research workshop on the occurrence, effects and fate of microplastic marine debris. September 9–11, 2008: NOAA Technical Memorandum NOS-OR&R30.

Atlas, R. M., and R. Bartha (1992): Hydrocarbon biodegradation and oil spill bioremediation. In *Advances in Microbial Ecology* 12, 287339. [Marshall K. C. (ed.)]. Plenum Press, New York, NY.

Barnes, D. K. A., F. Galgani, R. C. Thompson, and M. Barlaz (2009), Accumulation and fragmentation of plastic debris in global environments, *Philos. Trans. R. Soc. B*, 1985–1998.

Beasley, T. M., and H. V. Lorz (1986), A review of the Biological and geochemical behavior of Technetium in the marine environment, *J. Environ. Radioactiv.*, **3**, 1–22.

Bignert A., E. Boalt, S. Danielsson, J. Hedman, A. -K. Johansson, A. Miller, E. Nyberg, U. Berger, H. Borg, U. Eriksson, K. Holm, K. Nylund, and P. Haglund (2011), Comments Concerning the National Swedish Contaminant Monitoring Programme in Marine Biota, 2011. Report to the Swedish Nature Protection Agency no. 7, 2011.

Boehm, P. D., D. S. Page, J. S. Brown, J. M. Neff, and W. A. Burns (2004), Polycyclic aromatic hydrocarbons in mussels from Prince William Sound, Alaska, documented the return to baseline conditions, *Environ. Toxicol. Chem.*, **12**, 2916–2929.

Brretzen, P., and B. Salbu (2009), Bioavailability of sediment-associated and low-molecular-mass species of radionuclides/trace metals to the mussel *Mytilus edulis*, *J. Environ. Radioactiv.*, **100**, 333–341.

Browne, M. A., P. Crump, S. J. Niven, E. Teuten, A. Tonkin, T. Galloway, and R. Thompson (2011), Accumulation of Microplastic on Shorelines Worldwide: Sources and Sinks, *Environ. Sci. Technol.*, **45**, 9175–9179.

Codex Alimentarius. WHO/FAO. Food safety standards, http://www.codexalimentarius.org/.

Coe, J. M. and D. B. Rogers (eds.) (1997), *Marine debris: sources, impacts and solutions*. Springer, New York, 108.

Conover, R. J. (1971), Some relations between zooplankton and bunker C oil in Chedabucto Bay, following the wreck of the tanker Arrow, *J. Fish. Res. Board Can.*, **28**, 1327–1330.

Costello, M. J., M. Coll, R. Danovaro, P. Halpin, H. Ojaveer, and P. Miloslavich (2010), A census of marine biodiversity knowledge, resources, and future challenges, *PLoS ONE*, **5**, e12110.

Croll, D. A., C. W. Clark, J. Calambokidis, W. T. Ellison, and B. R. Tershy (2001), Effect of anthropogenic low-frequency noise on the foraging ecology of Balaenoptera whales, *Anim. Conserv.*, **4**, 13–27.

Derraik, J. G. B. (2002), The pollution of the marine environment by plastic debris: a review, *Mar. Pollut. Bull.*, **44**, 842–852.

de Wit, C. A., A. T. Fisk, K. E. Hobbs, D. C. G. Muir, G. W. Gabrielsen, R. Kallenborn, M. M. Krahn, R. J. Norstrom, and J. U. Skaare (2004) In: *AMAP assessment 2002: persistent organic pollutants in the Arctic* [de Wit C. A., A. T. Fisk, K. E. Hobbs and D. C. G. Muir (eds.)]. Arctic Monitoring and Assessment Programme, Oslo, Norway.

Dotinga, H. M., and A. G. O. Elferink (2000), Acoustic pollution in the oceans: the search for legal standards, *Ocean Development and International Law*, **31**, 151–182.

Elmgren, R., S. Hansson, U. Larsson, B. Sundelin, and P. Boehm (1983), The "Tsesis" oil spill: acute and long-term impact on the benthos, *Mar. Biol.*, **73**(1), 51–65.

Engås, A., S. Lokkeborg, E. Ona, and A. V. Soldal (1996), Effects of seismic shooting on local abundance and catch rates of cod (*Gadus morhua*) and haddock (*Melanogrammus aeglefinus*), *Can. J. Fish. Aquat. Sci.*, **53**, 2238–2249.

Erbe, C. (2002), Underwater noise of whale-watching boats and potential effects on killer whales (*Orcinus orca*), based on an acoustic impact model, *Marine Mammals Science*, **18**, 394–418.

EU, Commission of the European Communities (2001), *White paper: strategy for future chemicals policy*. (COM2001 88 final), Commission of the European Communities, Brussels, EU.

Fingas, M. F. (1995), A literature review of the physics and predictive modeling of oil spill evaporation, *J. Hazard. Mater.*, **42**, 157–175.

Fisher, W. S., L. M. Oliver, W. W. Walker, C. S. Manning, and T. F. Lytle (1999), Decreased resistance of eastern oysters (*Crassostrea virginica*) to a protozoan pathogen (*Perkinsus marinus*) after sublethal exposure to tributylltin oxide, *Mar. Environ. Res.*, **47**, 185–201.

Frantzis, A. (1998), Does acoustic testing strand whales? *Nature*, **392**, 29.

Galgani, F., D. Fleet, J. Van Franeker, S. Katsanevakis, T. Maes, J. Mouat, L. Oosterbaan, I. Poitou, G. Hanke, R. Thompson, E. Amato, A. Birkun, and C. Janssen (2010): Marine Strategy Framework Directive, Task Group 10 Report: Marine Litter. In: *JRC Scientific and Technical Reports* [Zampoukas N. (ed.)]. European Commission Joint Research Centre, Ispra.

Galgani, F., J. P. Leaute, P. Moguedet, A. Souplet, Y. Verin, A. Carpentier, H. Goraguer, D. Latrouite, B. Andral, Y. Cadiou, J. C. Mahe, J. C. Poulard, and P. Nerisson (2000), Litter on the sea floor along European coasts, *Mar. Pollut. Bull.*, **40**, 516–527.

GESAMP (2001), *A sea of troubles*, GESAMP, London.

GESAMP (2010): Proceedings of the GESAMP International Workshop on plastic particles as a vector in transporting persistent, bio-accumulating and toxic substances in the oceans. In: GESAMP Reports and Studies [Bowmer T. and P. J. Kershaw (eds.)]. MO/FAO/UNESCO-IOC/UNIDO/WMO/IAEA/UN/UNEP.

Glover, A. G., and C. R. Smith (2003), The deep-sea floor ecosystem: current status and prospects of anthropogenic change by the year 2025, *Environmental Conservation*, **30**, 219–241.

Goldberg, E. D. (1995), Emerging problems in the coastal zone for the twenty-first century, *Mar. Pollut. Bull.*, **31**(4–12), 152–158.

Grandjean, P., and P. J. Landrigan (2006), Developmental neurotoxicity of industrial chemicals. *Lancet*, **368**, 2167–2178.

Grandjean, P., L. K. Poulsen, C. Heilmann, U. Steuerwald, and P. Weihe (2010), Allergy and Sensitization during Childhood Associated with Prenatal and Lactational Exposure to Marine Pollutants, *Environ. Health Perspect.*, **118**(10), 1429–1433.

Gregory, M. R. (1978), Accumulation and distribution of virgin plastic granules on New Zealand beaches, *NZ J. Mar. Freshwat. Res.*, **12**, 339–414.

Gregory, M. R. (1996), Plastic 'scrubbers' in hand cleansers: A further (and minor) source for marine pollution identified, *Mar. Pollut. Bull.*, **32**, 867–871.

Gregory, M. R. (2009), Environmental implications of plastic debris in marine settings—entanglement, ingestion, smothering, hangers-on, hitch-hiking, and alien invasions, *Philos. Trans. R. Soc. B*, **364**, 2013–2026.

Grose, P., L. Mattson, and S. James (1977), *The Argo Merchant oil spill: a preliminary scientific report,* NOAA, US Dept Commence, Boulder.

Halpern, B. S., K. A. Selkoe, F. Micheli, and C. V. Kappel (2007), Evaluating and ranking the vulnerability of global marine ecosystems to anthropogenic threats, *Conserv. Biol.*, **21**, 1301–1315.

Halpern, B. S., S. Walbridge, K. A. Selkoe, et al. (2008), A global map of human impact on marine ecosystems, *Science*, **319**, 948–952.

Hardyniec, S., and S. Skeen (2005), Pile driving and barotrauma effects, *Journal of the Transportation Research Board*, 1941, 184–190.

Hartung, R. (1995): Assessment of the potential for long-term toxicological effects of the Exxon Valdez oil spill on birds and mammals. In: *Exxon Valdez Oil Spill: Fate and Effects in Alaskan Waters. ASTMSTP 1219.* [Wells P., J. N. Butler and J. S. Hughes (eds.)]. American Society for Testing and Materials, Philadelphia.

Hawkins, A. D. (1993): Underwater sound and fish behavior. In: *Behaviour of Teleost Fishes* (2nd edition). [Pitcher T. J. (ed.)]. Chapman & Hall, London, 129–169.

Heider, J., A. M. Spormann, H. R. Beller, and F. Widdel (1999), Anaerobic bacterial metabolism of hydrocarbons, *FEMS Microbiol. Rev.*, **22**, 459–473.

Hirai, H., H. Takada, Y. Ogata, R. Yamashita, K. Mizukawa, M. Saha, C. Kwan, C. Moore, H. Gray, D. Laursen, E. R. Zettler, J. W. Farrington, C. M. Reddy, E. E. Peacock, and M. W. Ward (2011), Organic micropollutants in marine plastics debris from the open ocean and remote and urban beaches, *Mar. Pollut. Bull.*, **62**, 1683–1692.

Hong, S., J. Candelone, C. C. Patterson, and C. F. Boutron (1994), Greenland ice evidence of hemispheric lead pollution two millennia ago by Greek and Roman civilizations, *Science*, **265**, 1841–1843.

ICES (2005), ICES Advisory Committee on Ecosystems (AGISC). Report of the Ad-Hoc Group on the impacts of sonar on Cetaceans and fish (AGISC) (2nd edition).

IMO Resolution A.982(24), 2005; http://www.imo.org/blast/blastDataHelper.asp?data_id=14373&filename=982.pdf.

IPCC (2007): Summary for Policy-makers, Climate Change 2007: Mitigation. Contribution of Working Group III to the Fourth Assessment Report of the IPCC. In: [Metz, B., O. R. Davidson, P. R. Bosch, R. Dave and L. A. Meyer (eds.)]. Cambridge University Press, Cambridge, United Kingdom and New York, NY, USA.

Islam, M., and M. Tanaka (2004), Impacts of pollution on coastal and marine ecosystems including coastal and marine fisheries and approach for management: a review and synthesis, *Mar. Pollut. Bull.*, **48**, 624–649.

ITOPF (2004), Oil spill effects on fisheries. ITOPF Technical Information Paper No. 3. www.itopf.org.

ITOPF (2008), The environmental impact of the Hebei Spirit Oil Spill, Taean, South Korea (7th December 2007), http://www.itopf.com/news-and-events/documents/HEBEISPIRIT-EnvironmentalImpact.pdf.

ITOPF (2011), ITOPF Handbook, www.itopf.org.

IWC/International Whaling Commission Scientific Committee (2004), *Annex K: Report of the Standing Working Group on Environmental Concerns,* 56 pp., Annual IWC meeting, Sorrento, Italy, 29 June-10 July 2004.

Jensen, S. (1972), The PCB story, *Ambio*, **1**(4), 123–131.

Jenssen, B. M. (2006), Endocrine-disrupting chemicals and climate change: a worst-case combination for Arctic marine mammals and seabirds? *Environ. Health Perspect.*, **114**, 76–80.

Jepson, P. D., M. Arbelo, R. Deaville, I. A. P. Patterson, P. Castro, J. R. Baker, E. Degollada, H. M. Ross, P. Herraez, A. M. Pocknell, F. Rodriguez, F. E. Howie, A. Espinosa, R. J. Reid, J. R. Jaber, V. Martin, A. A. Cunningham, and A. Fernandez (2003), Gas-bubble lesions in stranded cetaceans, *Nature*, **425**, 575–576.

Jernelöv, A., and O. Linden (1981), Ixtoc 1: A case study of the worlds' largest oil spill, *Ambio*, **10**(6), 299–306.

Jernelöv, A. (2011), The threats from Oil Spills: Now, then and in the future, *Ambio*, **39**, 353–366.

Johnson, C. B., and D. L. Garshelis (1995): Sea otter abundance and distribution, and pulp production in Prince William Sound following the Exxon Valdez oil spill. In: *"Exxon Valdez Oil Spill: Fate and Effects in Alaskan Waters. ASTM STP 1219.* [Wells P., J. N. Butler and J. S. Hughes (eds.)]. American Society for Testing and Materials, Philadelphia.

Kastelein, R. A., S. van der Heul, J. van der Veen, W. C. Verboom, N. Jennings, D. de Haan, and P. J. H. Reijnders (2007), Effects of acoustic alarms, designed to reduce small cetacean bycatch in gillnet fisheries, on the behaviour of North Sea fish species in a large tank, *Mar. Environ. Res.*, **64**, 160–180.

Kershaw, P., Katsuhiko, S. Lee, J. Leemseth, and D. Woodring (2011), Plastic debris in the ocean. In: Govers, T. and S. Beck (eds.), *UNEP year book: emerging issues in our environment*. UNEP, Nairobi, 78.

Ketten, D. R. (2004), Marine mammal auditory Systems: A summary of Audiometric and Anatomical Data and Implications for Underwater Acoustic Impacts, *Journal of Cetaceans Research and Management*, **7**, 37–39.

Khan, N. Y. (2007), Multiple stressors and ecosystem-based management in the Gulf, *Aquatic Ecosystem Health & Management*, **10**, 259–267.

Kingstone, P. (1999), Recovery of the marine environment following the Braer spill, Shetland. 1999 International Oil Spill Conference www.iosc.org/papers/01800.pdf#search¼"braer".

Lagardère, J. P. (1982), Effects of noise on growth and reproduction of *Crangon crangon* in rearing tanks, *Mar. Biol.*, **71**, 177–185.

Lamon, L., M. D. Valle, A. Critto, and A. Marcomini (2009), Introducing an integrated climate change perspective in POPs modeling and regulation, *Environ. Pollut.*, **157**, 1971–1980.

Langhamer, O., and D. Wilhelmsson (2009), Colonisation of fish and crabs of wave energy foundations and the effects of manufactured holes - A field experiment, *Mar. Environ. Res.*, **68**, 151–157.

Law, K. L., S. Moret-Ferguson, N. A. Maximenko, G. Proskurowski, E. E. Peacock, J. Hafner, and C. M. Reddy (2010). Plastic Accumulation in the North Atlantic Subtropical Gyre, *Science*, **329**, 1185–1188.

Linden, O., R. Elmgren, and P. Boehm (1979), The Tsesis oil spill: its impact on the coastal ecosystem of the Baltic Sea, *Ambio*, **8**(6) 244–253.

Linden, O., and T. Husain (2002): Impacts of wars: the Gulf War 1990-91. In: *Backhuys Publ. Leinde. The Gulf Ecosystem: Health and Sustainability* [Khan N. Y., M. Munawar and A. R. G. Price (eds.)].

Livingston, H. D., and R. F. Anderson (1983), Large particle transport of Plutonium and other Fallout Radionuclides to the deep ocean, *Nature*, **303**, 228–231.

Livingston, H. D., and P. P. Povinec (2000), Anthropogenic marine radioactivity, *Ocean Coast. Manag.*, **43**, 689–712.

MA, Millennium Ecosystem Assessment, http://www.maweb.org/en/Reports.aspx.

Macreadie, P. I., A. M. Fowler, and D. J. Booth (2011), Rigs-to-reefs: will the deep sea benefit from artificial habitat? *Frontiers in Ecology and the Environment*, **9**, 455–461.

Maria, E. J., F. P. Siringan, A. M. Bulos, and E. Z. Sombrito (2009), Estimating sediment accumulation rates in Manila Bay, a marine pollution hot spot in the Seas of East Asia, *Mar. Pollut. Bull.*, **59** (4–7), 164–174.

Marine Strategy Framework Directive (2008); http://eur-lex.europa.eu/LexUriServ/LexUriServ.do?uri=CELEX:32008L0056:EN:NOT.

Martins, G. M., R. C. Thompson, A. I. Neto, S. J. Hawkins, and S. R. Jenkins (2010), Enhancing stocks of the exploited limpet Patella candei d'Orbigny via modifications in coastal engineering, *Biol. Conserv.*, **143**, 203–211.

Mato, Y., T. Isobe, H. Takada, H. Kanehiro, C. Ohtake, and T. Kaminuma (2001), Plastic resin pellets as a transport medium for toxic chemicals in the marine environment, *Environ. Sci. Technol.*, **35**, 318–324.

McCauley, R. D., J. Fewtrell, and A. N. Popper (2003), High intensity anthropogenic sound damages fish ears, *J. Acoust. Soc. Am.*, **113**, 638–641.

McDonald, M. A., J. A. Hildebrand, and S. M. Wiggins (2006), Increases in deep ocean ambient noise in the Northeast Pacific west of San Nicolas Island, California, *J. Acoust. Soc. Am.*, **120**, 711–718.

McLeod, E., R. Moffit, A. Timmerman, R. Salm, L. Menviel, M. J. Palmer, E. R. Selig, K. S. Casey, and J. F. Bruno (2010), Warming seas in the coral triangle: coral reef vulnerability and management implications, *Coastal Management*, **38**, 518–539.

Miller, P. J. O., N. Biassoni, A. Samuels, and P. L. Tyack (2000), Whale songs lengthen in response to sonar, *Nature*, **405**, 903.

MMC (Marine Mammal Commission) Report to Congress (2007), Marine mammals and noise: A sound approach to research and management. Available from http://www.mmc.gov/sound/committee/pdf/soundFACAreport.pdf.

Moriyasu, M., R. Allain, K. Benhalima, and R. Claytor (2004), *Effects of seismic and marine noise on invertebrates: A literature review*, 47 pp., Canadian Science Advisory Secretariat, Ottawa, Canada.

Morton, A. B., and H. K. Symonds (2002), Displacement of Orcinus orca (L.) by high amplitude sound in British Columbia, *ICES J. Mar. Sci.*, **59**, 71–80.

Nakano, M., and P. P. Povinec (2003), Oceanic general circulation model for the assessment of the distribution of ^{137}Cs in the world ocean, *Deep-Sea Res. II*, **50**, 2803–2816.

Nedwell, J., and D. Howell (2003), *Assessment of sub-sea acoustic noise and vibration from offshore wind turbines and its impact on marine wildlife: initial measurements of underwater noise during construction of offshore windfarms, and comparison with background noise*, 68 pp., Collaborative Offshore Windfarm Research, London, UK.

Neff, J. M., and W. A. Stubblefield (1995): Chemical and toxicological evaluation of water quality following the Exxon Valdez oil spill. In: *Exxon Valdez Oil Spill: Fate and Effects in Alaskan Waters.* ASTM STP 1219. [Wells, P., J. N. Butler and J. S. Hughes (eds.)]. American Society for Testing and Materials, Philadelphia.

Negri, A. P., and M. O. Hoogenboom (2011), Water contamination reduces the tolerance of coral larvae to thermal stress, *PLoS ONE*, **6**, e19703, http://dx.doi.org/10.1371/journal.pone.0019703.

Nesheim, M. C., and A. L. Yaktine (2007), *Seafood choices: Balancing benefits and risks. Committee on nutrient relationships in seafood: selections to balance benefit and risks.* National Academy of Science, The national Academic press, Washington, DC. ISBN: 978-0-309-10218-6.

Nieukirk, S. L., K. M. Stafford, D. K. Mellinger, R. P. Dziak, and C. G. Fox (2004), Low-frequency whale and seismic airgun sounds recorded in the mid-Atlantic Ocean, *J. Acoust. Soc. Am.*, **115**(4), 1832–1843.

Noad, M. J., D. H. Cato, M. M. Bryden, M. N. Jenner, and K. C. S. Jenner (2002), Cultural revolution in whale songs, *Nature*, **408**, 537.

Noren, F. (2008), *Small plastic particles in Coastal Swedish waters*, 11 pp., KIMO Sweden, Lysekil.

Nowacek, D. P., L. H. Thorne, D. W. Johnston, and P. L. Tyack (2007), Responses of cetaceans to anthropogenic noise, *Mammal Review*, **37**, 81–115.

Noyes, P. D., M. K. McElwee, H. D. Miller, B. W. Clark, L. A. Van Tiem, K. C. Walcott, and E. D. Levin (2009), The toxicology of climate change: environmental contaminants in a warming world, *Environ. Int.*, **35**, 971–986.

NRC (2003), *Oil in the sea III, Inputs, fates, and effects.* National Research Council, National Academic Press, Washington, DC.

Nystuen, J. A., and D. M. Farmer (1987), The influence of wind on the underwater sound generated by light rain, *J. Acoust. Soc. Am.*, **82**, 270–274.

Olesiuk, P. F., L. M. Nichol, and J. K. B. Ford (2002), Effect of the sound generated by an acoustic harassment device on the relative abundance of and distribution of harbor porpoises (*Phocoena phocoena*) in Retreat Passage, British Columbia, *Marine Mammals Science*, **18**, 843–862.

Orpin, A. R., P. V. Ridd, S. Thomas, K. R. N. Anthony, P. Marchall, and J. Oliver (2004), Natural turbidity variability and weather forecasts in risk management of anthropogenic sediment discharge near sensitive environments, *Mar. Pollut. Bull.*, **49**, 602–612.

OSPAR (2009): Marine litter in the North-East Atlantic Region: Assessment and priorities for response. In: *UNEP Regional Seas*. London, 127.
Payne, R. J., W. B. Driskell, J. W. Short, and M. L. Larsen (2008), *Mar. Pollut. Bull.*, **56**, 2067–2081.
Phillips, D. J. H., and P. S. Rainbow (1993), *Biomonitoring of Trace Aquatic Contaminants*, Springer, Netherlands. ISBN: 978-0-412-53850-6.
Piola, R., and E. Johnston (2008), Pollution reduces native diversity and increases invader dominance in marine hard-substrate communities, *Divers. Distrib.*, **14**, 329–342.
Popper, N. A., and C. M. Hastings (2009), The effects of human-generated sound on fish. *Integrative Zoology*, **4**, 43–52.
Preston, A., and D. F. Jefferies (1969), The ICRP Critical Group Concept in Relation to the windscale sea discharges, *Health Phys.*, **16**, 33–46.
Ratcliffe, D. A. (1970), Changes attributable to pesticides in egg breakage frequency and eggshell thickness in some British birds, *J. Appl. Ecol.*, (7), 67–115.
Ritter, L., K. R. Solomon, J. Forget, M. Stemeroff, and C. O'Leary (1995), An Assessment Report on DDT, Aldrin, Dieldrin, Endrin, Chlordane, Heptachlor, Hexachlorobenzene, Mirex, Toxaphene, Polychlorinated Biphenyls, Dioxins and Furans. Prepared for the International Programme on Chemical Safety (IPCS).
Rockström, J., W. Steffen, and K. Noone (2009), Planetary Boundaries: Exploring the Safe Operating Space for Humanity, *Ecology and Society*, **14**, Article 32.
Russel, B. D., J. -A. I. Thompson, L. J. Falkenberg, and S. D. Connell (2009), Synergistic effects of climate change and local stressors: CO_2 and nutrient-driven change in subtidal rocky habitats, *Global Change Biology*, **15**, 2153–2162.
Ryan, P. G., C. J. Moore, J. A. van Franeker, and C. L. Moloney (2009), Monitoring the abundance of plastic debris in the marine environment, *Philos. Trans. R. Soc. B*, **364**, 1999–2012.
Samuel, Y., S. J. Morreale, C. W. Clark, C. H. Greene, and M. E. Richmond (2005), Underwater, low frequency noise in a coastal sea turtle habitat, *J. Acoust. Soc. Am.*, **117**, 1465–1472.
Sanders, H. L., J. F. Grassle, G. R. Hampson, L. S. Morse, S. Garner-Price, and C. C. Jones (1980), Anatomy of an oil spill: long-term effects from the grounding of the barge Florida off West Falmouth, Massachusetts, *J. Mar. Res.*, **38**, 265–380.
Sazykina, T. G., and I. I. Kryshev (1997), Current and potential doses from Arctic seafood consumption, *Sci. Total Environ.*, **202**, 57–65.
Sazykina, T. G., I. I. Kryshev, and A. I. Kryshev (1998), Doses to marine biota from radioactive waste dumping in the Fjords of Novaya Zemlya, *Radiat. Prot. Dosimetry*, **75**, 253–256.
Simpson, S. D., M. Meeka, J. Montgomery, R. McCauley, and A. Jeffs (2005), Homeward sound, *Science*, 308–221.
Slotte, A., K. Hansen, Dalen, and E. Ona (2004), Acoustic mapping of pelagic fish distribution and abundance in relation to a seismic shooting area off the Norwegian west coast, *Fish. Res.*, **67**, 143–150.
Snyder, B., and J. M. Kaiser (2009), Ecological and economic cost-benefit analysis of offshore wind energy, *Renewable Energy*, **34**, 1567–1578.
Sonne, C. (2010), Health effects from long-range transported contaminants in Arctic top predators: An integrated review based on studies of polar bears and relevant model species, *Environ. Int.*, **36**, 461–491.

Southward, A. J., and E. C. Southward (1978), Recolonization of rocky shores in Cornwall after use of toxic dispersants to clean up the Torrey Canyon spill. Presented at: Symposium on Recovery Potential of Oiled Marine Northern Environments; Halifax, Canada, *J. Fish. Res. Board Can.*, **35**, 682–706.

Spooner, M. F. (ed.) (1978). *The Amoco Cadiz oil spill.* Special edition of Marine Pollution Bulletin *9 (7).* Pergamon Press, Oxford and New York.

STAP (2011): Marine Debris as a Global Environmental Problem: Introducing a solutions based framework focused on plastic. In: *A STAP Information Document*, Global Environment Facility, Swedish Environmental Protection Agency, Washington, DC, 40. 2000. Hälsorisker med långlivade organiska miljögifter. Report 1521 (report in Swedish).

Suh, W. R. (2012), *The Hebei Spirit oil spill.* Proceedings from the International Conference on Oil Spill Risk Management 2011. World Maritime University, Malmö.

Swedish Environmental Protection Agency (2009), Sources, transport, reservoirs and fate of dioxins, PCBs and HCB in the Baltic Sea environment. Report 5912.

Teuten, E. L., J. M. Saquing, D. R. U. Knappe, M. A. Barlaz, S. Jonsson, A. Björn, S. J. Rowland, R. C. Thompson, T. S. Galloway, R. Yamashita, D. Ochi, Y. Watanuki, C. Moore, P. Viet, T. S. Tana, M. Prudente, R. Boonyatumanond, M. P. Zakaria, K. Akkhavong, Y. Ogata, H. Hirai, S. Iwasa, K. Mizukawa, Y. Hagino, A. Imamura, M. Saha, and S. Takada (2009), Transport and release of chemicals from plastics to the environment and to wildlife, *Philos. Trans. R. Soc. B*, **364**, 2027–2045.

Thomsen, F., K. Lüdemann, R. Kafemann, and W. Piper (2006), *Effects of offshore wind farm noise on marine mammals and fish*, 62 pp., Collaborative Offshore Windfarm Research, London, UK.

Thompson, R., C. Moore, A. Andrady, M. Gregory, H. Takada, and S. Weisberg (2005), New directions in plastic debris, *Science*, **310**, 1117.

Thompson, R. C., C. Moore, F. S. vom Saal, and S. H. Swan (2009), Plastics, the environment and human health: current consensus and future trends, *Philos. Trans. R. Soc. B*, **364**, 2153–2166.

Thompson, R. C., Y. Olsen, R. P. Mitchell, A. Davis, S. J. Rowland, A. W. G. John, D. McGonigle, and A. E. Russell (2004), Lost at sea: Where is all the plastic? *Science*, **304**, 838.

Tougaard, J., I. Ebbesen, S. Tougaard, T. Jensen, and J. Teilmann (2003), *Satellite tracking of harbour seals on Horns Reef. Use of the Horns Reef wind farm area and the North Sea*, 43 pp., National Environmental Research Institute, Roskilde, Denmark.

Tougaard, J., O. D. Henriksen, and L. A. Miller (2009), Underwater noise from three offshore wind turbines: estimation of impact zones for harbor porpoises and harbor seals, *J. Acoust. Soc. Am.*, **125**, 3766–3773.

Tyack, P. L. (2008), Implications for marine mammals of large scale changes in the marine acoustic environment, *J. Mammal.*, **89**, 549–558.

IMO Resolution A.720(17). http://www.imo.org/Pages/home.aspx.

UN, Report on Oceans and the Law of the Seas (2004), http://wwwupdate.un.org/Depts/los/general_assembly/general_assembly_reports.htm.

UNEP (United Nations Environment Programme) (2006). *Challenges to international waters; Regional assessments in a global perspective*, 120 pp., United Nations Environment Programme, London. ISBN: 91-89584-47-3.

The Stockholm Convention on Persistent Organic Pollutants (2001), www.pops.int/ United Nations Environment Programme (UNEP).

UNEP/AMAP (2011), Climate change and POPs: predicting the impacts. Report of the UENP/AMAP expert group. Secretariat of the Stockholm Convention, Geneva.

U.S. Environmental Protection Agency (1998), *Chemical hazard data availability study: what do we really know about the safety of high production volume chemicals?* Office of Pollution Prevention and Toxics, Washington, DC, USA.

UNSCEAR (1982), *Ionizing radiation: Sources and biological effects,* 773 pp., United Nations, New York.

Wahlberg, M., and H. Westerberg (2005), Hearing in fish and their reactions to sounds from offshore wind farms, *Mar. Ecol. Prog. Ser.*, **288**, 295–309.

Wania, F. (1998), The significance of long range transport of persistent organic pollutants by migratory animals. WECC Report 3/98.

Weilgart, L. S. (2007), The impact of anthropogenic ocean noise on cetaceans and implications for management, *Can. J. Zool.*, **85**, 1091–1116.

Wiens, J. A. (1995): Recovery of seabirds following the Exxon Valdez oil spill: An overview. In: *Exxon Valdez Oil Spill: Fate and Effects in Alaskan Waters.* ASTM STP 1219. [Wells, P., J. N. Butler and J. S. Hughes (eds.)]. American Society for Testing and Materials, Philadelphia.

Wilhelmsson, D., T. Malm, and M. Öhman (2006), The influence of offshore wind power on demersal fish, *ICES J. Mar. Sci.*, **63**(5), 775–784.

Wilhelmsson, D., T. Malm, R. Thompson, J. Tchou, G. Sarantakos, N. McCormick, S. Luitjens, M. Gullström, J. K. Patterson Edwards, O. Amir, and A. Dubi (2010), *Greening Blue Energy: Identifying and managing the biodiversity risks and opportunities of offshore renewable energy,* 102 pp., IUCN, Gland, Switzerland. ISBN: 978-2-8317-1241.

Williams, A. T., and D. T. Tudor (2001), Litter burial and exhumation: Spatial and temporal distribution on a cobble pocket beach, *Mar. Pollut. Bull.*, **42**, 1031–1039.

Wilson, O. B. J., S. N. Wolf, and F. Ingenito (1985), Measurements of acoustic ambient noise in shallow water due to breaking surf, *J. Acoust. Soc. Am.*, **78**, 190–195.

Wolfaardt, A. C., L. G. Underhill, R. Altwegg, J. Visagie, and A. J. Williams (2008), *Afr. J. Mar. Sci.*, **30**(2), 405–419.

CHAPTER

7

The Potential Economic Costs of the Overuse of Marine Fish Stocks

Ussif Rashid Sumaila, William W.L. Cheung†, A.D. Rogers‡*

*Fisheries Economics Research Unit, Fisheries Centre, The University of British Columbia, Vancouver, British Columbia, Vancouver, Canada

†Changing Ocean Research Unit, Fisheries Centre, The University of British Columbia, Vancouver, British Columbia, Canada

‡Department of Zoology, University of Oxford, Oxford, United Kingdom

FISH AND FISHERIES ARE IMPORTANT TO PEOPLE

At present, it is estimated that about 79.5 million tons (t) of marine fish are landed from capture fisheries annually, a decline from about 86.3 million t at a peak in landings in 1996 (FAO, 2010). However, it is estimated that about 11-26 million t per annum of fish are caught as illegal or unreported catches (Agnew et al., 2009). The value of marine fish at first landing has been estimated at about $84 billion US per annum (Sumaila et al., 2007), with an additional $10-24 billion lost to unreported and illegal fishing on an annual basis (Agnew et al., 2009).

These figures though do not reflect the overall economic value of marine capture fisheries. Fishing is an activity that depends on manufactured goods such as boats and nets and also the fish themselves are sold on up through a chain of supply at each step of which value is added, whether through packaging (e.g., canning), processing, marketing, transport, and management (Dyck and Sumaila, 2010). The total economic impact of fishing is about three times the value at first landing: $225-240 billion US per annum (Dyck and Sumaila, 2010). Of the overall global fish production from marine, inland, and aquaculture, about 81% (115 million t) was used for food, the rest (27 million t) being used not only for non-food purposes, mainly for reduction to fishmeal and fish oil, but also for ornamental purposes, culture, bait, and the pharmaceutical industry (FAO, 2010). Fish are also highly traded as commodities, with an annual rate of increase in value of 8.3% per annum from 1976 to 2008, and a concomitant increase in the proportion of fish production entering trade from 25% to 39% over the same period (FAO, 2010). China, Norway, and Thailand are the largest fish exporters with Japan, the United States, and the EU being the largest importers of fish (FAO, 2010). Exports reached a peak value of $102 billion US in 2008 but peaked in quantity in 2005 at 56 million t, dropping to 55 million t by 2008 (FAO, 2010). This may reflect not only a fall in production and demand for fishmeal but also the first impacts of rising food prices on consumer demand for fish (FAO, 2010). It is notable that, while developed states are the main importers of fish, exports of fish as a commodity are particularly important for developing countries and are higher in value than agricultural commodities such as rice, meat and sugar. For instance, while over 30% of fish caught are exported, only about 5% of rice produced is exported. Low-income, food-deficit countries are playing an increasing role in the fish-exporting industry (FAO, 2010).

The importance of fish as a component of diet and in relation to food security is not often considered. According to UN FAO, fish provide 20% of the intake of animal protein for 1.5 billion people and 15% for about 3 billion people (FAO, 2010), more than 40% of the world's population of 6.93 billion at the present time (US Census). In developing countries, small-island developing states and coastal areas, this figure can reach 90% of intake of animal protein (FAO, 2009). Some 44.9 million people are engaged in the fisheries sector at the present time (FAO, 2010), the vast majority of which are small-scale fishers (>90%; FAO, 2009). The majority of fishers and fish farmers are located in Asia (85.5%), followed by Africa (9.3%), Latin America and the Caribbean (2.9%), Europe (1.4%), North America (0.7%), and Oceania (0.1%; FAO, 2010).

However, as with the economic value of fisheries, there are linkages to other activities which mean that in developing countries alone more than 200 million people are dependent on small-scale fishing and many more may derive supplementary income from this activity (FAO, 2005). Fishing as a means of living may be particularly important in rural areas where there are few alternative areas of employment and has often been regarded as an employer of the "last resort" or as a "safety valve" for the poor (Coulthard et al., 2011). Because of this, fisheries have been perceived as a convenient means of generating employment and economic development in developing states, especially when coupled with the globalization of the trade in fish (Coulthard et al., 2011). This has resulted in organizations such as the World Bank and the UN FAO in funding the modernization of fisheries in the developing world.

Fisheries are clearly important, both economically and in terms of food security, especially when it is considered that the human population is increasing and is projected to rise to more than 9 billion by 2050, with almost all this growth occurring in developing states (UN-DESA, 2009). Taking the potential decline in wheat or rice production, the figure for population growth and the requirement for animal protein, an additional 75 million t of fish and invertebrates will have to be obtained from aquatic systems by 2050 (Rice and Garcia, 2011). This equates to a greater than 50% increase in fish production (Rice and Garcia, 2011). Aquaculture is playing an increasing role in meeting the demand for fish and from less than 1 million t in the 1950s to 52.5 million t now with an additional 15.8 million t of aquatic plants (FAO, 2010). It may soon exceed capture fisheries in the quantity of fish it produces for human consumption. However, given this projected increase in the requirement for fish and aquatic invertebrates, healthy fish stocks and the fisheries they support are likely to remain of global importance.

EVIDENCE OF OVERUSE OF FISH STOCKS

There is a general perception that overexploitation of marine biotic resources is a new problem that has increased in prevalence since the Second World War. However, the emergence of zooarchaeology coupled with a modern understanding of the ecological impacts of depletion of prey by humans in marine ecosystems has led to the conclusion that even ancient people had the ability to change the size structure and abundance of marine species (Erlandson and Rick, 2010). For example, it is likely that the hunting of sea otters by native peoples on the Channel Islands off the coast of California led to the release of prey species such as red abalone and mussels that in turn also decreased in size over the past 10,000 years as a result of resource depression (Erlandson and Rick, 2010).

Likewise, similar decreases in size, abundance, and range of molluscs, finfish, and pinnipeds have also been documented from the South Pacific Islands including New Zealand, although in some cases resolving the effects of environmental change on marine ecosystems compared with human predation are difficult (see Erlandson and Rick, 2010). This phase of human exploitation of marine biotic resources has been termed the "aboriginal" phase and is characterized by the use of primitive watercraft and fishing technologies (Jackson et al., 2001). During this phase of exploitation, humans were probably responsible for the extinction of terrestrial species (Martin, 2002), but there is no evidence of similar extinctions in marine species, probably because they existed in geographic (e.g., remote coastlines or islands) or habitat refugia (e.g., deep water) from hunters and gatherers (Erlandson and Rick, 2010).

While aboriginal fisheries continued in many parts of the world, there is evidence of a shift from subsistence fishing to artisanal or small-scale commercial fishing in Europe in medieval times. In the United Kingdom, assessments of archaeological bone assemblages reveal that prior to the end of the tenth century, catches were restricted to freshwater or migratory species of fish (cyprinids or eels; Barrett et al., 2004). During this period, fish were generally consumed by the fishers and family or indirectly by local elite households (Hoffmann, 2005). By the tenth century, there are records of fish being caught and sold to local markets (Hoffmann, 2005). This coincides with a switch in prey species in the United Kingdom to cod and herring, with flatfish showing an intermediate trend. It is thought that this change may have been driven partially by economic factors but was probably mainly a result of the depletion of freshwater fisheries, the increasing use of dams, and damage to inland waterways by more intensive agricultural practices, leading to the silting of streams (Barrett et al., 2004).

At this time, fish were stored in cages, tanks, or ponds or preserved by salting, smoking, or drying, although supplies were still restricted to coastal areas or towns close to fishing areas until the thirteenth century. Even at these early stages of commercial exploitation of fisheries, species such as salmon and sturgeon were forced into decline in many parts of Europe as a result of overexploitation and habitat alteration through modification of inland waterways and the silting up of estuaries, although climatic variability is likely to have also played a role (Hoffmann, 2005). Significant changes to fisheries occurred in the thirteenth and fourteenth centuries with the development of new preservation methods for fish like herring and the elaboration of transportation networks to transport these fish to more distant population centers (Hoffmann, 2005). This increasing commercialization of fisheries led to major exploitation of coastal spawning stocks of herring and other species and shifted offshore to more distant parts of the North Sea following the development of larger vessels and better processing techniques by the Dutch in the fourteenth century (Hoffmann, 2005). By the fifteenth century, the trade in fish had spread across Europe, with income contributing significantly to regional economies such as that of southwestern England (Kowaleski, 2000) and the number of exploited species of marine fish expanding (Lotze, 2007). This trend of increasing separation of fish consumers, located in urban centers, from increasingly capitalized fishing fleets with little incentive to preserve stocks (for local consumption) continued into colonial times. It is notable that as early as the thirteenth and fourteenth centuries, there is evidence of commercial collapse of inshore fisheries for marine finfish stocks of species like herring following heavy exploitation coupled with environmental change arising from human activities (use of land and freshwater changing nearshore ecosystems) or natural climatic variation.

In other parts of the world, the aboriginal phase of exploitation of marine ecosystems collided with colonial expansion from Europe generating a "colonial" phase of fishing (Jackson et al., 2001). In areas such as California, this was heralded by the arrival of colonial explorers, closely followed by missionaries and settlers. Despite exploitation for millennia by native peoples and alterations in the abundance or size of marine species, it was only during this second period of exploitation, the past 240 years in the case of California, that sea otters, whales, seals, bald eagles, and other species were driven to the brink of extinction (Erlandson and Rick, 2010). It is during the colonial phase of exploitation of marine species that the first recorded human-driven extinctions of marine species-occurred, an example being the elimination of Stellar's sea cow in the North Pacific in 1768 (Jackson et al., 2001). The geographic distribution of this species was probably reduced by aboriginal hunting, but the species survived in refugia on the Aleutian Islands until this time (Jackson et al., 2001). The Caribbean monk seal (*Monachus tropicalis*) was heavily exploited from the seventeenth century, although records of hunting go back as far as the fifteenth century (McClenachan and Cooper, 2008). Its exploitation, which was for food and for oil, used in machinery for sugar processing, expanded with the colonization of Caribbean Islands by European powers, although the species did not go extinct until 1952 (McClenachan and Cooper, 2008).

It is important to recognize that during this phase of human exploitation of marine biotic resources, significant changes occurred to marine ecosystems as a result of removal of target species (see below). For example, the exploitation of oyster reefs in estuaries by colonialists in North America began in the seventeenth century with exploitation of *Crassostrea virginica* in the Hudson River with the first recorded fisheries regulations being passed in the late seventeenth century and the first recorded collapses of these fisheries taking place in the early nineteenth century (Kirby, 2004). As oyster beds were depleted in estuaries close to urban centers, exploitation shifted to more distant estuaries moving southward with depletion following and continuing through into the twentieth century (Kirby, 2004). Similar patterns of exploitation are detectable in oyster populations from western North America (*Ostreola conchaphila*) and in eastern Australia (*Saccostrea glomerata*; Kirby, 2004).

The destruction of such reefs was completed when technological advances allowed the development of mechanical methods for harvesting in the late nineteenth century. Such vast oyster reefs exerted considerable top-down effects on estuarine ecosystems through their ability to filter phytoplankton and may be able to limit the effects of eutrophication, a common problem in estuaries in modern times (Jackson, 2001).

The industrial revolution led to the development of steam trawlers in the 1880s. Steam trawlers could fish further offshore for longer periods of time and deploy larger gear in deeper waters than sailing vessels (Thurstan et al., 2010). As early as 1885, outcries from line fishers in Britain lead to enquiries with regard to the damage to both fish stocks and marine ecosystems by trawling. These were inconclusive but lead to the establishment of systems for collecting fisheries data (Thurstan et al., 2010). From the late nineteenth century to the First World War, the domestic fleets in Britain switched from predominantly sailing vessels to steam-powered vessels and catch per unit fishing effort steadily declined (Thurstan et al., 2010).

Fishing was interrupted during the First World War, and following this period of recovery, there was an increase in catch per unit effort that lasted for several years (Thurstan et al., 2010). This was followed by a steady increase in catch per unit effort as British fishing fleets expanded the geographic coverage of fishing targeting areas off Africa and the Arctic (Thurstan et al., 2010; Kerby et al., 2012). This increase continued, with an interruption during the Second World War, until the early 1960s when trends reversed and catch per unit effort went into decline. Throughout this period, fishing technology advanced with the advent of diesel- and oil-powered fishing vessels, the development of sonar during the Second World War, advances in onboard processing and preservation of fish, and finally advances in navigation culminating in Global Positioning Systems (Kerby et al., 2012).

This picture is very much replicated for the global fishing fleet. Following the Second World War, most fishing effort was still concentrated in coastal waters, particularly off Europe, North America, and Japan. However, spatial coverage of global fishing effort rapidly expanded to cover most of the world's oceans by 2005 with an increase in overall fish catches continuing until 1996 when they peaked at 86.3 million t (Swartz et al., 2010). Catches thereafter have declined according to FAO statistics (see above). The expansion of the geographic extent of fishing has been accompanied by a 10-fold increase in global fishing effort since 1950, a figure that rises to 25-fold for Asia over the same period (Watson et al., 2012). Overall, the decline in global catch per unit effort suggests that a decrease in the biomass of many fished populations is likely to have declined by >50% (Watson et al., 2012).

Despite falling catches, many fisheries have been maintained by capacity-enhancing subsidies, especially for fuel, exacerbating the decline of target and bycatch populations (Sumaila et al., 2010). Overexploitation of large predatory fish species through targeted fishing and through depletion as a result of bycatch has been particularly severe with populations reported to have declined by >90% in many cases (e.g., Myers and Worm, 2003, 2005; Rosenberg et al., 2005). Such figures do not convey the overall global ecological and economic impact of fishing which has now extended into the deep ocean to depths as great as >2000 m and also into the high latitudes of Arctic and Antarctic waters (Morato et al., 2005; Swartz et al., 2010). The implications of overexploitation of fish extend from direct impacts on predators through the removal of prey (Cury et al., 2011), release of secondary predator populations where ecosystems are subject to top-down control by targeted fish populations (Worm and Myers, 2003; Myers et al., 2007), and large-scale ecosystem perturbations, especially where the effects of overfishing interact with other human impacts such as eutrophication, the introduction of alien species, and the impacts of climate change (Daskalov, 2002; Llope et al., 2011). In some cases, the latter include phase shifts from which ecosystem recovery is protracted or near-impossible because of impacts on ecosystem engineer species (Mumby et al., 2007; Alvarez-Filip et al., 2009; Jackson, 2010). Some fishing practices are also associated with direct destruction of habitat (e.g., trawling of deep-sea coral reefs; Althaus et al., 2009) or significant bycatch of species also with significant impacts at the ecosystem level and/or a direct threat of regional or global species extinction (Cavanagh and Gibson, 2007, Mediterranean sharks; Jaramillo-Legorreta et al., 2007, vaquita Gulf of California).

Studies have been hailed as indicating grounds for optimism that management efforts are resulting in recovery of fish stocks (e.g., Worm et al., 2009). It is true that recovery plans that involve a real reduction in fishing mortality do result in increased biomass of stocks (Wakeford et al., 2009; Murawski, 2010), even for species that are highly vulnerable to even low levels of fishing mortality (e.g., sharks; Ward-Paige et al., 2012). However, it must be pointed out that such cases often represent the most well-managed fisheries off the coasts of the world's most wealthy nations (see distribution of stocks studied in Worm et al., 2009), where there is substantial infrastructure for fisheries science and management including effective systems of monitoring control and surveillance coupled with strong enforcement of regulations (Mora et al., 2009). For many parts of the world, there is a lack of rigorous scientific assessment of stock status and fishing mortality, a lack of transparency in the translation of science to fisheries management policy, and an inability to enforce compliance to fisheries regulations (Mora et al., 2009). Even for wealthy States, levels of compliance with internationally agreed standards for fisheries management are not ideal and in many cases could be viewed as failed (Pitcher et al., 2009; Alder et al., 2010; Cullis-Suzuki and Pauly, 2010; Khalilian et al., 2010).

Ostrom and her collaborators provide a possible silver lining in terms of monitoring and control (Ostrom, 1990). This line of research suggests that the monitoring and enforcement needs can be accomplished through social norms and community institutions in a local community structure, which is more likely going to be the fisheries management regime faced in Asia, Africa, Latin America, and the Caribbean—which as we note where ~95% of the fishers and fish farmers are located. The implication here is that sustainability and healthy fish stocks may be achievable in small-scale fisheries, and in the developing world, with less sophisticated infrastructure for fisheries science and management.

CLIMATE CHANGE WILL EXACERBATE THE PROBLEM OF OVERUSE OF FISH STOCKS

Climate change is affecting ocean conditions and is expected to have large implications for marine ecosystems and fisheries. Overall, there is compelling evidence that the ocean has become warmer, with less sea-ice, more stratified and more acidic during the past century, with expansion of oxygen minimum zones in the twentieth century (IPCC, 2007). Given that global greenhouse gas emissions are already close to the high-end scenario considered by the Intergovernmental Panel on Climate Change (IPCC) Assessment Report Four (IPCC, 2007), it is expected that such changes will continue over the next few decades. Such ocean changes affect marine fisheries through changes in (1) resource availability, (2) fishing operations, (3) fisheries management and conservation measures, and (4) profits from fisheries (i.e., balance between price of fish products and fishing costs).

Effects of climate change on resource availability occur mainly through changes in distribution and productivity of fish stocks. Distributions of marine organisms are generally dependent on optimal environmental conditions (e.g., temperature, oxygen, food availability) and long-term changes in temperature and/or other ocean conditions often coincide with observed changes in species' distribution and fisheries (Sumaila and Cheung, 2010). Shifts in distributions of exploited species (generally poleward) that are attributed to ocean warming are increasingly reported in many regions (Simpson et al., 2011; Sumaila et al., 2011) and for commercially important species (Sumaila et al., 2011; Cheung et al., 2012a).

It is also projected that species distribution shifts will continue in the future, leading to increased diversity and fisheries potential in high-latitude regions and the opposite in the tropics (Blanchard et al., 2012; Cheung et al., 2008, 2011, 2012b; Hobday, 2010; Lindegren et al., 2010). Ocean primary production is projected to decrease by 2-13% by 2100 relative to 1860 under the SRES A2 scenario (Steinacher et al., 2010). In addition, because of the effects of warming and ocean dexoygenation on growth, maximum body size of fishes are projected to decrease by 14-24% by 2050 relative to present size under the SRES A2 scenario, potentially reducing their yield per recruit (Cheung et al., 2012c). Ocean deoxygenation and acidification are expected to reduce habitats for exploited marine organisms and fisheries yields in some regions (Cheung et al., 2011; Stramma et al., 2011). These may further reduce fisheries catch potential (Cheung et al., 2011). As a result, these changes are expected to cause large-scale redistribution of fisheries catch (Cheung et al., 2010), creating potential "winners" and "losers" through changes in fisheries potential of commercial species. Particularly, the tropics are identified as climate change "losers" in fisheries, because of the projected decrease in fisheries catch potential and low capacity to adapt to environmental changes (Allison et al., 2009; Cheung et al., 2010).

Climate change may affect fishing operations directly, through changes in weather conditions, and indirectly, through changes in distributions and behavior of exploited species (Cheung et al., 2012). Changes in frequency of extreme weather events may cause disruption to fishing activities and/or land-based fisheries-related infrastructure. Climate change may also affect fishing indirectly through its effect on behavior and activity of the exploited animals (e.g., distribution, movement pattern, etc.), affecting their catchability by fishing gears. Thus, fishing gears may need to be modified to adapt to these changes.

The range of climate change effects on marine ecosystems and fisheries may complicate fisheries and conservation policies, affecting their effectiveness. The shift in distribution of fish stocks and the increased uncertainty on their productivity can destabilize multinational fisheries agreement (Miller et al., 2013). Effectiveness of conservation plan may also be affected, for example, through changes in coverage of protected areas on vulnerable fish stocks (Jones et al., 2013). Particularly, at the moment, there is limited example that climate change is explicitly considered in fisheries management and marine conservation plans.

The economic impact of the above changes is estimated in Chapter 11. Here, we focus on the economic effects of overfishing of the world's marine fish stocks.

GLOBAL ECONOMIC LOSS DUE TO OVERFISHING

There is a growing literature on the potential economic loss due to overfishing. A prominent example of the World Bank's Sunken Billions Report (2010), which estimated that fisheries worldwide are losing US$ 50 billion a year due to overfishing. There are many economists publishing on this area (e.g., Dan Holland of NOAA Fisheries, James Sanchirico of University of California Davis, Robert Johnston of Clark University, and Eric Thunberg of NOAA). There are also a number of overviews such as the OECD 2010 report, The Economics of Rebuilding Fisheries: Workshop Proceedings (see http://www.oecd.org/greengrowth/fisheries/theeconomicsofrebuildingfisheriesworkshopproceedings.htm).

In the next paragraphs, we will present a summary of a number of recent works to illustrate and provide an estimate of the potential cost of overusing marine fish stocks. Sumaila et al. (2012) used estimates of catch loss reported in Srinivassan et al. (2010), defined as the difference between current landings and maximum sustainable yield (MSY) for those species that are considered to be overused. The MSY is applied in this analysis for practical and policy reasons, as it is the most commonly stipulated target or management reference point for many national legislations and international conventions.

For our present purposes, we assume that the estimated catch losses to overfishing reported by Srinivassan et al. (2010) may be fully regained after a period of rebuilding fisheries worldwide. To calculate potential catch losses, the authors used catch time series from the *Sea Around Us* project for 1066 taxa of fish and invertebrates in 301 Exclusive Economic Zones (EEZs), along with an empirical relationship they derived from catch data and stock assessments for 26 Northeast US species from the US National Oceanic and Atmospheric Administration (NOAA).

The log-linear relationship that they found between a species' mean maximum catch C_{max} from catch data and its MSY from stock assessment was robust ($R^2=0.84$, $p<0.001$) and has since been tested for 50 fully assessed stocks in the Northeast Atlantic, where variation in MSY accounted for 98% of the variability in C_{max} (Froese et al., 2012). Therefore, given the dearth of detailed stock assessments for the majority of species in the world's fisheries, Srinivasan et al. (2011) applied the relationship they derived (with a 50% prediction interval) to estimate MSY levels for all stocks they identified as overfished. By comparing with reported catch levels, they arrived at estimates of lost catch by mass, reporting that without overfishing, potential landings worldwide in the year 2000 may have been 9.1 million t higher than current landings (50% prediction interval: 3.6-19 million t higher).

To calculate the value of these potential landings under rebuilt global fisheries, Srinivassan et al. (2010) used a database of ex-vessel fish prices by Sumaila et al. (2007). For each taxon-EEZ pair designated as overfished, a price-per-ton p for the MSY was set by taking a weighted average of the actual prices corresponding to catches of the taxon within ±30%, ±50%, or ±100% of the estimated MSY level, in order of preference depending on data availability. This approach was used to account for the impacts of overfishing, and thus scarcity, on price levels.

There is debate among fisheries scientists as to the reliability of overfishing estimates based on catch trends rather than stock assessments, with some arguing that catch-based approaches are prone to overestimate depletion (Branch et al., 2011).

Based on Costello et al. (2012), who estimated the recovery time for 18 simulated fish species to be 11 years on average, with a range of 4-26 years depending on the species, Sumaila et al. (2012) assume a rebuilding period of 10 years ($t=0$-9) in their study. Following modeling work reported in UNEP's Green Economy Report (UNEP, 2011), the authors also assume that global fisheries landings decline linearly from ~80 to 50 million t per year from $t=0$-5 as fishing effort declines, but then rise linearly to the rebuilt level (~90 million t) by $t=9$. Once global fisheries have been rebuilt, this potential gain in resource rent would recur annually for the subsequent 40 years after rebuilding (Sumaila et al., 2012).

Fisheries economists use resource rent (i.e., what remains after fishing costs and subsidies are deducted from revenue) as an indicator of fisheries performance (Clark, 1990), although others argue that this is inadequate because it does not capture all the benefits derived from marine fisheries (Béné et al., 2010). Here, we adhere to using only resource rent as our indicator of economic performance, unlike in Sumaila et al. (2012), where they also report payments to labor (i.e., wages), and earnings to fishing companies as additional indicators of fisheries benefits.

As in Sumaila et al. (2012), we estimate R, resource rent adjusted for subsidies, as follows:

$$R = \mathrm{LV} - (C + S) \tag{7.1}$$

where LV represents the landed value of officially reported marine landings. The total variable cost of fishing is represented by C and subsidies are represented by S.

We compute the cost from overuse as the potential gain from rebuilding (P_{gains}) to be the value of the rebuilt resource rent (R_{rebuilt}) minus the value of current resource rent (R_{current}):

$$P_{\text{gains}_t} = R_{\text{rebuilt}_t} - R_{\text{current}_t} \tag{7.2}$$

where t represents time. We assume that globally, rebuilt fisheries will be successful in avoiding subsequent unsustainable increases in effort.

We then calculate, as in Sumaila et al. (2012), the present value of net gains from rebuilding global fisheries as follows:

$$\text{PV} = \sum_{t=0}^{49} \frac{P_{\text{gains}}}{(1+r)^t} \tag{7.3}$$

where PV is the present value of the net gain in resource rent, r is the prevailing rate of discount, and t represents time from present ($t \in [0, 49]$). A fixed discount rate of $r = 0.03$ (i.e., 3%) and compute the present value of net gains in resource rent for 50 years after rebuilding. A discount rate of 3% is generally lower than market rates as argued for by some environmental economists due to the central role of environmental resources in ensuring sustainable economies through time (Weitzman, 2001; Sumaila, 2004; Stern, 2006).

ESTIMATED ECONOMIC LOSSES

According to Srinivassan et al. (2010), global marine fisheries landings could increase to an average of 89 million t a year (range 83-99 million t; Table 7.1) if rebuilt, with a corresponding mean landed value of US$101 billion per year (range US$93-116 billion). This means that compared with current catches and landed values, the world could be losing up to 20 million t and US$30 billion.

Sumaila et al. (2012) find that by reducing fishing effort to what is considered optimal in terms of the capacity needed to land the MSY, eliminating harmful subsidies and putting in place effective management after rebuilding, resource rent from rebuilt global fisheries would be US$54 (US$39-77) billion per year (see Table 7.1). The *Sunken Billions* report of the World Bank (2009), which estimated economic rent without addressing the cost of reform and removing

TABLE 7.1 Key Economic Figures of Global Fisheries

		Rebuilt Fisheries		
Key Indicators, Annual Data (Unit)	**Current**	**Lower Bound**	**Mean**	**Upper Bound**
Catch (t)	80.2	82.7	88.7	99.4
Catch value (US$ billions)	87.7	92.6	100.5	116.3
Subsidies (US$ billions)	27.2	10.0	10.0	10.0
Rent net of subsidies[a] (US$ billions)	−12.5	39.0	54.0	77.0
Rent increase over current values (US$ billions)	–	51.2	66.4	89.4
NPV of resource rent increases (US$ billions)	–	665.2	972.0	1428.1
Transition costs[b] (US$ billions)	–	129.9	202.9	292.2
NPV net of transition costs (US$ billions)	–	535.3	769.1	1135.9

NPV, net present value.
[a] *The (resource) rent is the return to "owners" of fish stocks, which is the surplus from gross revenue after total cost of fishing is deducted and subsidies taken into account.*
[b] *Transition costs include the costs to society of reducing current fishing effort to levels consistent with maximum sustainable yield and the payments that governments may decide to employ to adjust capital and labor to uses outside the fisheries sector. Such payments may include vessel buyback programs and alternative employment training initiatives for fishers.*
Table adapted from Sumaila et al. (2012).

harmful subsidies, arrived at a potential resource rent of US$50 billion per year, using a different approach. The gain in resource rent from the current situation to a rebuilt global fishery would be US$66 (US$51-89) billion per year (Table 7.1).

The real cost to society of rebuilding fisheries, once the elimination of an estimated US$19 billion per year of harmful and ambiguous subsidies is taken into account Sumaila et al. (2010), is negative, implying that society as a whole will make money by engaging in rebuilding. However, fishing enterprises and fishers will lose profits and wages during rebuilding. Hence, to implement a rebuilding reform, governments may need to temporarily invest extra resources to mitigate these impacts. Sumaila et al. (2012)

therefore estimated the total amount that governments may need to invest in reform in order to rebuild world fisheries to be between US $130 and US$292 billion in present value, with a mean of US$203 billion. This latter point needs further research and the work reported in Hanna (2010 in OECD Proceedings) would be a good point of departure for such studies.

CONCLUSION

It is generally agreed that, as a result of the overuse of global fish stock, the world is losing a significant amount of economic benefits because our fisheries are not living up to their revenue potential; the total cost of fishing is too high and governments provide harmful subsidies to the sector, which results in a negative resource rent (i.e., economic loss to society) of about US$13 billion per year (Table 7.1). Sumaila et al. (2012) find that rebuilding overused global fish stocks would result in a gain in resource rent of US $66 billion per year, which, when discounted over the next 50 years using a 3% real discount rate, generates a present value of between US$660 and US$1430 billion (Table 7.1), i.e., between three and seven times the mean cost of fisheries rebuilding reform.

This result suggests that, even without accounting for the potential boost to recreational fisheries, processing, retail, and nonmarket values that would likely increase, there is a net economic benefit to be derived from rebuilding global fisheries. The overall ecological benefits of reducing the effects of overfishing and destructive fishing impacts are difficult to estimate in economic terms. However, socioeconomic benefits would include improved ecosystem resilience to environmental perturbations and greater security of ecosystem services such as coastal protection, nutrient recycling, provision of raw materials, and tourism. Hence, the current overused state of world fisheries is very costly.

References

Agnew, D. J., J. Pearce, G. Pramod, T. Peatman, R. Watson, J. R. Beddington, and T. J. Pitcher (2009), Estimating the worldwide extent of illegal fishing, *PLoS ONE*, **4**, e4570. http://dx.doi.org/10.1371/journal.pone.0004570.

Alder, J., S. Cullis-Suzuki, V. Karpouzi, K. Kaschner, S. Mondoux, W. Swartz, P. Trujillo, R. Watson, and D. Pauly (2010), Aggregate performance in managing marine ecosystems of 53 maritime countries, *Marine Policy*, **34**, 468–476.

Allison, E. H., A. L. Perry, M. -C. Badjeck, W. Neil Adger, K. Brown, D. Conway, A. S. Halls, G. M. Pilling, J. D. Reynolds, N. L. Andrew, and N. K. Dulvy (2009), Vulnerability of national economies to the impacts of climate change on fisheries, *Fish Fish.*, **10**, 173–196.

Althaus, F., A. Williams, T. A. Schlacher, R. J. Kloser, M. A. Green, B. A. Barker, N. J. Bax, P. Brodie, and M. A. Schlacher-Hoenlinger (2009), Impacts of bottom trawling on deep-coral ecosystems of seamounts are long-lasting, *Mar. Ecol. Prog. Ser.*, **397**, 279–294.

Alvarez-Filip, L., N. K. Dulvy, J. A. Gill, I. M. Côté, and A. R. Watkinson (2009), Flattening of Caribbean coral reefs: region-wide declines in architectural complexity, *Proc. R. Soc. Lond.* B, **276**, 3019–3025.

Barrett, J. H., A. M. Locker, and C. M. Roberts (2004), The origins of intensive marine fishing in medieval Europe: the English evidence, *Proc. R. Soc. Lond.* B, **271**, 2417–2421.

Béné, C., B. Hersoug, and E. H. Allison (2010). Not by rent alone: analysing the pro-poor functions of small-scale fisheries in developing countries, *Development Policy Review*, **28**, 325–358.

Blanchard, J., S. Jennings, R. Holmes, J. Harle, G. Merino, J. Allen, J. Holt, N. K. Dulvy, and M. Barange (2012), Potential consequences of climate change for primary production and fish production in large marine ecosystems, *Philos. Trans. R. Soc. B-Biol. Sci.*, **367**(1605), 2979–2989.

Branch, T., O. Jensen, D. Ricard, Y. Ye, and R. Hilborn (2011), Contrasting global trends in marine fishery status obtained from catches and from stock assessments, *Conserv. Biol.*, **25**, 777–786.

Cavanagh, R. D., and C. Gibson (2007), *Overview of the Conservation Status of Cartilaginous Fishes (Chondrichthyans) in the Mediterranean Sea,* IUCN, Gland, Switzerland and Malaga, Spain. vi + 42 pp.

Cheung, W. W. L., C. Close, V. Lam, R. Watson, and D. Pauly (2008). Application of macroecological theory to predict effects of climate change on global fisheries potential, *Mar. Ecol. Prog. Ser.*, **365**, 187–197.

Cheung, W. W. L., V. W. Y. Lam, J. L. Sarmiento, K. Kearney, R. E. G. Watson, D. Zeller, and D. Pauly (2010), Large-scale redistribution of maximum fisheries catch potential in the global ocean under climate change, *Global Change Biology*, **16**, 24–35.

Cheung, W. W. L., J. Dunne, J. L. Sarmiento, and D. Pauly (2011), Integrating ecophysiology and plankton dynamics into projected maximum fisheries catch potential under climate change in the Northeast Atlantic. *ICES J. Mar. Sci.*, **68**, 1008–1018.

Cheung, W. W. L., J. J. Meeuwig, M. Feng, E. Harvey, V. Lam, T. Langolis, D. Slawinski, C. Sun, and D. Pauly (2012a), Climate change induced *tropicalization of marine communities in Western Australia, Marine and Freshwater Research* **63**, 415–427.

Cheung, W. W. L., J. Pinnegar, G. Merino, M. C. Jones, and M. Barange (2012b), Review of climate change impacts on marine fisheries in the UK and Ireland, *Aquatic Conservation: Marine and Freshwater Ecosystems*, **22**, 368–388.

Cheung, W. W. L., J. L. Sarmiento, J. Dunne, T. L. Frolicher, V. W. Y. Lam, M. L. Deng Palomares, R. Watson, and D. Pauly (2012c), Shrinking of fishes exacerbates impacts of global ocean changes on marine ecosystems, *Nature Clim. Change*, http://dx.doi.org/10.1038/nclimate1691.

Clark, C. W. (1990), *Mathematical Bioeconomics: The Optimal Management of Renewable Resources,* Wiley, New York.

Costello, C., B. P. Kinlan, S. E. Lester, and S. D. Gaines (2012), *The economic value of rebuilding fisheries, OECD Food, Agriculture and Fisheries Working Papers, No. 55,* 68 pp., OECD Publishing, Paris.

Coulthard, S., D. Johnson, and J. A. McGregor (2011), Poverty, sustainability and human wellbeing: A social wellbeing approach to the global fisheries crisis, *Global Environmental Change,* **21**, 453–463.

Cullis-Suzuki, S., and D. Pauly (2010), Failing the high seas: A global evaluation of regional fisheries management organizations, *Marine Policy,* **34**, 1036–1042.

Cury, P. M., I. L. Boyd, S. Bonhommeau, T. Anker-Nilssen, R. J. M. Crawford, R. W. Furness, J. A. Mills, E. J. Murphy, H. Österblom, M. Paleczny, J. F. Piatt, J. -P. Roux, L. Shannon, and W. J. Sydeman (2011), Global seabird response to forage fish depletion—one-third for the birds, *Science,* **334**, 1703–1706.

Daskalov, G. M. (2002), Overfishing drives a trophic cascade in the Black Sea, *Mar. Ecol. Prog. Ser.,* **225**, 53–63.

Dyck, A. J., and U. R. Sumaila (2010), Economic impact of ocean fish populations in the global fishery, *J. Bioecon.,* **12**, 227–243.

Erlandson, J. M., and T. C. Rick (2010), Archaeology meets marine ecology: the antiquity of maritime cultures and human impacts on marine fisheries and ecosystems, *Annual Review of Marine Science,* **2**, 231–251.

FAO (2005): Increasing the contribution of small-scale fisheries to poverty alleviation and food security. In: *FAO Technical Guidelines for Responsible Fisheries.* No. 10, Food and Agriculture Organization of the United Nations, Rome, Italy, 79.

FAO (2009), *The State of World Fisheries and Aquaculture (SOFIA) 2008,* 176pp., Food and Agriculture Organization, Rome, Italy.

FAO (2010), *The State of World Fisheries and Aquaculture (SOFIA) 2010,* 197pp. Food and Agriculture Organization, Rome, Italy.

Froese, R., D. Zeller, K. Kleisner, and D. Pauly (2012), What catch data can tell us about the status of global fisheries. *Marine Biology* (Berlin) http://dx.doi.org/10.1007/s00227-012-1909-6.

Hobday, A. J. (2010), Ensemble analysis of the future distribution of large pelagic fishes off Australia, *Prog. Oceanogr.,* **86**, 291–301.

Hoffmann, R. C. (2005), A brief history of aquatic resource use in medieval Europe, *Helgol. Mar. Res.,* **59**, 22–30.

IPCC (2007): Summary for policymakers. In: *Climate Change 2007: The Physical Science Basis* [Solomon, S., D. Qin, M. Manning, Z. Chen, M. Marquis, K. B. Averyt, M. Tignor and H. L. Miller (eds.)]. Contribution of Working Group I to the Fourth Assessment Report of the Intergovernmental Panel on Climate Change, Cambridge University Press, Cambridge, United Kingdom, 18 pp.

Jackson, J. B. C. (2001), What was natural in the coastal oceans? *Proc. Natl. Acad. Sci. U. S. A.,* **98**, 5411–5418.

Jackson, J. B. C. (2010), The future of oceans past, *Philos. Trans. R. Soc. Lond.* B, **365**, 3765–3778.

Jackson, J. B. C., M. X. Kirby, W. H. Berger, K. A. Bjorndal, L. W. Botsford, B. J. Bourque, R. H. Bradbury, R. Cooke, J. Erlandson, J. A. Estes, T. P. Hughes, S. Kidwell, C. B. Lange, H. S. Lenihan, J. M. Pandolfi,

C. H. Peterson, R. S. Steneck, M. J. Tegner, and R. R. Warner (2001), Historical overfishing and the recent collapse of coastal ecosystems, *Science*, **293**, 629–638.

Jaramillo-Legorreta, A., L. Rojas-Bracho, R. L. Brownell, A. J. Read, R. Randall, R. R. Reeves, K. Ralls, and B. L. Taylor (2007), Saving the vaquita: immediate action, not more data, *Conserv. Biol.*, **21**, 1653–1655.

Jones, M. C., S. R. Dye, J. A. Fernandes, T. L. Frölicher, J. K. Pinnegar, R. Warren, and W. W. L. Cheung (2013), Predicting the Impact of Climate Change on Threatened Species in UK Waters, *PLoS ONE*, **8**, e54216.

Khalilian, S., R. Froese, A. Proelss, and T. Requate (2010), Designed for failure: a critique of the Common Fisheries Policy of the European Union, *Marine Policy*, **34**, 1178–1182.

Kerby, T., W. W. L. Cheung, and G. Engelhard (2012), The United Kingdom's role in North Sea demersal fisheries: a hundred year perspective, *Review in Fish Biology and Fisheries*, **22**(3), 621–634. http://dx.doi.org/10.1007/s11160-012-9261-y.

Kirby, M. X. (2004), Fishing down the coast: Historical expansion and collapse of oyster fisheries along continental margins, *Proc. Natl. Acad. Sci. U. S. A.*, **101**, 13096–13099.

Kowaleski, M. (2000), The expansion of south-western fisheries in late medieval England, *Economic History Review LIII*, **3**, 429–454.

Lindegren, M., C. Möllmann, A. Nielsen, K. Brander, B. R. MacKenzie, and N. C. Stenseth (2010), Ecological forecasting under climate change: the case of Baltic cod, *Proc. R. Soc. B Biol. Sci.*, **277**, 2121–2130.

Llope, M., G. M. Daskalov, T. A. Rouyer, V. Mihneva, K. -S. Chan, A. N. Grishin, and N. C. Stenseth (2011), Overfishing of top predators eroded the resilience of the Black Sea system regardless of the climate and anthropogenic conditions, *Global Change Biology*, **17**, 1251–1265.

Lotze, H. K. (2007), Rise and fall of fishing and marine resource use in the Wadden Sea, southern North Sea, *Fish. Res.*, **87**, 208–218.

Martin, P. S. (2002): Prehistoric extinctions: in the shadow of man. In Wilderness Political Ecology: Aboriginal Influences and the Original State of Nature. In: [Kay C. E. and R.T. Simmons (eds.)]. University of Utah Press, Salt Lake City, Utah, pp. 1–27.

McClenachan, L., and A. B. Cooper (2008). Extinction rate, historical population structure and ecological role of the Caribbean monk seal, *Proc. R. Soc. Lond.* B, **275**, 1351–1358.

Miller, K. A., G. R. Munro, U. R. Sumaila, and W. W. Cheung (2013), Governing Marine Fisheries in a Changing Climate: A Game-Theoretic Perspective, *Canadian Journal of Agricultural Economics*, **61**, 309–334.

Mora, C., R. A. Myers, M. Coll, S. Libralato, T. J. Pitcher, R. U. Sumaila, D. Zeller, R. Watson, K. J. Gaston, and B. Worm (2009), Management effectiveness of the world's marine fisheries, *PLoS Biol.*, **7**, e1000131, http://dx.doi.org/10.1371/journal.pbio.1000131.

Morato, T., R. Watson, T. J. Pitcher, and D. Pauly (2005), Fishing down the deep, *Fish Fish.*, **7**, 24–34.

Mumby, P. J., A. Hastings, and H. J. Edwards (2007), Thresholds and the resilience of Caribbean coral reefs, *Nature*, **450**, 98–101.

Murawski, S. A. (2010), Rebuilding depleted fish stocks: the good, the bad, and, mostly, the ugly, *ICES J. Mar. Sci.*, **67**, 1830–1840.

Myers, R. A., J. K. Baum, T. D. Shepherd, S. P. Powers, and C. H. Peterson (2007), Cascading effects of the loss of apex predatory sharks from a coastal ocean, *Science*, **315**, 1846–1850.

Myers, R. A., and B. Worm (2003), Rapid worldwide depletion of predatory fish communities, *Nature*, **423**, 280–283.

Myers, R. A., and B. Worm (2005), Extinction, survival or recovery of large predatory fishes, *Philos. Trans. R. Soc. Lond.* B, **360**, 13–20.

Ostrom, E. (1990), *Governing the Commons: The Evolution of Institutions for Collective Action*, Cambridge Univ. Press, New York.

Pitcher, T., D. Kalikoski, G. Pramod, and K. Short (2009), Not honouring the code, *Nature*, **457**, 658–659.

Rice, J. C., and S. M. Garcia (2011), fisheries, food security, climate change, and biodiversity: characteristics of the sector and perspectives on emerging issues, *ICES J. Mar. Sci.*, http://dx.doi.org/10.1093/icesjms/fsr041.

Rosenberg, A. A., W. J. Bolster, K. E. Alexander, W. B. Leavenworth, A. B. Cooper, and M. G. McKenzie (2005), The history of ocean resources: modelling cod biomass using historical records, *Frontiers in Ecology and the Environment*, **3**, 78–84.

Simpson, S. D., S. Jennings, M. P. Johnson, J. L. Blanchard, P. Schön, D. W. Sims, and M. J. Genner (2011), Continental Shelf-Wide Response of a Fish Assemblage to Rapid Warming of the Sea, *Curr. Biol.*, **21**, 1565–1570.

Srinivassan, U. T., W. W. L. Cheung, R. Watson, and U. R. Sumaila (2010), Food security implications of global marine catch losses due to overfishing, *J. Bioecon.*, **12**, 183–200.

Steinacher, M., F. Joos, T. L. Frölicher, L. Bopp, P. Cadule, V. Cocco, S. C. Doney, M. Gehlen, K. Lindsay, J. K. Moore, B. Schneider, and J. Segschneider (2010), Projected 21st century decrease in marine productivity: a multi-model analysis, *Biogeosciences*, **7**, 979–1005.

Stern, R. (2006), *Stern Review on the Economics of Climate Change*, Her Majesty's Treasury, London. ix + 579 p.

Stramma, L., E. D. Prince, S. Schmidtko, J. Luo, J. P. Hoolihan, M. Visbeck, D. W. R. Wallace, P. Brandt, and A. Körtzinger (2011), Expansion of oxygen minimum zones may reduce available habitat for tropical pelagic fishes, *Nature Climate Change*, **2**, 33–37.

Sumaila, U. R. (2004), Intergenerational cost benefit analysis and marine ecosystem restoration, *Fish Fish.*, **5**, 329–343.

Sumaila, U. R., A. D. Marsden, R. Watson, and D. Pauly (2007), A global ex-vessel fish price database: construction and applications, *J. Bioecon.*, **9**, 39–51.

Sumaila, U. R., and W. Cheung (2010), *Cost of adapting fisheries to climate change.* World Bank Discussion Paper Number 5. The International Bank for Reconstruction and Development/The World Bank, Washington, DC, U.S.A.

Sumaila, U. R., A. Khan, A. Dyck, R. Watson, R. Munro, P. Tydemers, and D. Pauly (2010), A bottom-up re-estimation of global fisheries subsidies, *J. Bioecon.*, **12**, 201–225.

Sumaila, U. R., W. W. L. Cheung, V. Lam, D. Pauly, and S. Herrick (2011), Climate change impacts on the biophysics and economics of world fisheries, *Nature Climate Change*, **1**, 449–456, http://dx.doi.org/10.1038/nclimate1301.

Sumaila, U. R., W. W. L. Cheung, A. Dyck, K. Gueye, L. Huang, V. Lam, D. Pauly, T. Srinivasan, W. Swartz, R. Watson, and D. Zeller (2012), Benefits of Rebuilding Global Marine Fisheries Outweigh Costs, *PLoS ONE*, **7**(7), e40542. http://dx.doi.org/10.1371/journal.pone.0040542.

Swartz, W., E. Sala, S. Tracey, R. Watson, and D. Pauly (2010), The spatial expansion and ecological footprint of fisheries (1950 to Present), *PLoS ONE*, **5**, e15143, http://dx.doi.org/10.1371/journal.pone.0015143.

Thurstan, R. H., S. Brockington, and C. M. Roberts (2010), The effects of 118 years of industrial fishing on UK bottom trawl fisheries, *Nature Communications*, **1**, http://dx.doi.org/10.1038/ncomms1013.

UN-DESA. (2009), Population Division of the Department of Economic and Social Affairs of the United Nations Secretariat, World Population Prospects: the 2008 Revision and World Urbanization Prospects: the 2009 Revision. http://esa.un.org/wup2009/unup/index.asp.

UNEP (2011), *Towards a Green Economy: Pathways to Sustainable Development and Poverty Eradication*, 631pp., UNEP, Nairobi, www.unep.org/greeneconomy.

Wakeford, R. C., D. J. Agnew, and C. C. Mees (2009), Review of institutional arrangements and evaluation of factors associated with successful stock recovery plans, *Rev. Fish. Sci.*, **17**, 190–222.

Ward-Paige, C. A., D. M. Keith, B. Worm, and H. K. Lotze (2012). Recovery potential and conservation options for elasmobranchs, *J. Fish Biol.*, **80**, 1844–1869.

Watson, R. A., W. W. L. Cheung, J. A. Anticamara, R. U. Sumaila, D. Zeller, and D. Pauly (2012), Global marine yield halved as fishing intensity redoubles, *Fish Fish.*, http://dx.doi.org/10.1111/j.1467-2979.2012.00483. x.

Weitzman, M. L. (2001), Gamma discounting, *Am. Econ. Rev.*, **91**, 260–271.

World Bank, FAO (2009), Sunken Billions: The Economic Justification for Fisheries Reform: Case Study Summaries. Permanent URL: http://go.worldbank.org/MGUTHSY7U0. Accessed April 15, 2009.

Worm, B., R. Hilborn, J. K. Baum, T. Branch, J. S. Collie, et al. (2009), Rebuilding global fisheries, *Science*, **325**, 578–585.

Worm, B., and R. A. Myers (2003), Meta-analysis of cod–shrimp interactions reveals top-down control in oceanic food webs, *Ecology*, **84**, 162–173.

CHAPTER

8

Impacts of Multiple Stressors

Julie Hall, Robert J. Diaz[†], Nicolas Gruber[‡], Dan Wilhelmsson[§]*

*NIWA, Kilbirnie, Wellington, New Zealand

[†]Department of Biological Sciences, Virginia Institute of Marine Science, College of William and Mary, Gloucester Point, Virginia, USA

[‡]Environmental Physics, Institute of Biogeochemistry and Pollutant Dynamics, Department of Environmental Sciences, Zurich, Switzerland

[§]Swedish Secretariat for Environmental Earth System Science, Stockholm, Sweden

INTRODUCTION

As the world's population increases, the intensity and spatial distribution of anthropogenic stressors impacting marine systems are expanding (Sanderson et al., 2002; Halpern et al., 2007a; Crain et al., 2008). Many of these stressors, such as increasing temperature, changes in ocean circulation and mixing, ocean acidification, hypoxia, increasing nutrients in coastal areas, overfishing, and pollution co-occur in time and space, resulting in almost all marine organisms and ecosystems becoming subject to the impacts of multiple stressors (Venter et al., 2006; Halpern et al., 2007b, 2008a,b). Our understanding of the impact of individual stressors is limited, but we have even less understanding of the impacts that a combination of these stressors will have on marine organisms and ecosystems and on the important services they provide to humanity. The need to understand the interactions and potentially cumulative or multiplicative effects of multiple stressors has been identified as one of the most important questions in marine ecology today (Zeidberg and Robison, 2007; Crain et al., 2008; Darling and Cote, 2008). This understanding would enable us to better evaluate and develop relevant management strategies for the impacts of multiple stressors on marine organisms and ecosystems, as the sustainability of ecosystem services will not be achieved by managing stressors at the individual level. Multiple stressors must be managed in an integrated way.

The occurrence of multiple stressors impacting marine systems is widespread. Halpern et al. (2008b) observed that no marine systems are unaffected by human impacts and calculated that 41% of the marine environment is strongly impacted by a combination of stressors. Marine stressors fall into two categories: Those that are forced by global processes and thus act globally such as increased temperature, decreased oxygen, and ocean acidification; and those that are a result of local and regional processes, and thus act primarily at a local to regional level. These include pollution, increased nutrient loads, coastal hypoxia associated with eutrophication, and overuse of marine resources. The global stressors exhibit substantial regional variability, while the local to regional stressors occur in many oceanic regions spanning the entire globe. The key difference between the global and regional/local stressors is the approach needed to mitigate them. Global stressors need to be addressed through global climate-policy-associated measures, while the regional/local stressors can be addressed through policy measures at this scale (Table 8.1).

Assessing the nature and impacts of multiple stressors requires looking at the problem from several perspectives. In this chapter,

TABLE 8.1 Links Between Scale of Impact and Stressors

Scale of Process	Stressor
Global	Warming Acidification Decreased oxygen
Aggregated from local/regional scales	Pollution Nutrient loading Overuse of marine resources Coastal hypoxia

we begin by looking at stressors that act on a global scale and, in a separate section, those that operate on smaller scales but have aggregate effects at the global scale. We then look at possible feedbacks and synergistic effects among the various stressors, in particular, examining what impacts these effects may have on management strategies. In this section, we look at examples of synergistic effects from the level of individual organisms to entire ecosystems.

GLOBAL-SCALE STRESSORS

The emission of carbon dioxide into the atmosphere as a consequence of the burning of fossil fuels and to a lesser degree by land-use change is the primary driver for all globally important stressors on marine systems (Doney, 2010; Gruber, 2011). Ocean acidification, warming, deoxygenation, as well as global change-induced changes in ocean circulation and mixing are all a consequence of the increase in anthropogenic CO_2 emissions. Given the long lifetime of CO_2 in the atmosphere, the anthropogenic emissions alter atmospheric CO_2 globally and, consequently, also alter Earth's entire climate system. Thus, these stressors act globally, but they have important local differences.

The first and most direct way in which the CO_2 emissions affect marine organisms and ecosystems is through ocean acidification. The ocean has taken up roughly 30% of the global anthropogenic CO_2 emissions since ~1800 (Sabine et al., 2004), thereby playing an important moderating effect on global climate change. But as the CO_2 enters the ocean, it forms a weak acid that reacts with the carbonate system in the ocean, shifting the acid-base balance toward a lower pH, i.e., it "acidifies" the ocean. Chemically, this set of equilibria is expressed as

$$CO_2 + H_2O \Leftrightarrow H^+ + HCO_3^- \Leftrightarrow 2H^+ + CO_3^{2-}$$

with the sum of all dissolved inorganic carbon species denoted as DIC, i.e., $DIC = CO_2 + HCO_3^- + CO_3^{2-}$ (see Chapter 2 for further details).

This does not mean that the ocean becomes acidic, however, as the ocean's pH is well above neutrality (e.g., the global mean pH of surface waters is about 8.1; Feely et al., 2008). The chemical changes associated with ocean acidification go well beyond the reduction in pH. Most of the CO_2 that enters the ocean will react with the carbonate ion (CO_3^{2-}) to form two bicarbonate ions (HCO_3^-). As a result, the carbonate ion concentration decreases, causing the saturation state of seawater with regard to mineral forms of $CaCO_3$ to decrease. This saturation state is commonly expressed in the form of the ratio of the *in situ* carbonate ion concentration over the carbonate ion concentration at saturation, expressed as Ω. Nearly, all surface waters are currently supersaturated with regard to the two most abundant forms of $CaCO_3$, i.e., the metastable form aragonite and more stable form calcite, i.e., $\Omega > 1$, for these compounds. However, if atmospheric CO_2 continues to increase, surface waters in several oceanic regions will become undersaturated ($\Omega < 1$) within the next 50 years. The decrease in saturation state is variable with the low latitudes showing about twice as large a decrease in Ω (Figure 8.1c). This difference is almost entirely due to the high latitudes having a Revelle (buffer) factor that is about twice as large as the low latitudes, and hence take up much less anthropogenic CO_2 from the atmosphere. The Revelle factor is a measure of how much the surface concentration of CO_2, often expressed by its equivalent partial pressure, is changing for a given change in the oceanic concentration of DIC (Zeebe and Wolf-Gladrow, 2001; Sarmiento and Gruber, 2006). Thus, the larger this factor, the less surface ocean DIC will be changing for a given change in the surface concentration of CO_2. Since the exchange of CO_2 across the air-sea interface is fast enough for surface ocean CO_2 to track well, the atmospheric increase in CO_2 causes the near surface concentration of anthropogenic CO_2 to be inversely related to the Revelle factor. While the high latitudes experience a smaller change, their initial preindustrial Ω is much lower than that in the low and mid-latitudes, so that it is the high latitudes that tend to become undersaturated first with regard to aragonite (Orr et al., 2005; Steinacher et al., 2009).

The second and more indirect way in which the CO_2 emissions affect marine organisms is through their impact on global climate. The accumulation of this greenhouse gas in the atmosphere is the primary cause for the observed warming and climate change over the past 50 years and will be the predominant force for

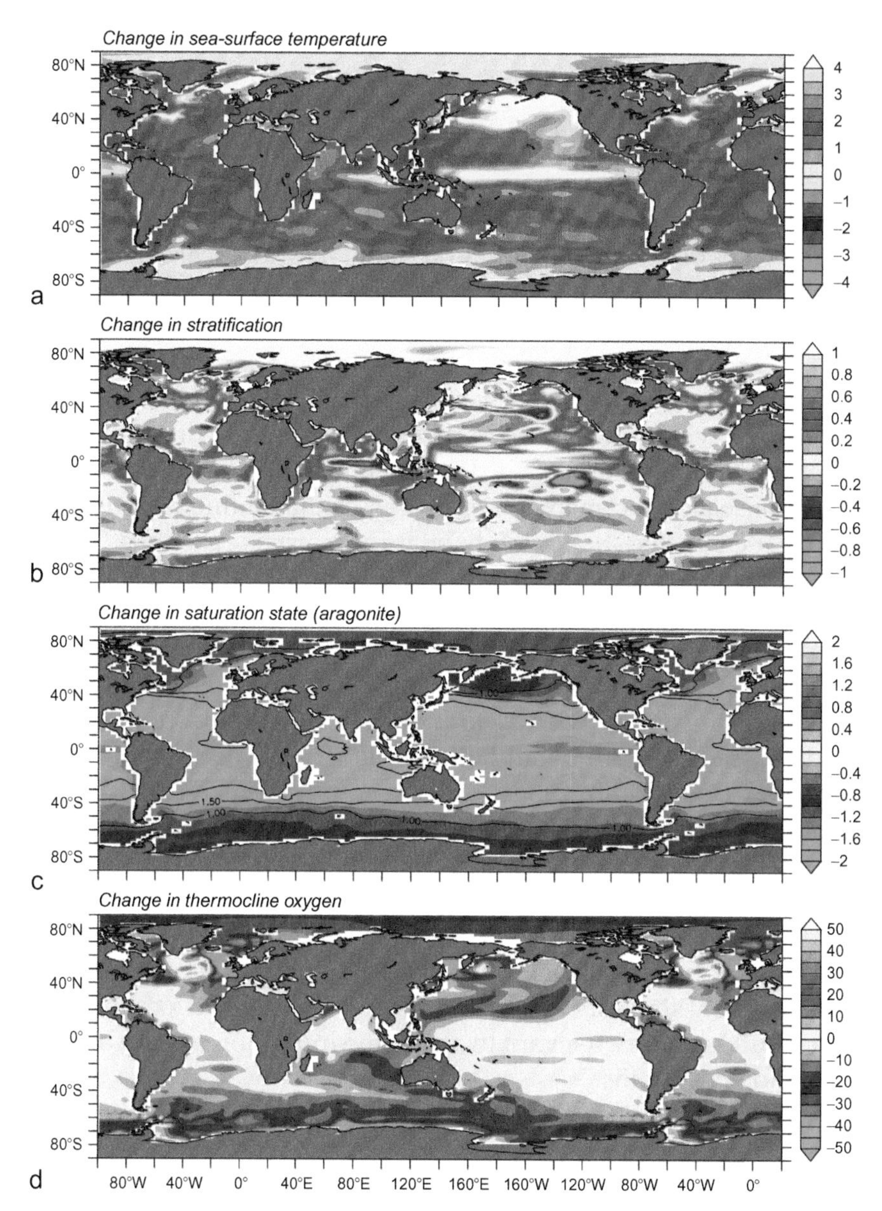

FIGURE 8.1 Global maps of model-simulated changes in ocean properties for 2099 (under the IPCC SRES A2 scenario) relative to the year 1850 (atmospheric CO_2 increased from 280 ppm in 1850 to 840 ppm in 2099). (a) Changes in sea-surface temperature, (b) change in upper ocean stratification, expressed as the density gradient between the upper ocean (0-50 m) and the upper thermocline (100-200 m); (c) change in the surface ocean saturation state with respect to aragonite—also shown as contours is the absolute saturation state for 2099; (d) change in the mean concentration of oxygen in the thermocline (200-600 m). Results are from the NCAR CSM 1.4 model (Frölicher et al., 2009; Steinacher et al., 2009).

additional warming and climate change in the next 50-100 years (Bernstein et al., 2007). The ocean tends to moderate this warming by taking up a substantial amount of heat from the atmosphere (e.g., Levitus et al., 2005; Lyman et al., 2010), but this heat uptake warms the ocean, changes its currents and mixing, and increases stratification. Additional physical changes occur in the ocean due to global change-induced alteration of the hydrological cycle, leading to altered pattern of net evaporation and precipitation, causing changes in salinity and consequently changes in stratification and ocean circulation as well. These physical changes alter the supply of nutrients to the surface ocean, impact the depth of the surface mixed layer, change the paths and intensity of ocean currents, affect the strength of upwelling and downwelling, and change the distribution of sea ice and many other processes that define the physicochemical characteristics of the habitat for marine organisms. These physical changes, particularly ocean warming and the associated increase in vertical stratification, lead to a decrease in the oceanic oxygen concentration in the ocean's interior, a process termed ocean deoxygenation (Keeling et al., 2010). We distinguish this process from coastal hypoxia to highlight the difference in cause. As discussed in more detail below, this latter process is a consequence of coastal eutrophication, driven by the increased supply of nutrients into the ocean. In contrast, ocean deoxygenation is essentially a consequence of global warming.

These chemical and physical changes occur globally (Figure 8.1), and their future magnitude will be primarily determined by the evolution of future CO_2 emissions. At the same time, the imprint of these global stressors will have a strong regional and local nature, which will be highly relevant to assess the impact on marine organisms and ecosystems.

Global warming of the surface ocean will tend to be maximal in the low to mid-latitudes, while the very high latitudes tend to warm less (Figure 8.1a), particularly if they remain covered by sea ice such as the regions around Antarctica. The magnitude of warming in the Arctic depends very sensitively on the magnitude of sea-ice loss in that region, a process that differs strongly between current model projections. Otherwise, the magnitude of the predicted surface ocean warming varies between the different models largely in response to their climate sensitivity. The results shown in

Figure 8.1 from the NCAR CSM 1.4 model tend to have a relatively small climate sensitivity (Meehl et al., 2000). Models with a higher climate sensitivity, but still within the currently accepted range, can give a warming of nearly twice the amplitude shown here.

As the warming is maximal at the surface and decreases with increasing depth, this change increases upper-ocean stratification nearly everywhere (Figure 8.1b). Exceptions are places where shifting currents alter the upper-ocean water-density structure in a more complex manner or where changes in surface salinity may offset the warming-induced stratification. In other places, particularly the Arctic, the opposite is the case. There, the decrease in salinity substantially enhances the temperature-induced increase in upper-ocean stratification.

Compared to the other stressors, the predicted ocean deoxygenation is spatially the least homogenous stressor (Figure 8.1d). While the global mean decrease of dissolved oxygen is relatively modest (a few percent relative to the mean ocean O_2 concentration of $\sim$180 μmol kg^{-1} (Sarmiento and Gruber, 2006) or a few μmol kg^{-1}), the local decreases are predicted to be up to 50 μmol kg^{-1} by the end of the century. The pattern of deoxygenation also differs substantially between models. Most models show the largest decrease in oxygen in the mid- to high latitudes (Keeling et al., 2010), i.e., in regions that have, in general, relatively high dissolved oxygen concentrations. The NCAR CSM 1.4 model used for Figure 8.1d shows nearly no change in the thermocline of the low latitudes, where oxygen is generally already low. In contrast, other models predict a substantial decrease of oxygen in these regions, leading to a substantial extension of the oxygen minimum zones (Cocco et al., in preparation), a result that tends to be supported by surveys of these zones over the past 50 years (Stramma et al., 2008).

An important caveat when considering the model-based projections shown in Figure 8.1 is the low spatial resolution of the underlying Earth System Model (between about 2° and 4°). As a result, a number of critical processes and regions cannot be properly resolved in such models. One prominent example is the Eastern Boundary Upwelling Regions, which have been shown to have a very low saturation state with regard to aragonite (low Ω) (Feely et al., 2008) and are set to progress toward a state of widespread undersaturation within the next few decades (Gruber et al., 2012).

Furthermore, these regions also tend to have naturally low oxygen concentrations in their thermoclines, a condition that appears to have intensified in recent decades, at least in the California Current System off the coast of the North America (Bograd et al., 2008; Chan et al., 2008). This coastal deoxygenation needs to be viewed separately from coastal hypoxia as the former is driven by changes in climate.

This coalescence of the different global drivers creates a number of hotspots, among which the Eastern Boundary Upwelling Regions stand out (Figure 8.2), but also the high-latitude North Pacific, the Arctic, and the Southern Ocean deserve attention.

LOCAL- AND REGIONAL-SCALE STRESSORS

Two key local stressors whose effects aggregate to global scales are hypoxia and pollutants. Impacts of hypoxia (or low dissolved oxygen environments—also known as dead zones) associated with human activities are limited to local and regional scales but are globally distributed (see Figure 4.2, Chapter 4). It also takes a combination of multiple stressors for a hypoxic area to develop. The

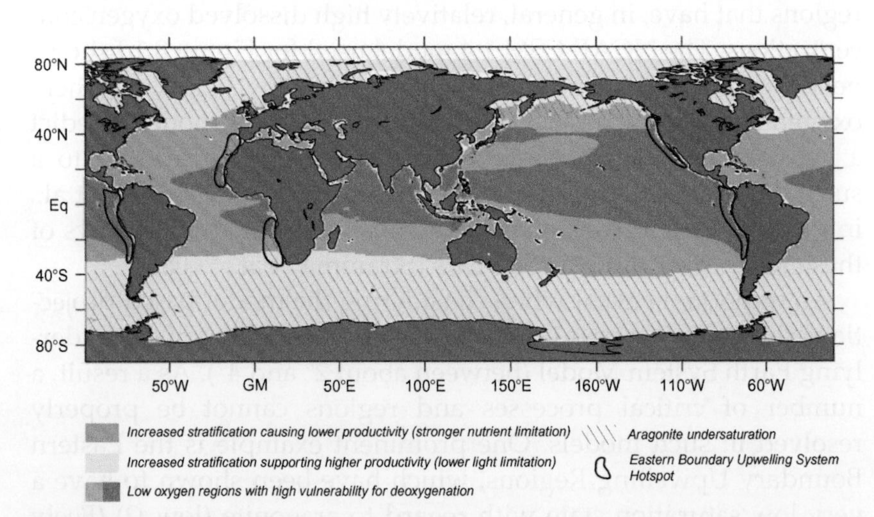

FIGURE 8.2 Global map showing regions of particular vulnerability to the three main stressors, i.e., ocean warming, acidification, and deoxygenation. *From Gruber (2011), figure 3. © The Royal Society.*

principal stressors that can lead to coastal hypoxia are water-column stratification that isolates bottom waters from oxygen-rich surface waters and accumulation of excess organic matter on the seafloor due to eutrophication. The global stressors ocean deoxygenation and ocean warming can accelerate and worsen coastal hypoxia by lowering the initial oxygen concentration of water that is transported toward the coasts and by increasing stratification.

Hypoxia affects a wide range of aquatic systems with varying frequency, seasonality, and persistence. While there always have been naturally occurring hypoxic habitats, such as deep fjords and some coastal upwelling regions, anthropogenic activities related primarily to organic and nutrient enrichment from sewage/industrial discharges and from runoff from agricultural lands have led to increases in hypoxia in both freshwater and marine systems. A consequence of this nutrient enrichment has been a rapid rise in the areas affected by hypoxia over the past 50 years (Diaz and Rosenberg, 2008). By the early 1900s, dissolved oxygen was a topic of interest in research and management; and by 1920s, it was recognized that a lack of oxygen was a major hazard to fishes and other higher organisms. It was not obvious, however, that oxygen would become critically low in coastal systems until the 1970s and 1980s when large areas of low dissolved oxygen started to appear with associated mass mortalities of invertebrate and fishes. From the middle of the twentieth century to today, there have been drastic changes in dissolved oxygen concentrations and dynamics in many marine coastal waters. Prime examples would be the northwest continental shelf of the Black Sea, which is now recovering from hypoxia, the Baltic Sea, the Gulf of Mexico continental shelf off Louisiana and Texas, and the East China Sea (Diaz and Rosenberg, 2008).

There are now close to 600 documented cases of large-scale hypoxia around the globe, primarily from North America and Europe where reporting water quality is routine. Because of a lack of reporting, there is little known about hypoxia from much of Asia, Indonesia, or the Pacific. If the strong association between human population density and hypoxia seen in North America and Europe holds for other parts of the globe, then there are likely well over 1000 hypoxic areas globally. This leads us to the conclusion that no other environmental variable of such ecological importance to coastal ecosystems as dissolved oxygen has changed so drastically, in such a short time. Increasing stress from expanding agricultural and industrial outputs to meet the demands of the expanding human population will only increase areas of coastal hypoxia and may also impact aquaculture site selection or production (Diaz et al., 2012).

Pollution of the marine environment occurs in many different forms, such as input of chemicals, radioactivity, solid waste, human-induced sedimentation, energy (i.e., heat and noise), as well as in the form of oil spills and what is more and more referred to as "biological pollution" (pathogens, parasites, and invasive species). Pollutants mainly enter the marine environment through discharge and runoff of material and effluents from land (mainly via rivers), offshore, and shipping activities and as fallout from the atmosphere. Coastal areas and habitats are generally under higher pressure from pollution than the high seas (e.g., MA, 2005; Halpern et al., 2008b).

The main sources and types of pollution of concern for the marine ecosystems vary with region (e.g., UNEP, 2006; Halpern et al., 2007a). The ultimate impact of pollution of a certain type and magnitude, further, varies with the physical conditions of the environment, such as water circulation and temperature, and, obviously, with the susceptibility of the marine organisms and habitats present (Halpern et al., 2007b; Costello et al., 2010). Impacts of pollution on marine organisms range from disturbance and hampered reproduction to mortality and population declines, which in turn can lead to quality and productivity losses at ecosystem scales.

Chemicals, such as persistent organic pollutants and metals, can be toxic for humans and wildlife and impair physiological functions. This can, for example, lead to immune dysfunctions and reproductive and neurological disorders among humans, marine organisms, and seabirds (e.g., Ratcliffe, 1970; Jensen, 1972; Ritter et al., 1995; Grandjean et al., 2010). Persistent chemicals can be transported over long ranges by ocean currents to remote and nonindustrialized areas. Due to their chemical properties, a number of substances can accumulate in marine organisms in concentrations several orders of magnitude higher than in the surrounding water (bioaccumulation), and chemicals can become more and more concentrated while passing through the food chain (biomagnification).

For example, polar bears are among the animals that have the highest concentrations of toxic organic pollutants (de Wit et al., 2002).

Solid waste, oil spills, and elevated sediment inputs can kill, harm, or inhibit growth and reproduction of marine animals as well as degrade habitats at larger scales (Derraik, 2002; Orpin et al., 2004; Halpern et al., 2007a; Teuten et al., 2009). Close to noise sources, animals can be killed by the sound pressure, and at larger distances, noise can cause animals to avoid areas or hamper their ability to communicate, feed, and navigate (e.g., Weilgart, 2007). The effects of radionuclides in the marine environment are largely related to health hazards for humans, although it seems that this is a cause of concern only in specific regions (Sazykina et al., 1998; Livingston and Povinec, 2000).

While pollution can have serious impacts on species and habitats in many areas and regions, the global marine ecosystems may on average be relatively resilient to single types of pollution of typical magnitudes (Halpern et al., 2007a; Costello et al., 2010). However, the long-term combined effects of, e.g., chemicals (i.e., "cocktail effects") and the cumulative effects of noise, as well as synergies between different types of pollutants, are virtually unknown. Pollution impacts may undermine the resilience of ecosystems to other stressors such as elevated sea-surface temperatures and ocean acidification (Jenssen, 2006; Khan, 2007; Piola and Johnston, 2008; Clark et al., 2009; Noyes et al., 2009; Russel et al., 2009; McLeod et al., 2010; Negri and Hoogenboom, 2011).

FEEDBACKS AND SYNERGISTIC EFFECTS

As shown in earlier chapters, stressors such as increasing temperature, ocean acidification, hypoxia, increasing nutrients in coastal areas, and pollution have impacts on biogeochemical cycles, the physiology and populations of marine organisms, and the structure and function of marine ecosystems. In areas of the ocean where there are strong impacts of multiple stressors, understanding the impacts of multiple stressors on marine organisms, populations, and ecosystems is critical to the management of marine ecosystems. There are three types of biological response to multiple stressors: additive, antagonistic, and synergistic. The additive response occurs when the impact of the multiple stressors is a direct addition of the stressors acting alone. An antagonistic response is where the impact of the multiple stressors is less than the addition of the stressors acting alone, and a synergistic response is where the impact of the multiple stressors is greater than the additive impact (Figure 8.3).

In an analysis which considered 112 population studies of marine organisms which used a range of stressors, organisms subjected to two stressors showed that 23% had an additive impact, 42% were antagonistic, and 35% were synergistic (Darling and Cote, 2008). This was similar to an analysis by Crain et al. (2008) which considered 171 (78% single-species studies) studies of community or population responses and reported 26%, 38%, and 36%, respectively, for additive, antagonistic, and synergistic interactions. They also showed that when a third stressor was introduced, the occurrence of synergistic interactions doubled to 66%. The Crain et al. (2008) analysis also showed that synergistic interactions varied both within and among stressor pairs, suggesting that population dynamics and ecosystem traits play an important role in response to stressors. Factors such as trophic level, response type, ecosystem type, starting abundance, reproductive rate, intensity of the stressor, and length of the experiment may all impact the interactions. Trophic level was shown to have an effect on the interaction

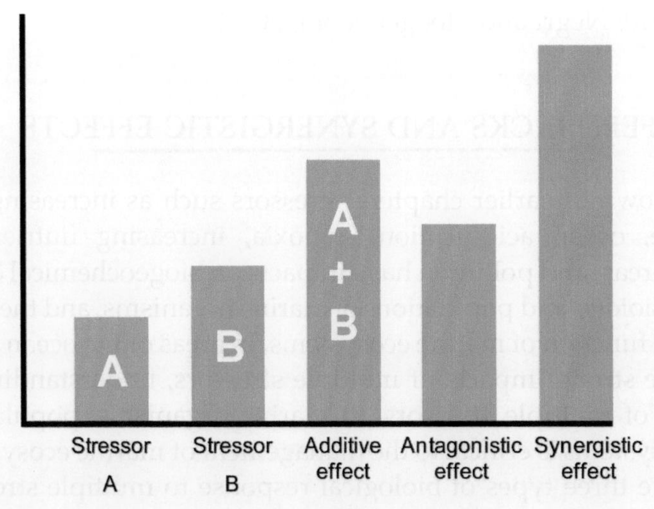

FIGURE 8.3 Potential interactions of multiple stressors.

with studies of algae showing significant antagonistic impacts and studies on animals showing significant synergistic impacts. These results indicate that the impacts of multiple stressors could be more negative for organisms at higher trophic levels which may reflect the sensitivity of more complex food webs to disruption (Crain et al., 2008).

The impact of multiple stressors may also depend on whether a species or community-level response is being measured. Species-level impacts may be reduced or increased through species interactions such as species diversity and redundancy, trophic complexity, ecological history, and ecosystem type (Vinebrooke et al., 2004). Crain et al. (2008) analysis found that species-level impacts were primarily synergistic and community-level impacts were primarily antagonistic indicating that studies of a single species may not be a good predictor of community-level responses. The prediction of impacts of multiple stressors on marine ecosystems is complex and influenced by a wide range of factors including species diversity, openness of the system, environmental variability, temporal patterns of stressor occurrence (simultaneous versus sequential), intensity of the stressor, and the variable used to measure the response (Breitburg et al., 1998; Relyea and Hoverman, 2006).

The potential for synergistic impacts of multiple stressors is one of the largest uncertainties in the prediction of the impacts of anthropogenic activities (Sala and Knowlton, 2006; Sutherland et al., 2006). From a community/ecosystem perspective, finding out that 77% of the multiple stressor interactions were nonadditive (Darling and Cote, 2008) is of concern as antagonistic and synergistic responses can alter species interactions and food-web structure and function in unpredictable ways. This calls into question the common ecosystem modeling approach of using additive functions to represent multiple stressors and highlights the need for nonadditive multiple stressor interactions to be considered in ecosystem models where anthropogenic impacts on ecosystems are being considered.

Experiments to investigate the impacts of multiple stressors in marine ecosystems have used a wide range of measurements to assess the impacts including physiological and organism-level measures. Physiological studies of multiple stressors focus on how processes within the organism are impacted. For example,

O'Donnell et al. (2009) measured the expression of a central molecular gene (chaperone *hsp70*) as a bioindicator to assess changes in responses of sea urchin larvae stressed by elevated pCO_2, decreased pH, and temperature changes. The larvae raised at elevated pCO_2 showed reduced expression of the thermally induced gene which suggests that successful larval development at elevated pCO_2 resulted in changes in the ability to respond to thermal stress (Figure 8.4). Taken in a broader perspective, this result indicates that environmental conditions during free-floating larval stages may impact the number of urchins and potentially disrupt community composition (O'Donnell et al., 2009).

Another example of the physiological measurements being used to assess the impact of increased pCO_2 and temperature was undertaken by Metzger et al. (2007), when they considered the impacts of these stressors on the ability of the blood of the edible crab (*Cancer pagurus*) to carry oxygen. They found that elevated CO_2 resulted in a large shift in the thermal tolerance limits of *C. pagurus* to cooler

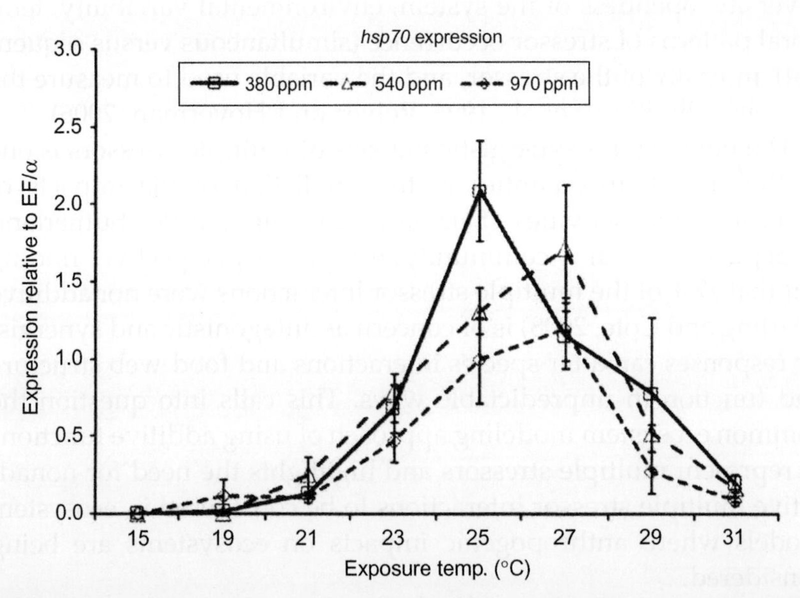

FIGURE 8.4 qPCR analysis of the expression of molecular gene (chaperone *hsp70*) in response to 1 h exposure to elevated temperature. Values along the abscissa represent temperature of exposure. Larvae were raised at 15 °C. Expression was normalized by expression of EF1α and grouped according to the pCO_2 at which the chamber was aerated. *Error bars* are ±1 standard error (O'Donnell et al., 2009).

temperatures and suggested a narrowing of the thermal tolerance window with increasing atmospheric CO_2 concentrations. These physiological studies indicate that there is a potential for the narrowing of the species distribution ranges due to the narrowing of the species thermal window under increased atmospheric CO_2.

Understanding organism and population responses to multiple stressors is important to understanding the anthropogenic impacts on marine ecosystems. Coral reefs have emerged as vulnerable ecosystems to stressors associated with climate change. They are among the most diverse ecosystems on earth and are estimated to contain approximately one-third of all described marine species (Reaka-Kudla, 2001). The occurrence of mass bleaching of coral reefs has been shown to occur more commonly in response to a combination of stressors acting simultaneously and is often synergistic (Baker et al., 2008). The impacts of globally acting stressors, such as increasing temperature and decreasing pH, have been shown to have a significant impact on coral reefs worldwide. Experiments to assess impacts of increased pCO_2 and increased temperature on the bleaching, productivity, and calcification rates of three groups of reef-building corals representing some of the most common and functionally important benthic organisms on coral reefs found negative responses in three species (*Porolithon onkodes, Acropora intermedia, and Porites lobata*; Anthony et al., 2008) (Figure 8.5).

The *P. onkodes* coral species was very sensitive, with decreasing productivity and calcification with increasing CO_2 which was exacerbated by increased temperature. The productivity response of the *A. intermedia* and *P. lobata* was variable; *A. intermedia* showed maximum productivity at an intermediate pH followed by a significant decline, whereas *P. lobata* showed a continuous decrease in productivity with decreasing pH. The calcification of both species was less responsive to pH changes than was bleaching and productivity. These results indicate that future predictions of bleaching in response to increased temperature must also take into account the additional effects of ocean acidification and suggest that any potential adaptation of corals to thermal stress may be offset by the impacts of decreased pH in the ocean.

Coral reefs primarily occur in coastal regions where anthropogenic impacts often manifest at the local level. A significant number of local/regional-scale stressors such as increased nutrients,

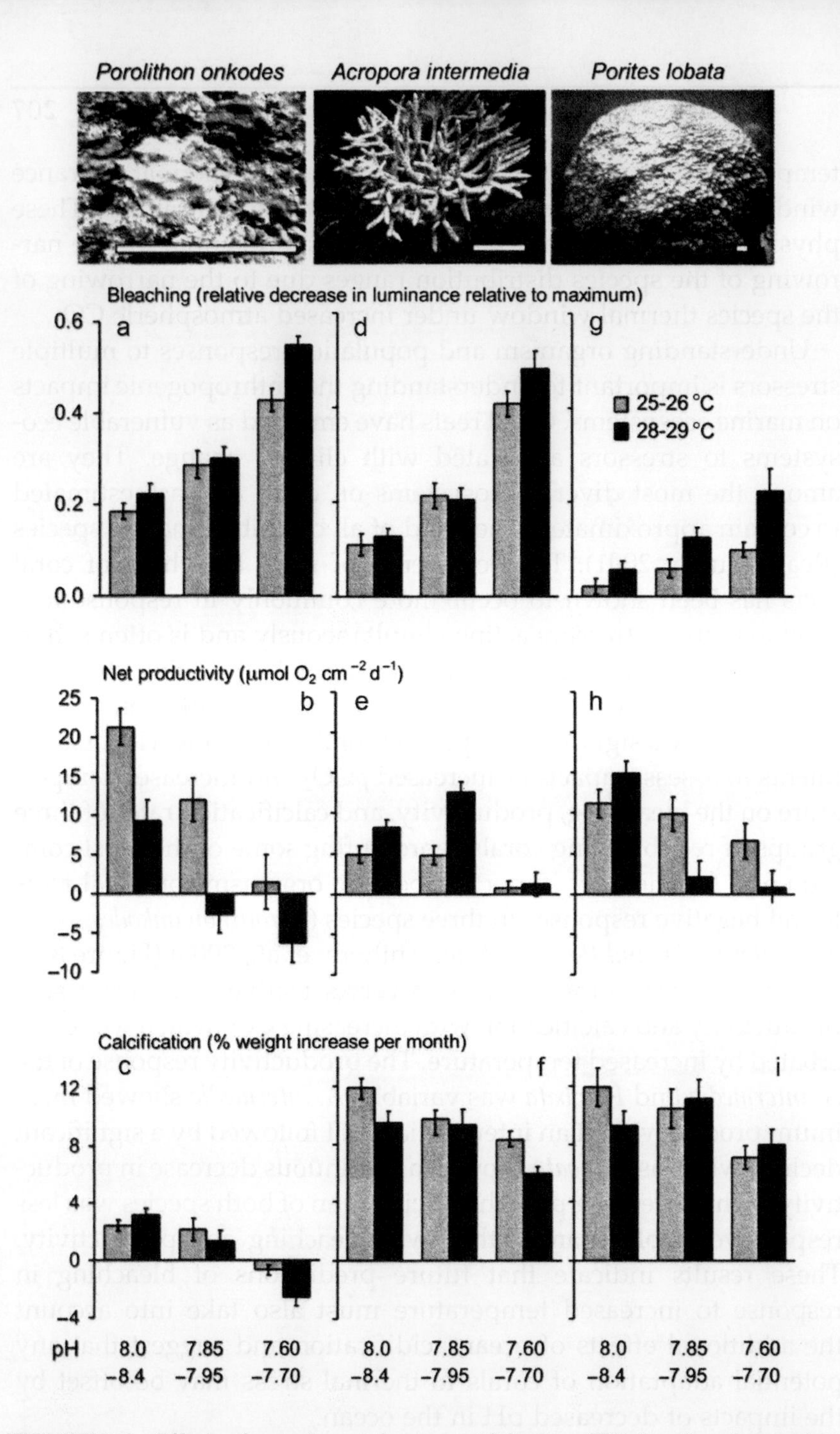

FIGURE 8.5 Effects of experimental ocean acidification (CO_2 level) and warming on three key performance variables of three major coral reef builders: (a-c) crustose coralline algae (CCA, *Porolithon onkodes*), (d-f) branching *Acropora* (*A. intermedia*), and (g-i) massive *Porites* (*P. lobata*). Gray and black bars show low- and high-temperature treatments, respectively. Data are means ± SEM of $n = 15/25$ specimens for each combination of CO_2 and temperature. Levels of CO_2 represented the present-day control condition (380 ppm atmospheric CO_2) and projected scenarios for high categories IV (520-700 ppm) and VI (1000-1300 ppm) by the IPCC (Anthony et al., 2008). *From O'Donnell et al. (2009), figure 3. © Springer.*

hypoxia, sediments, and pollutants have been shown to impact coral reefs. The potential for these local/regional-scale stressors to interact with global-scale stressors such as increased temperature and ocean acidification is high. For example, impacts of increased temperature and pollutants (Cu concentrations) at the critical life stage of metamorphosis for coral larvae reduce successful metamorphosis in two species (Negri and Hoogenboom, 2011). The combined effects for *Acropora tenuis* were additive above 29 °C and were synergistic for both species above 31 °C with *Acropora millepora* being more sensitive (Figure 8.6).

This study demonstrates synergistic effects at environmentally relevant temperatures and Cu concentration at a critical life history stage for coral species. Another example of the interaction of global and local/regional stressors is the impact of increased temperatures and exposure to commonly used herbicides. A study of the coral

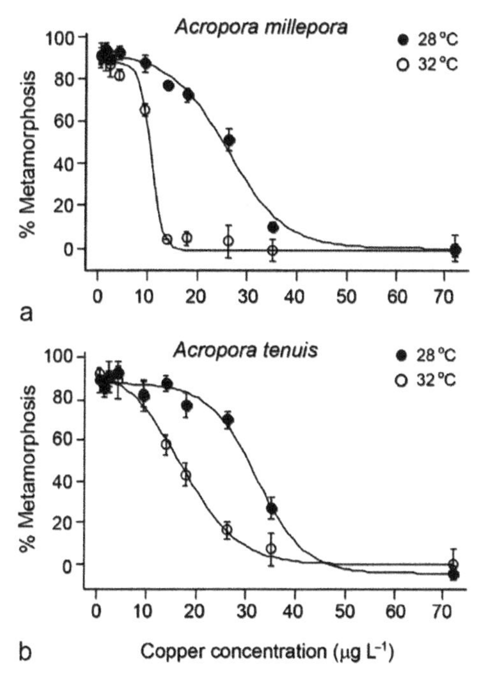

FIGURE 8.6 Relationship between larval metamorphosis and copper concentration for *Acropora millepora* (a) and *Acropora tenuis* (b). Points represent the mean and standard error of measured %metamorphosis at each of two temperatures. Lines are the best-fit nonlinear regressions of Equation (1) (from Negri and Hoogenboom, 2011) to each set of data. *Source: http://dx.doi.org/10.1371/journal.pone.0019703.g001.*

A. millepora exposed to three agricultural herbicides at environmentally relevant concentrations showed that there was a significant impact on both photosynthetic efficiency and photoinhibition (Negri et al., 2011). The herbicides had a negative impact on the thermal stress measurement at 31 and 32 °C with the impact being additive for photosynthetic efficiency. At high temperatures, the impact was synergistic with two of the herbicides, diuron and atrazine. A similar impact was not observed with the crustose coralline algae, *Neogoniolithon fosliei*. The experimental results showed that reducing the diuron concentration by 1 μg L^{-1} above 30 °C would protect photosynthetic efficiency by the equivalent of a 1.8 °C temperature increase and a 1 °C change for photoinhibition. The key finding of this study is that the control of local- to regional-scale stressors (herbicides in this case) has the potential to mitigate impacts of global stressors.

Coral bleaching has also been shown to be strongly related to water quality measured as both chlorophyll *a* and dissolved inorganic nitrogen concentration (Wagner et al., 2010). This is consistent with experimental studies by Nordemar et al. (2003) and Schloder and D'Croz (2004) who showed that increased nitrate concentrations exacerbated temperature stress in corals. The suggested mechanism for this is that the increased nitrogen results in increased photosynthetic pigments within the corals which make them darker and more sensitive to both increased temperature and high irradiance (Fabricius, 2006). Wooldridge (2009) has also shown that for inshore reef areas in the Great Barrier Reef system susceptible to high land runoff, a reduction in nutrient concentrations would result in an increase of approximately 2-2.5 °C, the upper thermal bleaching limit. These results all support the conclusion of Dodge et al. (2008) who suggested that understanding water quality was critical to understanding the resilience of coral reef to bleaching events. The examples given above indicate that the outlook for reefs impacted by multiple stressors is exceptionally serious (Veron et al., 2009) and management interventions will be required to save reefs and increase their resilience to future change (Veron et al., 2009).

There are other important ecosystems which are subject to both global and local/regional stressors including a range of coastal benthic ecosystems. The local/regional impacts include increased eutrophication (increased nutrient concentrations) and increased hypoxic areas from the decomposition of excess organic matter. The increase in hypoxia in the coastal oceans as a result of anthropogenic activities impacts benthic organisms with the potential for greater impacts with increases in water temperature (Chapter 4). The interaction of increased hypoxia and temperature will likely increase vulnerability from both physical factors such as stratification (Chapter 3) and biological factors such as temperature-dependent metabolism. All this will lead to a reduced area of habitat for benthic and pelagic species (Stramma et al., 2008). Vaquer-Sunyer and Duarte (2010) conducted a meta-analysis of experimental assessments of over 100 species of benthic organisms which showed that the survival range and median lethal concentration of oxygen increased with increased temperature. Crustaceans were the most sensitive to hypoxia and their thresholds for mortality were the most sensitive to increased temperature. They also showed that, for benthic organisms that are unable to move away from the hypoxia, survival time under hypoxia decreased by 25% with a 1.8 °C rise in temperature and by 47% with a 4 °C rise in temperature. For the oyster *Crassostrea virginica*, this meant a reduction in survival time from 20 to 14 days with a 1.8 °C temperature increase and a decrease to a survival time of 10 days with a 4 °C increase in temperature. These results imply that synergistic effects between hypoxia and increased temperature will result in an increasing vulnerability of benthic fauna to hypoxia in a warmer ocean.

An example of the impacts of increased nutrients in the coastal region coupled with decreased pH is the impact of these stressors on coastally occurring macroalgal species. Russel et al. (2009) undertook a series of experiments to investigate the impacts of these two stressors on marine coastal communities by measuring the response of the calcifying algae, *Lithophyllum* sp., and the turf-forming algae *Feldmannia* spp. in mesocosm experiments. Both decreased pH and increased nutrients had a negative impact on the calcifying algae; in comparison, the decrease in pH had a positive impact on the turf algae and the increased nutrient had no impact (Figure 8.7).

When the ability of these species to colonize an area was measured at current nutrient concentrations and pH, the calcifying algae colonized 21% of the area but there was no colonization at

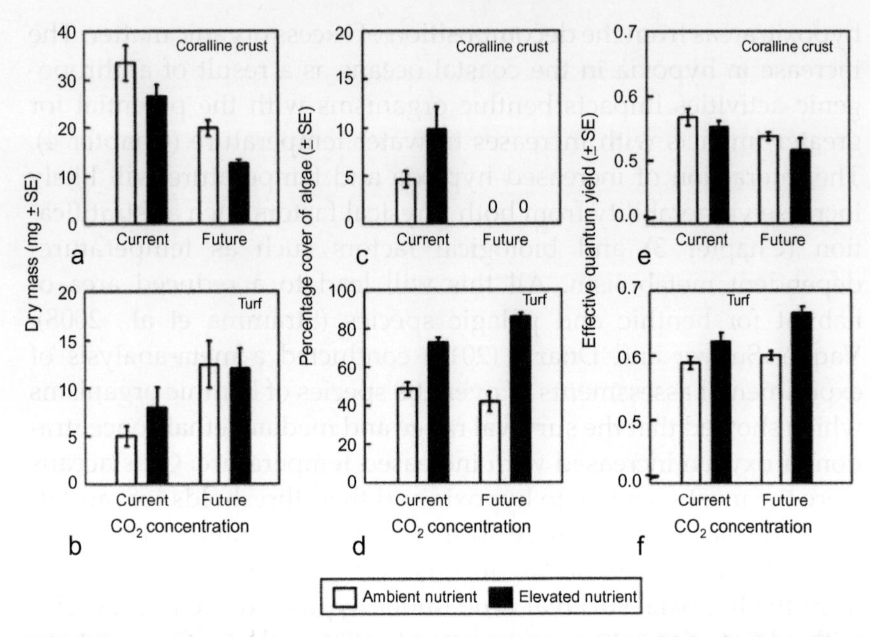

FIGURE 8.7 The dry mass (a, b) and effective quantum yield (e, f) of coralline crusts (a, e) and turfs (b, f) on the natural rock substrate and percentage cover of crusts (c) and turfs (d) that recruited to unoccupied substrate when exposed to different concentrations of CO_2 (current vs. future) and nutrients (ambient vs. elevated). "0" in (c) show 0% cover (recruitment) (Russel et al., 2009).

elevated nutrient or decreased pH, whereas turf algae colonize under all conditions. When photosynthetic yield was measured, there was a decrease for the calcifying algae with decreased pH and no impact of the increased nutrients. For the turf algae, there was a positive response to both perturbations with the impact being synergistic with a 41% increase over the additive impact. These results have significant implications for the Australian temperature coastal region where currently 80% of the hard substrate is dominated by calcifying algae and associated ecosystems (Irving et al., 2004). The calcifying algae form the foundation of a structurally complex and productive habitat with over 1000 species (Goodsell and Connell, 2002) which supports commercial harvest of species such as abalone (Shepherd and Daume, 1994). In contrast, the turf ecosystems represent a substantial change in ecological function including reduced productivity (Mann, 1973) and ecosystem services (Vitousek et al., 1997).

The combination of global and local/regional stressors does not only occur in coastal regions. In the Eastern Tropical Pacific, the combination of decreased pH, increased temperature, and hypoxia has the potential to have a significant impact on one of the top predators in the region. The jumbo squid (*Dosidicus gigas*) grows up to 2 m in length and 50 kg in weight and has a high O_2 demand and blood with a low O_2-carrying capacity. These animals are suggested to live chronically "on the edge of O_2 limitation" and are not considered to be well poised to adapt to changes in O_2 supply and demand. Rosa and Seibel (2008) found that decreased pH and increased temperature would reduce the jumbo squid's ability to enter near surface waters and an increasingly shallower hypoxic zone may decrease the depth the squid can go to at night resulting in a vertical compression of habitat. This habitat compression may alter the behavior and feeding ecology and subsequently impact growth and reproduction of the jumbo squid (Prince and Goodyear, 2006). This in turn has the potential to impact not only the fishery for jumbo squid but also the ecosystem within the Eastern Tropical Pacific as the jumbo squid are an important component of the diet of birds, fish, and mammals (Davis et al., 2007).

An example of multiple global stressors at the ecosystems level is shown in a study which considers the impacts of multiple stressors on fish population and fisheries catches. An estimate of the impact of decreased pH and O_2 content has been undertaken in a modeling study of the fisheries in the northeast Atlantic. Cheung et al. (2011) modeled 120 species of bottom-living fish and invertebrates which make up 95% of the fisheries catch for this region and showed that the estimated catch potential was reduced by 20-30% under the SRES A1B scenario to 2050 (Figure 8.8). When the model was extended to include the impacts on phytoplankton species composition, there was a further 10% reduction in estimated catch potential as the size structure of the phytoplankton population can have significant ecological impacts. The authors note that the magnitude of the predicted changes is uncertain, given the data and sensitivity of the models.

In this chapter, we have shown that multiple stressors do occur throughout the world's oceans at varying levels of intensity and we presented a range of examples of the impacts of multiple global and local/regional stressors on marine organisms and ecosystems.

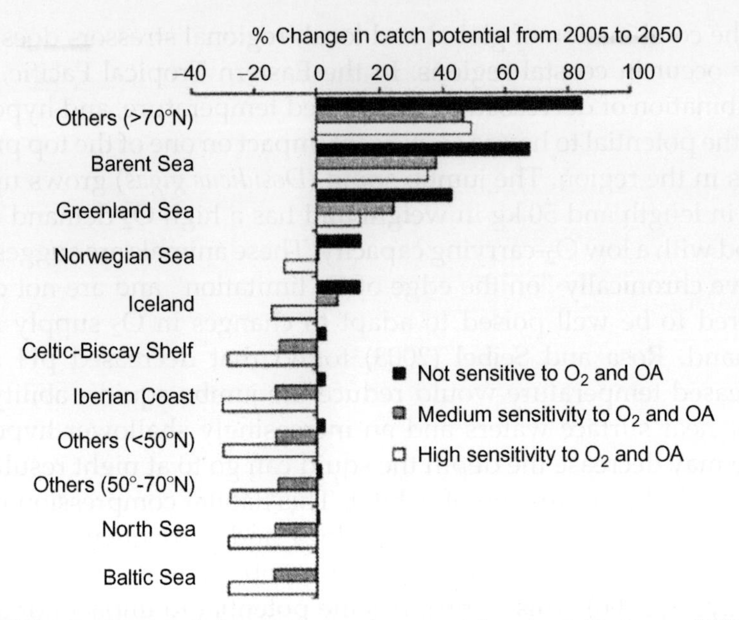

FIGURE 8.8 Projected changes in maximum catch potential between 2005 and 2050 (10-year average) in the LMEs in the Northeast Atlantic with high sensitivity (open bar), medium sensitivity (gray bar), and insensitive (black bar) to changes in oxygen content and pH (Cheung et al., 2011).

What has been shown to be critical to our thinking as we look to manage the impacts of the global stressors is that there are local actions that can be taken which have the potential to reduce the impacts of global stressors. In the case of coral reefs, the reduction of locally distributed pollutants such as nutrients, Cu, and herbicides will decrease the susceptibility of coral reef to bleaching at high temperatures and the susceptibility to the impacts of decreasing pH, showing that management interventions will for a time increase reef resilience (Veron et al., 2009). This is demonstrated in Figure 8.9 where the impact of increased nutrients from land-use changes alters the threshold temperature for coral bleaching.

In the case of coastal temperate ecosystems, the reduction of nutrients has been shown to potentially decrease the impact of decreasing pH on the structure and function of coastal temperate ecosystems. These examples highlight that the management of local stressors has the potential to mitigate at least temporally the impact of global stressors. The wide range of synergistic impacts of multiple stressors also indicates that a precautionary approach to the

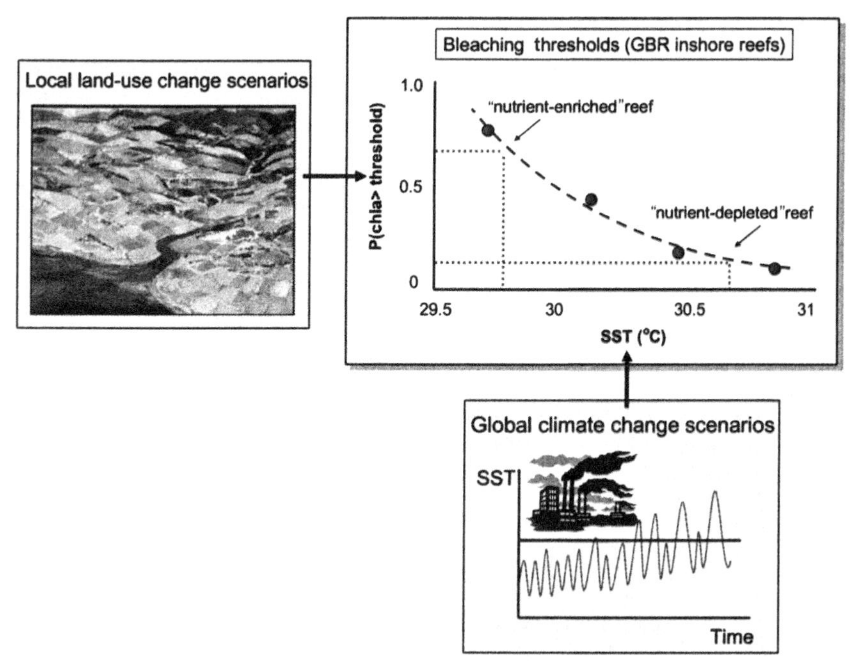

FIGURE 8.9 Conceptual modeling framework that enables the impact of "local" (viz. water quality improvement) and "global" (viz. CO_2/temperature reduction) management strategies to be assessed in terms of their *joint* (conditional) potential to reduce the future likelihood of mass coral bleaching on the GBR (Wooldridge, 2009).

impacts of multiple stressors is critical until we extend our understanding and ability to predict the impacts of multiple stressors on marine organisms and ecosystems.

CONCLUDING REMARKS

Our understanding of the feedbacks and interactions between the main threats to the global oceans described in the other chapters in this book is unfortunately very rudimentary. We are only at the beginning of the process of determining how the complex interplay between these processes works and much more research is necessary. Nonetheless, we are far enough along in our understanding to draw some preliminary conclusions:

- Multiple stressors tend to amplify each other in a substantial number of cases

- Looking at the response of single species of organisms is not a good way to estimate how communities of species and larger ecosystems will respond to multiple stressors
- Coral ecosystems are particularly vulnerable to multiple stressors
- Local measures (such as reducing pollution or nutrient levels) can help ameliorate the effects of global-scale ocean stressors such as warming or acidification

The fact that in many cases multiple stressors amplify each other requires us to change the way we manage and make decisions about marine resources. We no longer have the luxury of thinking only about local issues or about one threat at a time. We need to devise a management system that works across scales from local to global and that allows us to optimize our use of marine resources in a sustainable way, given several simultaneous threats. This challenge is a difficult one, but far from impossible to achieve. Beyond acquiring a better understanding of the interactions and feedbacks between multiple stressors, we need to reduce the level of compartmentalization in our management and decision-making infrastructure for marine resources. Creating a cross-scale governance system for marine resources that actually works is an urgent challenge.

References

Anthony, K. R., D. I. Kline, G. Diaz-Pulido, S. Dove, and O. Hoegh-Guldberg (2008), Ocean acidification causes bleaching and productivity loss in coral reef builders, *Proceedings of the National Academy of Sciences*, **105**, 17442–17446.

Baker, A. C., P. W. Glynn, and B. Riegl (2008), Climate change and coral reef bleaching: an ecological assessment of long term impacts, recovery trends and future outlook, Estuarine, *Coastal Shelf Science*, **80**(4), 435–471.

Bernstein, L, et al. (2007): IPCC, 2007: climate change 2007: synthesis report. Contribution of working groups I. In: *II and III to the Fourth Assessment Report of the Intergovernmental Panel on Climate Change*. Intergovernmental Panel on Climate Change, Geneva, http://www.ipcc.ch/ipccreports/ar4-syr.htm.

Breitburg, D. L., J. W. Baxter, C. A. Hatfield, R. W. Howarth, C. G. Jones, G. M. Lovett, and C. Wigand (1998): Understanding effects of multiple stressors: ideas and challenges. In: *Successes, Limitations and Frontiers in Ecosystem Science* [Pace M. L. and P. M. Groffman (eds.)]. Springer Press, New York, 416–431.

Bograd, S. J., C. G. Castro, E. D. Lorenzo, D. M. Palacios, H. Bailey, W. Gilly, and F. P. Chavez (2008), Oxygen declines and the shoaling of the hypoxic boundary in the California Current, *Geophys. Res. Lett.*, **35**, L12607, http://dx.doi.org/10.1029/2008GL034185.

Chan, F., J. A. Barth, J. Lubchenco, A. Kirincich, H. Weeks, W. T. Peterson, and B. A. Menge (2008), Emergence of anoxia in the California Current large marine ecosystem, *Science*, **319**, 920, http://dx.doi.org/10.1126/science.1149016.

Cheung, W. W. L., J. Dunne, J. L. Sarmiento, and D. Pauly (2011), Integrating ecophysiology and plankton dynamics into projected maximum fisheries catch potential under climate change in the Northeast Atlantic, *Marine Science*, http://dx.doi.org/10.1093/icejms/fsr012.

Clark, C. W., Ellison, W. T., Southall, B. L., Hatch, L., Van Parijs, S. M., Frankel, A., Ponirakis, D. (2009) Acoustic masking in marine ecosystems: intuitions, analysis, and implication. *Mar. Ecol. Prog. Ser.*, **395**, 201–222.

Costello, M. J., M. Coll, R. Danovaro, P. Halpin, H. Ojaveer, and P. Miloslavich (2010), A census of marine biodiversity knowledge, resources, and future challenges, *PLoS ONE*, **5**, e12110.

Crain, M. C., K. Kroeker, and B. S. Halpern (2008), Interactive and cumulative effects of multiple human stressors in marine systems, *Ecol. Lett.*, **11**, 1304–1315.

Darling, E. S., and I. M. Cote (2008), Quantifying the evidence for ecological synergies, *Ecol. Lett.*, **11**, 1278–1286.

Davis, R. W., N. Jaquet, D. Gendron, U. Markaida, G. Bazzino, and W. Gilly (2007), Diving behaviour of sperm whales in relation to behaviour of a major prey species, the jumbo squid, in the Gulf of California, Mexico, *Mar. Ecol. Prog. Ser.*, **333**, 291–302.

Derraik, J. G. B. (2002), The pollution of the marine environment by plastic debris: a review, *Mar. Poll. Bull.*, **44**, 842–852.

de Wit, C. A., A. T. Fisk, K. E. Hobbs, D. C. G. Muir, G. W. Gabrielsen, R. Kallenborn, M. M. Krahn, R. J. Norstrom, and J. U. Skaare (2002): Persistent organic pollutants in the Arctic. In: *AMAP assessment 2002: Arctic Monitoring and Assessment Programme* [de Wit C. A., A. T. Fisk, K. E. Hobbs and D. C. G. Muir (eds.)]. Oslo, Norway.

Diaz, R., N. N. Rabalais and D. L. Breitburg (2012), Agriculture's Impact on Aquaculture: Hypoxia and Eutrophication in Marine Waters. OECD, Committee for Agriculture and the Environment Policy Committee. COM/TAD/CA/ENV/EPOC(2010)16/FINAL. 46 p. Accessed April 1, 2013 http://www.oecd.org/tad/sustainable-agriculture/49841630.pdf

Diaz, R. J., and R. Rosenberg (2008), Spreading dead zones and consequences for marine ecosystems, *Science*, **321**, 926–928.

Dodge, R. E., C. Birkeland, M. Hatziolos, J. Kleypas, S. R. Palumbi, O. Hoegh-Guldberg, R. Van Woesik, J. C. Ogden, R. B. Aronson, B. D. Causey, and F. Staub (2008), A call to action for coral reefs, *Science*, **322**, 189–190.

Doney, S. C. (2010), The Growing Human Footprint on Coastal and Open-Ocean Biogeochemistry, *Science*, **328**(1512), 1185198, http://dx.doi.org/10.1126/science.

Fabricius, K. E. (2006), Effects of irradiance, flow, and colony pigmentation on the temperature microenvironment around corals: Implications for coral bleaching? *Limnol. Oceanogr.*, **51**, 30–37.

Feely, R. A., C. L. Sabine, M. Hernandez-Ayon, D. Ianson, and B. Hales (2008), Evidence for upwelling of corrosive 'acidified' water onto the continental shelf, *Science*, **320**, 1490–1492, http://dx.doi.org/10.1126/science.1155676.

Frölicher, T. L., F. Joos, G. -K. Plattner, M. Steinacher, and S. C. Doney (2009), Natural variability and anthropogenic trends in oceanic oxygen in a coupled carbon cycle-climate model ensemble, *Global Biogeochem. Cy.*, **23**, GB1003, http://dx.doi.org/10.1029/2008GB003316.

Goodsell, P. J., and S. D. Connell (2002), Can habitat loss be treated independently of habitat configuration? Implications for rare and common taxa in fragmented landscapes, *Mar. Ecol-Prog. Ser.*, **239**, 37–44.

Grandjean, P., L. K. Poulsen, C. Heilmann, U. Steuerwald, and P. Weihe (2010), Allergy and Sensitization during Childhood Associated with Prenatal and Lactational Exposure to Marine Pollutants, *Environ. Health Perspect.*, **118**(10).

Gruber, N. (2011), Warming up, turning sour, losing breath: Ocean biogeochemistry under global change, *Phil. Trans. R. Soc. A*, **369**, 1980–1996, http://dx.doi.org/10.1098/rsta.2011.0003.

Gruber, N., C. Hauri, Z. Lachkar, D. Loher, T. L. Frölicher, and G. K. Plattner (2012), Rapid progression of ocean acidification in the California Current System, *Science*, **337**, 220–223, http://dx.doi.org/10.1126/science.1216773.

Halpern, B. S., K. L. McLeod, A. A. Rosenberg, and L. B. Crowder (2007a), Managing for cumulative impacts in ecosystem-based management through ocean zoning, *Ocean Coast. Manage.*, **51**, 203–211.

Halpern, B. S., K. A. Selkoe, F. Micheli, and C. V. Kappel (2007b), Evaluating and ranking the vulnerability of global marine ecosystems to anthropogenic threats, *Conserv. Biol.*, **21**, 1301–1315.

Halpern, B. S., K. L. McLeod, A. A. Rosenberg, and L. B. Crowder (2008a), Managing for cumulative impacts in ecosystem-based management through ocean zoning, *Ocean Coast. Manage.*, **51**, 203–211.

Halpern, B. S., S. Walbridge, K. A. Selkoe, C. V. Kappel, F. Micheli, C. D'Agrosa, J. F. Bruno, K. S. Casey, C. Ebert, H. E. Fox, R. Fujita, D. Heinemann, H. S. Lenihan, E. M. P. Madin, M. T. Perry, E. R. Selig, M. Spalding, R. Steneck, and R. Watson (2008b), A Global Map of Human Impact on Marine Ecosystems, *Science*, **319**, 948–952.

Irving, A. D., S. D. Connell, and B. M. Gillanders (2004b), Local complexity in patterns of canopy-benthos associations produces regional patterns across temperate Australasia, *Mar. Biol.*, **144**, 361–368.

Jensen, S. (1972), The PCB story, *Ambio*, **1**(4), 123–131.

Jenssen, B. M. (2006), Endocrine-disrupting chemicals and climate change: a worst-case combination for Arctic marine mammals and seabirds? *Environ. Health Persp.*, **114**, 76–80.

Keeling, R. F., A. Kortzinger, and N. Gruber (2010), Ocean Deoxygenation in a Warming World, *Annual Review of Marine Science*, **2**, 199–229, http://dx.doi.org/10.1146/annurev.marine.010908.163855.

Khan, N. Y. (2007), Multiple stressors and ecosystem-based management in the Gulf, *Aquatic Ecosystem Health & Management*, **10**, 259–267.

Levitus, S., J. Antonov, and T. Boyer (2005), Warming of the world ocean, Geophys. *Res. Lett.*, **32**, L02604, http://dx.doi.org/10.1029/2004GL021592.

Livingston, H. D., and P. P. Povinec (2000), Anthropogenic marine radioactivity, *Ocean Coast. Manage.*, **43**, 689–712.

Lyman, J. M., S. A. Good, V. V. Gouretski, M. Ishii, G. C. Johnson, M. D. Palmer, D. M. Smith, and J. K. Willis (2010), Robust warming of the global upper ocean, *Nature*, **465**, 334–337, http://dx.doi.org/10.1038/nature09043.

MA, Millennium Ecosystem Assessment, http://www.maweb.org/en/Reports.aspx.

Mannn, K. H. (1973), Seaweeds: Their productivity and strategy for growth, *Science*, **182**, 975–981.

McLeod, E., R. Moffit, A. Timmerman, R. Salm, L. Menviel, M. J. Palmer, E. R. Selig, K. S. Casey, and J. F. Bruno (2010), Warming seas in the coral triangle: coral reef vulnerability and management implications, *Coast. Manage.*, **38**, 518–539.

Meehl, G. A., W. D. Collins, B. A. Boville, J. T. Kiehl, T. M. L. Wigley, and J. M. Arblaster (2000), Response of the NCAR Climate System Model to Increased CO_2 and the Role of Physical Processes, *J. Climate*, **13**, 1879–1898.

Metzger, R., F. J. Sartoris, M. Langenbuch, and H. O. Portner (2007), Influence of elevated CO_2 concentrations on thermal tolerance of the edible crab Cancer pagurus, *J. Therm. Biol.*, **32**, 144–151.

Negri, A. P., and M. O. Hoogenboom (2011), Water contamination reduces the tolerance of coral larvae to thermal stress, *PLoS ONE*, **6**, e19703, http://dx.doi.org/10.1371/journal.pone.0019703.g001.

Negri, A. P., F. Flores, T. Rothig, and S. Uthicke (2011), Herbicides increase the vulnerability of corals to rising sea surface temperature, *Limnol. Oceanogr.*, **56**(2), 471–485.

Nordemar, I., M. Nystrom, and R. Dizon (2003), Effects of elevated seawater temperature and nitrate enrichment on the branching coral *Porites cylindrical* in the absence of particulate food, *Mar. Biol.*, **142**, 669–677.

Noyes, P. D., M. K. McElwee, H. D. Miller, B. W. Clark, L. A. Van Tiem, K. C. Walcott, and E. D. Levin (2009), The toxicology of climate change: environmental contaminants in a warming world, *Environ. Int.*, **35**, 971–986.

O'Donnell, M. J., L. M. Hammond, and G. E. Hofmann (2009), Predicted impact of ocean acidification on a marine invertebrate: elevated CO_2 alters response to thermal stress in sea urchin larvae, Mar Biol, **156**, 439–446.

Orpin, A. R., P. V. Ridd, S. Thomas, K. R. N. Anthony, P. Marchall, and J. Oliver (2004), Natural turbidity variability and weather forecasts in risk management of anthropogenic sediment discharge near sensitive environments, *Mar. Poll. Bull.*, **49**, 602–612.

Orr, J. C., V. J. Fabry, O. Aumont, L. Bopp, S. C. Doney, R. A. Feely, A. Gnanadesikan, N. Gruber, A. Ishida, F. Joos, R. M. Key, K. Lindsay, E. Maier-Reimer, R. Matear, P. Monfray, A. Mouchet, R. G. Najjar, G. -K. Plattner, K. B. Rodgers, C. L. Sabine, J. L. Sarmiento, R. Schlitzer, R. D. Slater, I. J. Totterdell, M. -Fr. Weirig, Y. Yamanaka, and A. Yool (2005), Anthropogenic ocean acidification over the twenty-first century and its impact on calcifying organisms, *Nature*, **437**, 681–686, http://dx.doi.org/10.1038/nature04095.

Piola, R., and E. Johnston (2008), Pollution reduces native diversity and increases invader dominance in marine hard-substrate communities, *Diversity and Distributions*, **14**, 329–342.

Prince, E. D., and F. P. Goodyear (2006), Hypoxia-based habitat compression of tropical pelagic fishes, *Fish. Oceanogr.*, **15**, 451–464.

Ratcliffe, D. A. (1970), Changes attributable to pesticides in egg breakage frequency and eggshell thickness in some British birds, *J. Appl. Ecol.*, **7**, 67–115.

Reaka-Kudla, M. L. (2001), Known and unknown biodiversity, risk of extinction and conservation strategy in the sea. In: *Waters in Peril* [Leah Bendell-Young and Patricia Gallaugher (eds.)]. Springer, New York, 19–33.

Relyea, R., and J. Hoverman (2006), Assessing the ecology in eco-toxicology: a review and synthesis in freshwater systems, *Ecol. Lett.*, **9**, 1157–1171.

Ritter, L., K. R. Solomon, J. Forget, M. Stemeroff, and C. O'Leary (1995), An Assessment Report on DDT, Aldrin, Dieldrin, Endrin, Chlordane, Heptachlor, Hexachlorobenzene, Mirex, Toxaphene, Polychlorinated Biphenyls, Dioxins and Furans, Prepared for the International Programme on Chemical Safety (IPCS).

Rosa, R., and B. A. Seibel (2008), Synergistic effects of climate-related variables suggest future physiological impairment in a top oceanic predator, *PNAS*, **105**(52), 20776–20780.

Russel, B. D., J. -A. I. Thompson, L. J. Falkenberg, and S. D. Connell (2009), Synergistic effects of climate change and local stressors: CO_2 and nutrient-driven change in subtidal rocky habitats, *Glob. Change Biol.*, **15**, 2153–2162.

Sabine, C. L., R. A. Feely, N. Gruber, R. M. Key, K. Lee, J. L. Bullister, R. Wanninkhof, C. S. Wong, D. W. R. Wallace, B. Tilbrook, F. J. Millero, T. -H. Peng, A. Kozyr, T. Ono, and A. F. Rios (2004), The oceanic sink for anthropogenic CO_2, *Science*, **305**, 367–371, http://dx.doi.org/10.1126/science.1097403.

Sala, E., and N. Knowlton (2006), Global marine biodiversity trends, *Annu. Rev. Environ. Resour.*, **31**, 93–122.

Sanderson, E. W., M. Jaiteh, M. A. Levy, K. H. Redford, A. V. Wannebo, and G. Woolmer (2002), The human footprint and the last of the wild, *Bioscience*, **52**, 891–904.

Sarmiento, J. L., and N. Gruber (2006), *Ocean Biogeochemical Dynamics*, 503 pp., Princeton University Press, Princeton, Woodstock.

Sazykina, T. G., I. I. Kryshev, and A. I. Kryshev (1998), Doses to marine biota from radioactive waste dumping in the Fjords of Novaya Zemlya, *Radiat. Prot. Dosimetry*, **75**, 253–256.

Schloder, C., and L. D'Croz (2004), Responses of massive and branching coral species to the combined effects of water temperature and nitrate enrichment, *J. Exp. Mar. Biol. Ecol.*, **313**, 255–268.

Shepherd, S. A., and S. Daume (1994): Ecology and survival of juvenile abalone in a crustose coralline habitat in South Australia. In: *Survival Strategies in Early Life Stages of Marine Resources* [Watanabe Y. et al (eds.)]. Balkema, Rotterdam, 297–313.

Steinacher, M., F. Joos, T. L. Frölicher, G. -K. Plattner, and S. C. Doney (2009) Imminent ocean acidification projected with the NCAR global coupled carbon cycle-climate model, *Biogeosciences*, 6, 515–533 (http://dx.doi.org/10.5194/bg-6-515-2009).

Stramma, L., G. C. Johnson, J. Sprintall, and V. Mohrholz (2008), Expanding oxygen-minimum zones in the tropical oceans, *Science*, **320**, 655–658, http://dx.doi.org/10.1126/science.1153847.

Sutherland, W. J., S. Armstrong-Brown, P. R. Armsworth, T. Brereton, J. Brickland, C. D. Campbell, D. E. Chamberlain, A. I. Cooke, N. K. Dulvy, N. R. Dusic, M. Fitton, R. P. Freckleton, H. C. J. Godfray, N. Grout, H. J. Harvey, C. Hedley, J. J. Hopkins, N. B. Kift, J. Kirby, W. E. Kunin, D. W. Macdonald, B. Marker, M. Naura, A. R. Neale, T. Oliver, D. Osborn, A. S. Pullin, M. E. A. Shardlow, D. A. Showler, P. L. Smith, R. J. Smithers, J. -L. Solandt,

J. Spencer, C. J. Spray, C. D. Thomas, J. Thompson, S. E. Webb, D. W. Yalden, and A. R. Watkinson (2006), The identification of 100 ecological questions of high policy relevance in the UK, *J. Appl. Ecol.*, **43**, 617–627.

Teuten, E. L., J. M. Saquing, D. R. U. Knappe, M. A. Barlaz, S. Jonsson, A. Bjorn, S. J. Rowland, R. C. Thompson, T. S. Galloway, R. Yamashita, D. Ochi, Y. Watanuki, C. Moore, P. H. Viet, T. S. Tana, M. Prudente, R. Boonyatumanond, M. P. Zakaria, K. Akkhavong, Y. Ogata, H. Hirai, S. Iwasa, K. Mizukawa, Y. Hagino, A. Imamura, M. Saha, and H. Takada (2009), Transport and release of chemicals from plastics to the environment and to wildlife, *Philos. T. Roy. Soc. B*, **364**, 2027–2045.

UNEP (United Nations Environment Programme) (2006), *Challenges to international waters; Regional assessments in a global perspective*, 120 pp., United Nations Environment Programme, ISBN: 91-89584-47-3.

Vaquer-Sunyer, R., and C. M. Duarte (2010), Temperature effects on oxygen thresholds for hypoxia in marine benthic organisms, Global Change Biology, "Accepted Article".

Venter, O., N. N. Brodeur, L. Nemiroff, B. Belland, I. J. Dolinsek, and J. W. A. Grant (2006), Threats to endangered species in Canada, *Bioscience*, **56**, 903–910.

Veron, J. E. N., O. Hoegh-Guldberg, T. M. Lenton, J. M. Lough, D. O. Obura, P. Pearce-Kelly, C. R. C. Sheppard, M. Spalding, M. G. Stafford-Smith, and A. D. Rogers (2009), *Mar. Pollut. Bull.*, **58**, 1428–1436.

Vinebrooke, R. D., K. L. Cottingham, J. Norberg, M. Scheffer, S. I. Dodson, S. C. Maberly, and U. Sommer (2004), Impacts of multiple stressors on biodiversity and ecosystem functioning: the role of species co-tolerance, *Oikos*, **104**, 451–457.

Vitousek, P. M., H. A. Mooney, J. Lubchenco, and J. M. Melillo (1997), Human domination of earth's ecosystems, *Science*, **277**, 494–499.

Wagner, D. E., P. Kramer, and R. van Woesik (2010), Species composition, habitat, and water quality influence coral bleaching in southern Florida, *Mar. Ecol. Prog. Ser.*, **408**, 65–78.

Weilgart, L. S. (2007), The impact of anthropogenic ocean noise on cetaceans and implications for management, *Can. J. Zool.*, **85**, 1091–1116.

Wooldridge, S. A. (2009), Water quality and coral bleaching thresholds: Formalising the linkage for the inshore reefs of the Great Barrier Reef, Australia, *Mar. Pollut. Bull.*, **58**, 745–751.

Zeidberg, L. D., and B. H. Robison (2007), Invasive range expansion by the Humboldt squid, *Dosidicus gigas*, in the eastern North Pacific, *Proc. Natl. Acad. Sci. U.S.A.*, **104**, 12946–12948.

Zeebe, R. E., and D. A. Wolf-Gladrow (2001), *CO_2 in seawater: equilibrium, kinetics, isotopes*, 346 pp., Elsevier, Amsterdam.

CHAPTER

9

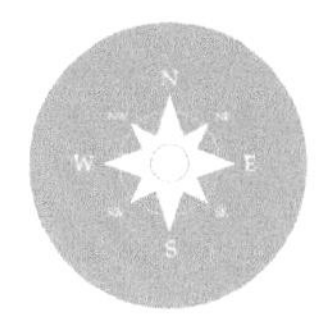

Tipping Points, Uncertainty, and Precaution: Preparing for Surprise

Kevin J. Noone, Frank Ackerman*[†]

*Department of Applied Environmental Science, Stockholm University, Stockholm, Sweden

[†]Synapse Energy Economics, Cambridge, Massachusetts, USA

INTRODUCTION

Chapters 2-8 present the current state of the science regarding causes and consequences of several major threats to the global oceans. In many cases, these threats involve gradual, predictable changes, leading to a clear picture of the most likely outcomes. Year after year, the world is growing warmer, sea levels are rising, storms are intensifying, the oceans are acidifying, anoxic zones are spreading, and many forms of pollution are on the increase. That picture is an ominous one, involving serious losses and economic costs. The likely, avoidable costs of five categories of ocean-related climate impacts, as described in Chapter 10, will amount to more than $1 trillion a year by the end of this century. A full calculation of the likely, avoidable costs of impacts discussed throughout this book would be much greater (although the necessary data are unfortunately not available for that calculation).

Yet the gradual, likely changes are not the whole story. There are also abrupt changes, involving less likely and less predictable impacts. If pushed far enough, many natural systems reach thresholds or tipping points at which the behavior of the system suddenly changes.

The oceans have many potential tipping points, as described in many of the other chapters in this book. As one example, the North Atlantic cod fishery, a reliable mainstay of Canada's eastern provinces for centuries, suddenly collapsed in the early 1990s (Figure 9.1). Many studies suggest that there is a temperature threshold above which coral reefs will not survive, with dire consequences for islands, coastal communities, and fisheries that depend on the ecosystem services (components of nature, directly enjoyed, consumed, or used to yield human well-being (Boyd and Banzhaf, 2007)) of living reefs. Ocean acidification will eventually lead to undersaturation of aragonite, at which point the oceans will become corrosive to some mollusks and other calcifying organisms.

A final example is crucial for sea-level rise: the Greenland ice sheet is not considered likely to collapse in this century, but it is

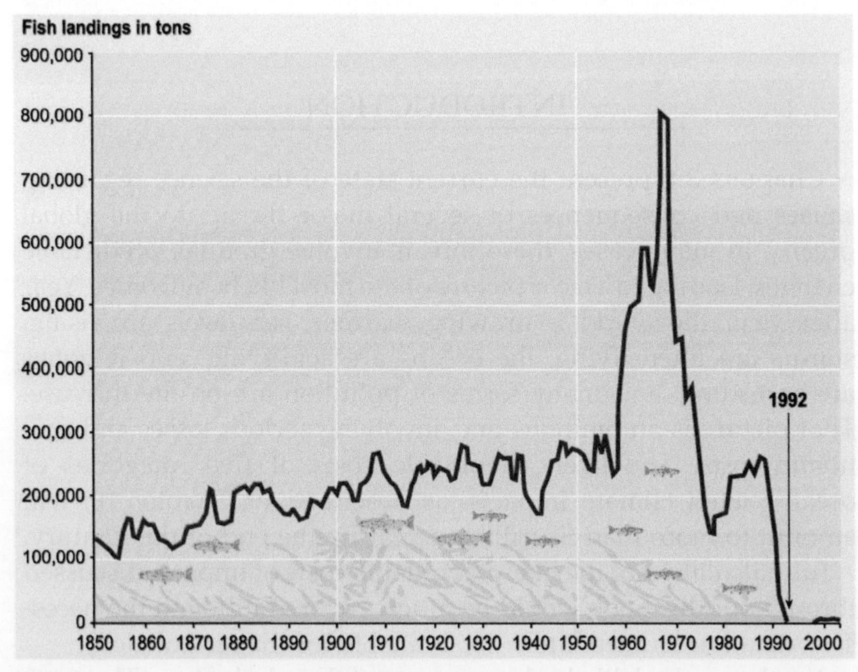

FIGURE 9.1 Collapse of North Atlantic cod stocks. *Reprinted from Reid et al. (2005), figure 11. © World Resources Institute.*

all too clear that ice melts or slides off the land into the oceans more rapidly in warmer weather. Based on oceanic sediment measurements, we know that the world did not have major ice sheets in Antarctica or Greenland between 40 and 65 million years ago when atmospheric CO_2 concentrations were above about 500 parts per million (ppm) (Zachos et al., 2001) (we have now reached 400 ppm). Stable ice sheets in the northern hemisphere did not form until about 8 million years ago when CO_2 concentrations were below about 300 ppm. We have good evidence of at least two stable states for the planet: one with major ice sheets and one without. We do not know exactly what combination of conditions could cause the system to flip from one state to the other in the future, nor do we know how long such a transition would take. However, as CO_2 concentrations rise and the resulting global warming raises temperatures, complete loss of the Greenland ice sheet (causing an eventual 7 m of sea-level rise) becomes less and less unlikely.

Preparing for multiple marine threats and uncertainties is a daunting task, and one for which we are currently not very well equipped. Prediction of impacts and tipping points is difficult, in part, because individual threats interact in complex, nonlinear fashions, as seen in Chapter 8. Coral reefs are threatened by ocean warming, acidification, and pollution; fisheries are potentially harmed by almost all the threats we have examined; islands and coastal communities are at risk from sea-level rise, intensifying storms, and the loss of ocean-based livelihoods. Feedbacks among these threats, cases in which the whole is worse than the sum of the parts, are frequent and inescapable.

Prediction is also difficult because we are dealing with potential tipping points outside our collective experience. We have never, in all the millennia of human experience, killed all the coral reefs before or melted a block of ice the size of Greenland; no one can be sure exactly how or when these disasters will occur. We are walking within an uncertain distance from the edge of a cliff, in a fog that obscures our vision.

How should we make policy decisions, responding to both predictable, gradual changes—such as those to which monetary values can be assigned—and abrupt, low-probability changes associated with tipping points and catastrophic losses? Unfortunately, economic and policy debate often narrows its focus to the former category, particularly when relying on cost-benefit analysis. For lack of available numerical estimates, cost-benefit analysis, in effect, often assigns a value or a probability of zero to the most ominous, but uncertain, threats that we face.

WHY BUY INSURANCE?

In other areas of life, individuals and institutions adopt much more sophisticated approaches to uncertainty. We routinely buy insurance to cover rare calamities such as residential fires, car accidents, or the death of young parents, even though these events are extremely unlikely to occur.

The number of residential fires that U.S. fire departments responded to in 2010 was less than 0.3% of the number of housing units in the country.[1] At that rate, each home would average one fire every 350 years; by far the most likely number of residential fires experienced in anyone's lifetime (or a building's lifetime) is zero. Yet fire insurance coverage is widespread; few people would choose to play the excellent odds of saving money by cancelling their policies.

Life insurance is frequently bought by young parents, to protect their children's future in the event of the parents' death. In the United States in 2007, the average annual rate of death from all causes was under 0.3% until age 46.[2] Thus parents of newborn children typically faced risks of death at least as low as the risk of residential fires. Again, despite the excellent odds in favor of survival, voluntary purchases of life insurance are widespread.

In short, people buy insurance against personal catastrophes that are known to have probabilities in the tenths of a percent per year.

[1] Fire departments responded to 369,500 residential fires in 2010 (http://www.nfpa.org, "The U.S. Fire Problem"), while there were 130.6 million housing units in the country (U.S. Census Bureau, *2012 Statistical Abstract*, Table 982, http://www.census.gov/compendia/statab/cats/construction_housing/housing_units_and_characteristics.html).

[2] U.S. Department of Health and Human Services, Centers for Disease Control and Prevention, *National Vital Statistics Reports*, Volume 59, no. 9 (2011), Table 1, "Life table for the total population: United States, 2007," available at ftp://ftp.cdc.gov/pub/Health_Statistics/NCHS/Publications/NVSR/59_09/Table01.xls.

The traditional (but arbitrary) scientific standard of 95% confidence is insufficient: we have much more than 95% confidence—indeed, more than 99% confidence—that an individual household will not need fire insurance, and an individual young parent will not need life insurance, this year.

This is not to say that every risk, no matter how small, deserves attention, as caricatures of the "precautionary principle" have suggested.[3] It is possible to suffer personal or property damage from a meteorite falling on you, but this appears to be six to eight orders of magnitude less likely (i.e., about 1 million to 100 million times less likely) than a residential fire.[4] Meteorite insurance, if sold separately, would be unlikely to appeal to many customers. At a low enough level of likelihood, risks are routinely ignored; but at probabilities of only tenths of a percent per year, major risks motivate purchases of insurance.

What tipping points and catastrophic losses in the oceans are as likely as the events that individuals insure themselves against? Should we adopt policies that amount to collective insurance against such losses? The insurance paradigm suggests an important alternative to cost-benefit analysis of the most likely and most easily quantified impacts. It is, moreover, consistent with the way that individuals act on their own: if good decisions were based only on most likely outcomes, no one would ever buy insurance.

The analogy to private insurance is an informative but inexact model for public policy. The analogy breaks down in two ways. First, we all know something, and insurance companies know a great deal, about the probability of fires, premature deaths, and meteorite impacts. No one has comparable knowledge of the likelihood of dangerous tipping points in the oceans and other major ecosystems—although we do know that business as usual makes them increasingly probable. Second, because private risks are well understood and manageable in size, it is possible to buy insurance for them. Global-scale risks—like wholesale changes in marine

[3] For example, "If we take costly steps to address all risks, however improbable they are, we will quickly impoverish ourselves" (Sunstein, 2005, p. 25).

[4] One estimate puts the annual average for the United States and Canada as a whole at 0.8 meteorite impacts causing damage to buildings and 0.0055 impacts on people (Halliday et al., 1985).

ecosystems caused by human activities—are a different story. There is no galactic insurance company offering coverage for catastrophic environmental losses on the only planet we have got. In practice, the insurance approach calls for public policy, often at a national and international level, that represents self-insurance, taking action to prevent threats rather than seeking compensation after the fact.

MAKING DECISIONS IN THE DARK

How can we make decisions in the dark, in the absence of knowledge of the probabilities of reaching dangerous tipping points? It may help to start by distinguishing the multiple meanings of uncertainty. Economists frequently refer to the distinction made long ago by Frank Knight, in which "risk" refers to events with known probabilities, while "uncertainty"—sometimes called "Knightian uncertainty"—refers to events with unknown probabilities (Knight, 1921). For a modern commentary and clarification of Knight's approach, see Runde (1998). Other disciplines have at times used different terminologies for the same concepts.[5]

In an examination of the unknowns related to global change, Schneider et al. (1998) extend Knight's categories to add a third level beyond uncertainty, dubbed "surprise":

- *Risk* is the condition in which the set of possible events or outcomes, and the probability that each will occur, is known;
- *Uncertainty* is the condition in which the possible events or outcomes are known (factually or hypothetically), but the probabilities that each will occur are not known or are highly subjective estimates;
- *Surprise* is the condition in which the event or outcome is not known or expected.

Surprise, in this definition, resembles Donald Rumsfeld's famous "unknown unknowns," defined as "things we don't know we don't

[5] In the field of risk assessment, Knightian risk, with known probabilities, is sometimes referred to as "known uncertainty," while Knightian uncertainty is called "unknown uncertainty" (Daneshkhah, 2004).

know."[6] As the former U.S. Secretary of Defense observed, "And if one looks throughout the history of our country and other free countries, it is the latter category [unknown unknowns] that tend to be the difficult ones." Strategic planning for surprises is, indeed, fraught with peril, as demonstrated by Rumsfeld's disastrously wrong assertion that the United States could win a quick and inexpensive military victory in Iraq in 2003.

The definition of surprise, as unknown or unexpected outcomes, is difficult to use in practice. Schneider et al. (1998) make a further distinction between strict surprise (or truly unknown unknowns) and "imaginable surprise," in which the event or outcome departs from the expectations of the observing community. That is, an event occurs that was thought to have very low probability.

Exploring ever-deeper levels of uncertainty, however, may not be crucial for public policy purposes. A recent analysis of ecosystem management with an unknown threshold finds that there is a complex relationship between uncertainty and precautionary behavior (Brozović and Schlenker, 2011). At a very low level of uncertainty, outcomes are so predictable that there is no need for precaution. At a very high level of uncertainty, precautionary policies have such a small probability of success that the expected benefits do not justify the costs; that is, if we are extremely uncertain about the location of a dangerous threshold, then there is not much chance that any specific policy leads to avoiding it. Precautionary expenditures are the optimal policy response only at an intermediate level—when there is enough uncertainty to make us worried, but not enough to make us give up in despair.

There is a long-standing but limited literature on decision making under Knightian uncertainty. An important economic theorem, coauthored by Nobel laureates Arrow and Hurwicz (1972), addresses decision making under conditions of what they called "ignorance," when the possible outcomes are known, but nothing is known about their probabilities; in the Schneider et al. terminology, this is decision making under uncertainty, not surprise. Arrow and Hurwicz proved that under these conditions, the most efficient approach to public policy is based solely on knowledge of the extremes of the range of possibilities. Nothing is added by attempts to find the midpoint, average, or best point estimate.

[6] http://www.defense.gov/transcripts/transcript.aspx?transcriptid=2636.

Published in an obscure location and presented in dense technical terms, the Arrow-Hurwicz analysis was for many years known only to a handful of specialists. Its significance for environmental policy was recognized and popularized by Woodward and Bishop (1997). (For a nontechnical discussion of the Arrow-Hurwicz result and summary of its proof, see Ackerman (2008) Chapter 4 and Appendix A.)

The Arrow-Hurwicz approach implies that decision making under uncertainty depends solely on the best and worst possible outcomes. This approach has been formalized, in general terms, as the "α-maxmin" model, where α represents the relative weight given to the best versus worst cases (see Ghirardato et al. (2004) for an axiomatic development and Farber (2011) for nontechnical discussion).

Other analyses in a similar framework have demonstrated that if society is risk-averse, or if it is desirable to preserve a diversity of options in the face of uncertainty, then only the worst case matters (Gilboa and Schmeidler, 1989; Kelsey, 1993). In short, under reasonable assumptions, such as risk aversion, the best decision under uncertainty is based primarily or entirely on information about the worst-case outcome—just as suggested by the precautionary principle (Farber, 2011).

Some attempts have been made to quantify this perspective in climate economics modeling, using the so-called minimax regret criterion (Ackerman and Stanton, 2013, Chapter 7). Suppose that a policy choice is made under uncertainty, and then the uncertainty is resolved. The "regret" associated with the policy is the difference between the value of the actual outcome and the best outcome that could have been obtained with perfect foresight. The maximum regret for a policy is the greatest regret associated with it under any state of the world, or the worst-case outcome for that policy. The minimax regret criterion picks the policy with the smallest maximum regret or the least-bad worst case.

Such approaches, while logical in theory, confront the immense practical challenge of specifying the range of possibilities to be considered. Under conditions of serious uncertainty, there is often disagreement about the credible best-case and worst-case outcomes. The relevant worst case may not be the literally worst possible event; in the context of individual insurance decisions, discussed above, risks of residential fires or premature deaths of young parents are clearly relevant, but risks of meteorite damage probably are not.

PEERING INTO THE FUTURE

Exploration of the relevant best and worst possible environmental futures is often addressed through scenarios that trace multiple possible paths of evolution. The IPCC's scenarios for greenhouse gas emissions provide a well-known but limited example. The SRES scenarios (A1, A2, B1, and B2) represented an early attempt; the RCP scenarios, two of which are used in Chapter 10, offer a newer, improved version. Yet the IPCC scenarios, both new and old, stop short of describing the outcomes of greatest concern. Tipping points and catastrophic losses might well result from the increasing temperatures implied by these emission scenarios, but the IPCC scenarios themselves do not include such outcomes. All of the IPCC scenarios—old and new—describe continuous, incremental changes into the future. At best, they may span the range of plausible emissions trajectories, from runaway growth of greenhouse gases under business-as-usual in RCP8.5 to extremely rapid emission reduction in RCP2.6.

There is a need for more comprehensive scenarios, to understand the range of possible futures and to prepare for the unpleasant surprises that are unfortunately imaginable. Given the complexity of the natural and human systems that are involved, creation of useful, appropriate scenarios is an art as well as a science. Scenario analysis and planning exercises have been conducted by many institutions—including some that might seem to be unlikely candidates for environmental concern. We will look at examples of two of these—Royal Dutch Shell and the U.S. military—to explore how major organizational entities prepare for surprise.

SHELL GAME

Royal Dutch Shell, a multinational oil corporation, has been using scenarios analysis for almost a half century as a way to gain a deeper understanding of global development, changes in the world's energy system, and how these developments affect the corporation. The Shell scenarios are not based on detailed economic modeling. Rather, they are internally consistent pictures of the evolving world situation that Shell uses to explore how best to navigate in an uncertain future. A recent version of their scenarios, released in 2011, uses two contrasting stories of future trends through 2050—called *Scramble* and *Blueprints*—to span the range of potential global developments.[7]

In the *Scramble* scenario, nations focus on their own needs and aims. States enter into bilateral or regional agreements aimed at local resource development, with progress driven as much by political opportunism as by rational focus. Demand-side policies are not meaningfully pursued until supply limitations are acute. Climate and other environmental issues are not seriously addressed until major environmental challenges force policy responses. These events compel late and severe responses that result in energy price volatility. There is significant growth in the use of coal and biofuels. CO_2 concentrations rise well above 550 ppm by 2050, and an increasing fraction of economic activity in the later years of the scenario is directed toward preparing for the impacts of climate change.

In the world of the *Blueprints* scenario, concern about lifestyles and economic prospects forges new alliances, promoting action in both developed and developing countries. These actions first take root locally as cities or regions take the lead in planning and implementing new initiatives. Over time, these local initiatives are progressively linked and harmonized at national levels. Emissions management policies are successful in limiting growth in atmospheric carbon dioxide. Rapid increases in energy efficiency are achieved, coupled with the emergence of mass-market electric vehicles. A decoupling of world GDP and energy growth occurs before 2050.

These two contrasting scenarios allow Shell to map out possible decision pathways for the future development of the corporation. Not being predictions of the future, they cannot be used to decide about making a particular strategic decision on a given date in the future. Rather, they allow strategic planners to prepare a suite of decision pathways that could be adapted to respond to specific contingencies as they arise. Creating a preparedness for surprise is part of the process.

[7] *Signals & Signposts: Shell Energy Scenarios to 2050*, Shell International BV, The Hague, The Netherlands, http://www.shell.com/home/content/aboutshell/our_strategy/shell_global_scenarios/signals_signposts/.

PENTAGON PLANNING

Planning for worst-case scenarios is central to military and security policies. Costly, or at least time-consuming, precautions against unlikely security risks are familiar to everyone who travels by plane: if predictable, average, or most likely outcomes were all that mattered, there would be no need for airport screening of passengers or luggage. Military planning is all about worst cases; on an average day, no one needs an army.

A precautionary framework underlies two U.S. military analyses of climate change risks. Both appeared at times (2003 and 2007) when much of the U.S. government was denying the existence or minimizing the importance of climate change. Yet for Pentagon planning, climate change posed serious threats that could not be ignored.

In a 2003 report to the Department of Defense,[8] Peter Schwartz and Doug Randall wrote:

> The purpose of this report is to imagine the unthinkable—to push the boundaries of current research on climate change so we may better understand the potential implications on United States national security. . . .
>
> We have created a climate change scenario that although not the most likely, is plausible, and would challenge United States national security in ways that should be considered immediately.

In their scenario, a major, abrupt climate upheaval causes large temperature changes across the globe (both increases and decreases in temperature, at different locations). Persistent, decade-long droughts occur, coupled with intensifying winter storms. These climatic changes result in food shortages, decreased availability and quality of freshwater, and disrupted access to energy supplies due to storms and other natural disasters. Such abrupt changes could make some regions of the world unable to sustain their current populations. Competition for scarce and necessary resources would then become a significant driver of conflict worldwide, with dire implications for national security in the United States and elsewhere.

[8] Peter Schwartz and Doug Randall, "An Abrupt Climate Change Scenario and Its Implications for United States National Security" (2003), http://www.gbn.com/consulting/article_details.php?id=53.

Even more remarkable is the 2007 report from a group of 11 retired three-star and four-star admirals and generals.[9] They found that the projected impacts of climate change pose a serious threat to national security, acting as a "threat multiplier" in volatile regions of the world and adding to tensions even in stable regions. The retired admirals and generals—several of whom were skeptical about climate change when they started work on the report—spoke in terms that echo, from a military perspective, the precautionary concerns of environmental advocates.

Discussing uncertainty about the extent of climate change, the report explained that

> Military leaders see a range of estimates and tend not to see it as a stark disagreement, but as evidence of varying degrees of risk. They don't see the range of possibilities as justification for inaction.

Along similar lines, General Gordon Sullivan said that

> We never have 100 percent certainty... If you wait until you have 100 percent certainty, something bad is going to happen on the battlefield.

Like the Shell scenarios, the Pentagon analyses are explorations of the impacts of abrupt change. They paint pictures of worst case, but entirely possible outcomes; they are describing imaginable surprises. These scenarios, however, are not based on any theoretical or numerical model; rather, they are the result of interviews with climate experts, analysis of market and political trends, and examination of the roots of instability and conflict. They intentionally use extreme possibilities, not best guesses or median estimates, of what may happen in the future. Neither of the Shell or the Pentagon approaches are predictive; they are not used to tell in advance that a particular event will happen at a particular point in the future. Rather, they are used to frame a set of possible futures, and to imagine and analyze how decisions made today could play out in and perhaps modify these possible futures. In this sense, they are diagnostic tools for decision support rather than predictive ones.

[9] The CNA Corporation, *National Security and the Threat of Climate Change* (2007), http://www.cna.org/reports/climate.

SAFE STANDARDS AND PLANETARY BOUNDARIES

The analysis needed for scenario development is complicated by the difficulty of identifying critical risks in advance. Climate scientists have identified a number of potential tipping points—that is, points at which a small change in conditions leads to a large, lasting change in the global climate or major ecosystems (Lenton et al., 2008). Many of these involve the oceans, including loss of major ice sheets, weakening or collapse of African and South Asian monsoons, and disruption of the Atlantic thermohaline circulation or the El Niño-Southern Oscillation. With sufficient data, it may be possible in some cases to detect early warning signs of critical tipping points, but this is an untested new approach where more research is needed (Lenton, 2011).

Policy responses to uncertainty about tipping points, discontinuities, and irreversible losses have included the "safe minimum standards" and "tolerable windows" frameworks. Safe minimum standards were originally proposed in response to extinction risks and threats to biodiversity. Such policies would identify and protect the minimum level of environmental resources necessary to ensure survival of a species or ecosystem service, overriding cost-benefit analyses of rival uses of those resources (Farmer and Randall, 1998; Palmini, 1999). The tolerable windows approach recasts the climate policy debate, specifying standards or "guardrails" for tolerable climate evolution, such as a maximum increase in global average temperature or a minimum level of per capita food production; policies that are incompatible with those guardrails are then ruled out (Petschel-Held et al., 1999).

The "Planetary Boundaries" analysis by Rockström et al. (2009) provides a recent example of a similar approach, examining uncertainty about dangerous thresholds in complex, coupled global socioenvironmental issues. The authors identify nine global boundaries beyond which we increase the risk of a planetary-scale change in how the Earth system functions, as shown in Figure 9.2:

1. Climate change
2. Ocean acidification
3. Stratospheric ozone depletion
4. Global cycling of nitrogen and phosphorus

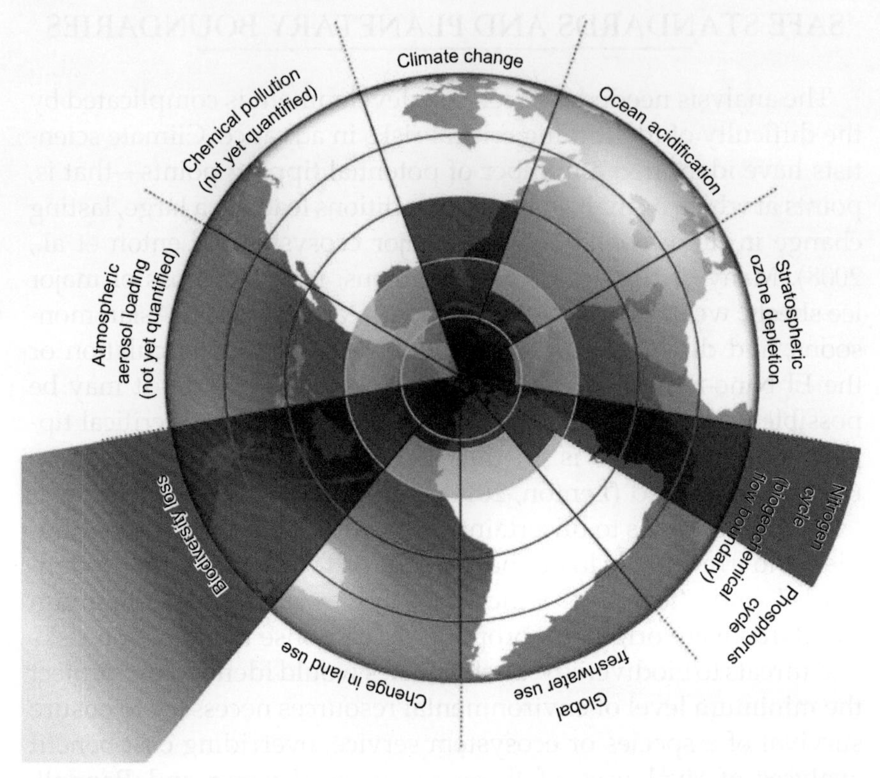

FIGURE 9.2 Illustration of the planetary boundaries. Green circles represent the boundaries; red wedges represent current levels of exploitation of resources. *Reprinted from* Nature.

5. Global freshwater use
6. Land use change
7. Biodiversity loss
8. Atmospheric aerosol (particulate) loading
9. Chemical pollution

The authors were able to provide quantitative estimates for seven of the boundaries; boundaries for atmospheric aerosol loading and chemical pollution remain undetermined.

The boundaries have several defining characteristics:

- They are associated with a large-scale change in how planetary systems function, often in the form of a threshold or tipping point

- There is a zone of uncertainty for each of the boundaries; the boundary value itself is set at the lower end of this zone of uncertainty
- They are tightly coupled (e.g., the primary driver for both the climate and ocean acidification boundaries is human CO_2 emissions; biodiversity, land, water, nitrogen, and phosphorus use are all tied through agricultural activities)
- Each boundary has a "control variable" for which humans have become a dominant factor
- They are normatively defined, relative to a preferred state—namely, the relatively stable Holocene period (roughly the last 10,000 years) in which our societies have developed
- The processes controlling the boundaries operate on timescales over which ethical considerations and political action are relevant (e.g., years to decades)
- A "safe operating space" can be created within the boundaries

These boundaries are nonnegotiable in the sense that they are all hard-wired into how the Earth system works. The ocean acidification boundary is perhaps the best example of this aspect. This boundary is expressed in terms of a chemical equilibrium—the point at which a particular form of calcium carbonate (aragonite) made by many marine organisms becomes soluble. It is impossible to renegotiate a chemical equilibrium.

Applications of multicriteria analysis and optimization are not new in the marine planning domain (Brown et al., 2001; Himes, 2007). The Planetary Boundaries framework offers a different perspective on multicriteria analysis from a scenarios perspective, bringing in the concept of abrupt transitions between multiple stable environmental states and bridging the terrestrial, marine and atmospheric domains. While still a relatively new concept, it has nonetheless already been incorporated into decision and planning support in the private (Shell, 2011) and policy sectors (United Nations Secretary General's High-level Panel on Global Sustainability, 2012).

It is informative to view the threats to the global oceans described in the previous chapters through the prism of the Planetary Boundaries framework. An example is shown in Figure 9.3, illustrating the complex interactions among multiple ocean threats and outcomes.

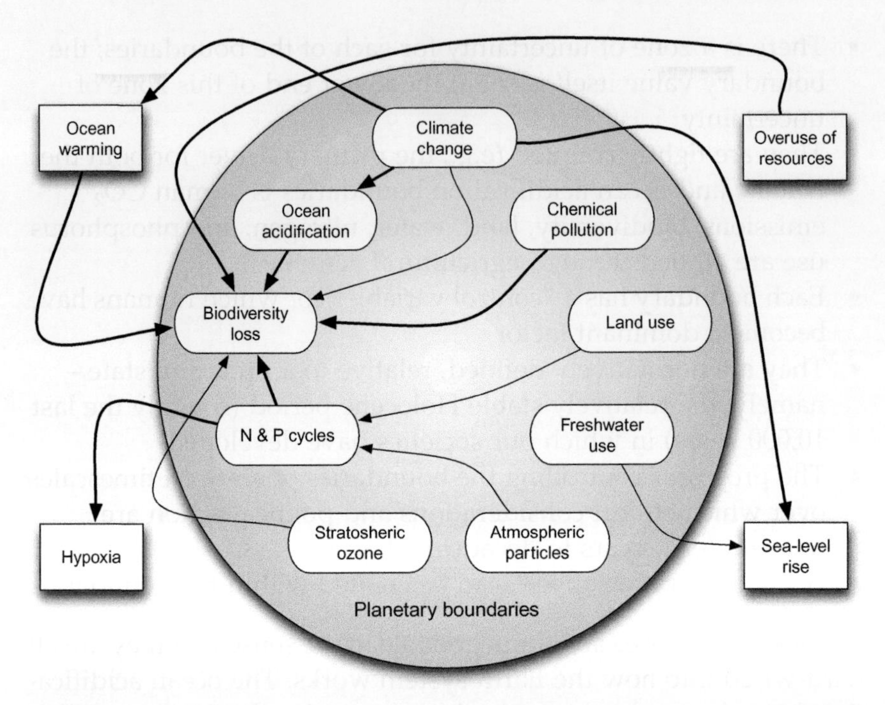

FIGURE 9.3 Examples of links between threats to the global oceans and planetary boundaries.

The round boxes within the large circle are the nine planetary boundaries from Rockström et al. Some of the threats described in our earlier chapters overlap with these planetary boundaries; others appear in the square boxes outside the circle. The arrows in the diagram show some of the linkages that are involved in this dense web of interactions; all of the planetary boundaries are connected with the marine impacts discussed in this book. For example, climate change caused by increasing atmospheric CO_2 concentrations drives sea-level rise and ocean warming, which in turn affects marine biodiversity loss. Increasing CO_2 concentrations also drive ocean acidification, which also negatively impacts marine biodiversity. Additional arrows, representing further connections, could undoubtedly be added to Figure 9.3.

The point of this discussion is not to suggest that the Planetary Boundaries concept *per se* should be applied to marine issues. Rather, we need to develop tools that aid us in developing strategic plans that work across scales and for several variables simultaneously. Trying to solve one problem at a time while ignoring the others is not an option; neither is only pursuing local management initiatives without accounting for the aggregate effects of local policies.

Section IV of Chapter 12 presents suggestions for how to prepare for surprise in marine planning. This is already being done in many individual areas such as contingency planning for oil spills (Qiao et al., 2002) and through sophisticated mathematical models of decision chains in oil spill management (Bassey and Chigbu, 2012). What we advocate here is a more holistic, risk-based approach that is applicable across scales from local to global.

The problem of decision making under uncertainty, or planning for surprises, is inescapable in major environmental policy problems such as protecting the world's oceans. Frameworks for addressing uncertainty are available, stretching far beyond the limited formulas of cost-benefit analysis. Yet practical application of these frameworks requires us to picture in detail the best and worst cases, to explore the outer bounds of possible futures. The ominous scientific findings described in earlier chapters provide a warning of what could happen if we do not change our ways. What is needed now is the collective will to respond to that warning.

References

Ackerman, F. (2008), *Poisoned for Pennies: The Economics of Toxics and Precaution*, Island Press, Washington, DC.

Ackerman, F., and E. A. Stanton (2013), *Climate Economics: The State of the Art*. Chapter 7. Routledge.

Arrow, K. J., and L. Hurwicz (1972): An Optimality Criterion for Decision-Making Under Ignorance. In: Uncertainty and Expectations in Economics: Essays in Honor of G. L. S. Shackle [Carter C. F. and J. L. Ford (eds.)]. Augustus M. Kelly, Clifton, NJ.

Bassey, K. J., and P. E. Chigbu (2012), On optimal control theory in marine oil spill management: A Markovian decision approach. *Eur. J. Oper. Res.*, **217**(2), 470–478, http://dx.doi.org/10.1016/j.ejor.2011.09.036.

Boyd, J., and S. Banzhaf (2007), What are ecosystem services? The need for standardized environmental accounting units, *Ecol. Econ.*, **63**(2–3), 616–626, http://dx.doi.org/10.1016/j.ecolecon.2007.01.002.

Brown, K., W. N. Adger, E. L. Tompkins, P. Bacon, D. Shim, and K. Young (2001), Trade-off analysis for marine protected area management, *Ecol. Econ.*, **37**, 417.

Brozovic´, N., and W. Schlenker (2011), Optimal Management of an Ecosystem with an Unknown Threshold, *Ecol. Econ.*, **70**(4), (February), 627–640, http://dx.doi.org/10.1016/j.ecolecon.2010.10.001.

Daneshkhah, A. R. (2004), *Uncertainty in Probability Risk Assessment: A Review.* Working Paper. BEEP Working Paper: Bayesian Elicitation of Experts' Probabilities, University of Sheffield, Sheffield, UK.

Farber, D. A. (2011), Uncertainty, *Georgetown Law J.*, **99**(4) (April), 901–959.

Farmer, M. C., and A. Randall (1998), The Rationality of a Safe Minimum Standard, *Land Econ.*, **74**(3), 287–302.

Ghirardato, P., F. Maccheroni, and M. Marinacci (2004), Differentiating Ambiguity and Ambiguity Attitude, *J. Econ. Theory*, **118**(2) (October), 133–173, http://dx.doi.org/10.1016/j.jet.2003.12.004.

Gilboa, I., and D. Schmeidler (1989), Maxmin Expected Utility with Non- unique Prior, *J. Math. Econ.*, **18**(2), 141–153.

Halliday, I., A. T. Blackwell, and A. A. Griffin (1985), Meteorite Impacts on Humans and on Buildings, *Nature*, **318**(6044) (November 28), 317, http://dx.doi.org/10.1038/318317a0.

Himes, A. H. (2007), Performance Indicator Importance in MPA Management Using a Multi-Criteria Approach, *Coast. Manage.*, **35**(5), 601–618, http://dx.doi.org/10.1080/08920750701593436.

Kelsey, D. (1993), Choice Under Partial Uncertainty, *Int. Econ. Rev.*, **34**(2), 297–308.

Knight, F. H. (1921), *Risk, Uncertainty and Profit*, Harper & Row, New York.

Lenton, T. M. (2011), Early Warning of Climate Tipping Points, *Nature Climate Change*, **1**(4) (June 19), 201–209, http://dx.doi.org/10.1038/nclimate1143.

Lenton, T. M., H. Held, E. Kriegler, J. W. Hall, W. Lucht, S. Rahmstorf, and H. J. Schellnhuber (2008), Tipping Elements in the Earth's Climate System, *P. Natl. Acad. Sci. U. S. A.*, **105**(6) (February 12), 1786–1793, http://dx.doi.org/10.1073/pnas.0705414105.

Palmini, D. (1999), Uncertainty, Risk Aversion, and the Game Theoretic Foundations of the Safe Minimum Standard: a Reassessment, *Ecol. Econ.*, **29**, 463–472.

Petschel-Held, G., H. -J. Schellnhuber, T. Bruckner, F. L. Toth, and K. Hasselmann (1999). The Tolerable Windows Approach: Theoretical and Methodological Foundations, *Climatic Change*, **41**(3–4), 303–331, http://dx.doi.org/10.1023/A:1005487123751.

Qiao, B., J. C. Chu, P. Zhao, Y. Yu, and Y. Li (2002), Marine oil spill contingency planning, *J. Environ. Sci. (China)*, **14**(1), 102–107.

Reid, W. V., et al. (2005), *Ecosystems and Human Well-being: Synthesis, in Millennium Ecosystem Assessment, 2005, edited*, 137 pp., Island Press, Washington, D.C.

Rockström, J., W. Steffen, K. Noone, F. Åsa Persson, S. Chapin, E. F. Lambin, T. M. Lenton, et al. (2009), A Safe Operating Space for Humanity, *Nature* **461**(7263), 472–475, http://dx.doi.org/10.1038/461472a.

Runde, J. (1998), Clarifying Frank Knight's Discussion of the Meaning of Risk and Uncertainty, *Cambridge J. Econ.* **22**(5) (September), 539–546, http://dx.doi.org/10.1093/cje/22.5.539.

Schneider, S. H., B. L. Turner, and H. M. Garriga (1998), Imaginable Surprise in Global Change Science, *Journal of Risk Research*, **1**(2) (April), 165–185, http://dx.doi.org/10.1080/136698798377240.

Shell, I. B. (2011), *Signals & Signposts: Shell Energy Scenarios to 2050Rep.*, 78 pp., Shell International BV, The Hague, The Netherlands.

Sunstein, C. (2005), *Laws of Fear: Beyond the Precautionary Principle,* Cambridge University Press, New York.

United Nations Secretary General's High-level Panel on Global Sustainability (2012), *Resilient People, Resilient Planet: A future worth choosing Rep.,* 94 pp., United Nations, New York, NY USA.

Woodward, R. T., and R. C. Bishop (1997), How to Decide When Experts Disagree: Uncertainty-Based Choice Rules in Environmental Policy, *Land Econ.,* **73**(4), 492–507.

Zachos, J., M. Pagani, L. Sloan, E. Thomas, and K. Billups (2001), Trends, Rhythms, and Aberrations in Global Climate 65 Ma to Present, *Science,* **292**(5517) (April 27), 686–693, http://dx.doi.org/10.1126/science.1059412.

CHAPTER

10

Valuing the Ocean Environment

Frank Ackerman

Synapse Energy Economics, Cambridge, Massachusetts, USA

INTRODUCTION

Some questions are too big for meaningful answers.

What are the world's oceans worth? Everything, perhaps, since life as we know it would be impossible without the oceans. But this is not a helpful way to frame the economics of the ocean environment. Some, but not all, valuable functions of oceans would survive under the worst environmental assaults that could be imagined. What we really want to know is: how much of the value of the oceans is at risk?

What is the cost of the environmental damage that could be done to the oceans? Or, looking at the same question from a positive perspective, what is the value of protecting the oceans from that damage by controlling climate change, acidification, and other environmental threats?

Even this question is challenging and still too vast for a complete response; it will require careful definition and framing to calculate a useful, partial answer. This chapter builds on the analyses of specific threats in earlier chapters, and on the climate economics and science literature, to develop monetary valuations of selected major impacts on ocean ecosystems and services.

This chapter begins with several boundary conditions for the quantitative analysis, followed by a review of important past studies that have estimated values for ocean ecosystems. It then turns to new calculations, estimating the avoidable impacts of climate change for five specific categories of services provided by the oceans:

- Fishing
- Tourism and recreation
- Moderation of extreme events, including
  - sea-level rise (SLR)
  - storm damages
- Carbon absorption by the oceans

WHAT IS NOT INCLUDED

It is essential to be clear about what is and what is not included in the calculations presented later in this chapter. Several important boundaries to our analysis need to be emphasized at the outset.

Much more extensive economic research is available for damages related to climate change than for other environmental impacts. As a result, our estimates are focused on the comparison between high and low impact climate scenarios. The effects of climate change include both the relatively predictable, expected effects of global warming, changes in precipitation, and SLR, and the risks of catastrophic changes, with lower (but nonzero) probability and disastrous costs. Recent research has greatly expanded our knowledge of both predictable climate impacts and catastrophic risks (Ackerman and Stanton, 2013).

Much of the urgency and controversy surrounding climate change reflects the possibility of tipping points and the risk of catastrophic, irreversible outcomes. Economic analysis of catastrophic risk is discussed in Chapter 9, while this chapter addresses the more predictable effects of climate change. The results presented in this chapter are not meant to be a complete, stand-alone valuation of climate risks and damages as a whole. Rather, Chapters 9 and 10 should be read together as complementary aspects of an economic analysis of climate impacts.

Only commodities with meaningful prices are included in our calculations. This excludes profound and irreplaceable meanings of nature that are too valuable to have prices. Entire communities, livelihoods, and ways of life depend on the natural environment—as discussed throughout this book. Human lives, health, the existence of other species, and unique habitats and ecosystems are immensely valuable but do not have meaningful prices; they are literally priceless (Ackerman and Heinzerling, 2004). Considerable foolishness has resulted from economists' attempts to make up artificial prices for priceless values.

It is not a new idea to suggest that some values are beyond prices. More than 200 years ago, the German philosopher Immanuel Kant distinguished between exchange value, or price, and intrinsic value, or dignity:

> In the kingdom of ends everything has either *price* or *dignity*. Whichever has price can be replaced by something else which is *equivalent*; whatever, on the other hand, is above all price, and therefore admits of no equivalent, has a *dignity*. ***Kant (1785)***

To determine whether something has a price or a dignity, consider whether you would be offended by an offer to buy it. If you recently bought the last ticket on a whale-watching trip and someone else offers to buy the ticket from you for twice its price, you might not accept the offer, but you would probably not take offense. Compare this to studies that have claimed that the existence of humpback whales, as a species, is "worth" $18 billion.[1] Suppose that some super-rich individual offers to pay $36 billion for the right to hunt and kill all the humpback whales in the ocean. This again is twice the "price," but it is a completely different, clearly offensive offer.

The moral of this story is simple: a ticket to go on a whale-watching trip is a commodity with a price; the existence of whales is an irreplaceable value, with a dignity. Economic analysis can offer a meaningful price for whale-watching trips, but not for whales. Invoking such considerations does not resolve every

[1] For details, see Ackerman and Heinzerling (2004). The $18 billion estimate was for U.S. willingness to pay for the existence of humpbacks; the conclusion would be qualitatively unchanged if it were replaced by a global estimate.

question of policy evaluation; in some cases, rival irreplaceable values are at stake on both sides of a debate. What level of protection should be provided to endangered species such as tigers or sharks, which occasionally kill people? It is unlikely, however, that any economic formula or methodology would provide much guidance on such questions.

Aggregate economic valuation of costs and benefits also tends to ignore inequality of income and resources. Damages to richer people look bigger in monetary terms, because they are surrounded by more valuable property. If someone who has built a mansion on the beach in Florida and someone who lives in a seaside village in Bangladesh both have their homes destroyed by storms, the Florida loss will involve many more dollars—but homelessness in Bangladesh may represent the greater human tragedy. The largest losses of human life from storms are frequently in South Asia; the largest losses of property are typically in the United States or other high-income countries.

An adequate analysis of this important issue was far beyond the scope of this chapter. As a result, the economic estimates presented below are global aggregate values and may not represent the true human costs of climate impacts.

Another boundary condition is that only impacts that can be affected by policy will be included; some important damages to the oceans can no longer be changed by any policy that could be adopted today. For example, overfishing has led to decimation of stocks of Atlantic cod and bluefin tuna, losses that have had a real, measurable cost. But those losses have already happened; they are part of the baseline for any analysis of current and future options. Therefore, a calculation of the costs of those past losses is not relevant to choices being made today, except as a cautionary tale for management of other valuable species. (Policies adopted today could lead to more or less extensive recovery of depleted fish populations, which is a different matter.)

In the case of climate change, there are not only past damages but also some future damages that can no longer be avoided. Climate change is a slow-moving crisis, with substantial inertia; even if carbon emissions stopped tomorrow, the changes already under way would lead to significant additional warming and SLR, among other effects. Moreover, there is no realistic scenario in which emissions will stop tomorrow. Even with an extremely ambitious and successful response to the climate threat, there will be several decades more of significant (though declining) carbon emissions. Those transitional decades will add to the concentration of greenhouse gases in the atmosphere and in the oceans, contributing to the irreducible level of climate change.

The avoidable portion of future climate damages is the distance between our fears and our hopes: between climate impacts under a scenario of only limited, slow-moving abatement, where emissions continue to grow for many decades, versus a rapid abatement scenario, where emissions are reduced rapidly and effectively, starting at once and approaching zero within this century. To guide policy decisions, economic calculations should measure the avoidable portion of climate damages or the difference between the two scenarios. Comparison to twentieth-century norms, or to a hypothetical world without climate change, is unfortunately no longer relevant. The *status quo* is no longer an option; the environment of the past century is not one of the available choices for the future.

CLASSIC STUDIES OF THE VALUE OF OCEAN ENVIRONMENTS

Some aspects of the ocean environment have been studied much more extensively than others; leading topics in economic valuation studies to date include fishing, tourism, coral reefs, storms, and SLR (each will be discussed in connection with the detailed estimates below). In addition, there have been at least two well-known investigations of the value of ocean environments in general. Both have been quoted widely as demonstrating the large magnitudes of the values at stake. Neither, unfortunately, provides an appropriate methodology for our calculations.

The first was a 1997 study by Robert Costanza and a dozen coauthors, seeking to estimate the economic value of all the world's ecosystems (Costanza et al., 1997). It produced a controversial estimate of $33 trillion in 1994 U.S. dollars, greater than the value of world GDP at the time. Although it covered 16 different biomes and 17 different ecosystem services, the calculation was dominated by the two ocean biomes and just a few natural services, as seen in Table 10.1.

TABLE 10.1 Ocean Ecosystem Values in Costanza et al. (1997)

	Billions of 1994 U.S. dollars						
	Gas Regulation	**Nutrient Cycling**	**Food Production**	**Cultural**	**Recreation**	**All Other**	**Total**
Open ocean	1262	3918	498	2523	–	181	8381
Coastal	–	11,406	288	192	254	427	12,568
All other	79	1751	600	299	561	9029	12,319
Global total	1341	17,075	1386	3015	815	9636	33,268

Source: Calculated from Costanza et al. (1997), table 2.

The two ocean biomes were the open ocean (excluding continental shelf regions) and coastal zones (including estuaries, other coastal ecosystems, coral reefs, and continental shelf regions). These two ocean biomes accounted for 63% of the total economic value of all world ecosystems in their calculations. In terms of ecosystem services, nutrient cycling alone accounted for more than half of the global total; it was predominantly provided by oceans, particularly concentrated in coastal zones. The second most valuable service was "cultural," accounting for almost 10% of global value and primarily attributed to the open oceans. The article describes this service as "providing opportunities for noncommercial uses" and lists as examples "aesthetic, artistic, educational, spiritual, and/or scientific values of ecosystems."[2] Other valuable ocean services, in this analysis, included gas regulation (CO_2 absorption), food production, and recreation.

The Constanza et al. study has drawn much-deserved attention to the substantial economic value of ecosystem services. Its methodology and results, however, have proved more controversial, with debate surrounding the calculation of total rather than marginal values, and details of the specific estimates (among many others,

[2] Costanza et al. (1997), Table 1, p. 254.

Pearce, 1998; Serafy, 1998; Toman, 1998). As seen above, most of the headline number, the $33 trillion total for global ecosystem services, depends on judgments about the value of nutrient cycling and cultural services of the oceans—categories for which it appears virtually impossible to develop rigorous estimates.

The second comprehensive overview was The Economics of Ecosystems and Biodiversity (TEEB), a study which ran from 2007 through 2010.[3] The TEEB study was hosted by the United Nations Environment Program (UNEP), with financial support from the European Commission, several European governments, and Japan. Its goal was to analyze the economic significance of biodiversity and ecosystem services. In addition to providing many thoughtful statements about the difficulties and limitations of valuation, TEEB surveyed hundreds of publications estimating the values of particular ecosystem services and assembled a database of estimates that met the project's quality standards (de Groot et al., 2010). All estimates were converted into 2007 "international dollars"—that is, in purchasing power parity (PPP) terms[4]—and measured as international dollars per hectare per year for a given biome and ecosystem service.

TEEB identified 11 biomes, of which 4 are relevant to oceans: open ocean (excluding continental shelf regions, islands, and reefs), coral reefs, coastal systems (continental shelf regions, and all coastal areas except reefs and wetlands), and coastal wetlands (tidal marshes and mangroves). Table 10.2 shows the total value per hectare and the principal categories of value for each biome.[5]

[3] See www.teebweb.org for more information.

[4] PPP calculations adjust for the fact that nontraded goods and services tend to have lower prices in lower-income countries, so the purchasing power of income in such countries is greater than a conventional calculation would suggest. By definition, the United States is the benchmark so that U.S. income per capita is the same in PPP or market exchange rate terms. The use of international (PPP) dollars in the TEEB database means that estimates reported in developing-country currencies are treated as being larger than they appear to be at market exchange rates.

[5] Minor value categories are omitted so that the categories shown do not add up to the totals for each biome. TEEB reported separately, but did not include in these totals, value categories for which only one study could be found; the main totals, reported here, are based on two or more estimates in every category.

TABLE 10.2 Valuation of Selected Ocean Ecosystem Services, from TEEB Database

	2007 PPP $/Ha/Year		Number of Studies
	Mean	**Median**	
Open ocean			
All values	49	49	6
Climate regulation	30	30	2
Food production	15	15	2
Coral reefs			
All values	105,126	18,327	96
Tourism	68,453	883	32
Conservation of genetic diversity	11,697	1196	9
Moderation of extreme events	6149	1071	13
Aesthetic information	14,759	14,759	2
Coastal systems			
All values	27,948	27,845	27
Nutrient cycling	19,979	27,421	3
Tourism	7065	245	6
Coastal wetlands			
All values	47,542	11,276	96
Waste treatment, water purification	33,966	6926	4
Moderation of extreme events	3294	2387	13
Habitat/nursery service	3800	362	21
Conservation of genetic diversity	2539	491	4

Source: Excerpted from de Groot et al. (2010).

In this database, values per hectare are low in the open ocean, orders of magnitude lower than in reefs and coastal areas; on the other hand, most of the world's 36 billion hectares of ocean are in the "open ocean" (i.e., outside the continental shelf region, which is in "coastal systems"). Equally important is the small number of studies valuing the open ocean; this area has been little researched.

Most of the ocean-related studies identified by TEEB focused on valuation of coral reefs or coastal wetlands (including mangroves). For tourism, in particular, the values were very site specific, with some extraordinarily high estimates—as suggested by the large ratio of mean to median values for both tourism entries in Table 10.2.

TEEB cautions against simple application of these estimates to other analyses. And indeed, they are not directly applicable for our purposes. Like the Costanza group's estimates, the TEEB numbers are valuations for ecosystem services as a whole, not for the fraction of these services that is at risk from avoidable environmental damage. Nonetheless, TEEB provides a helpful look at the state of the literature in ecosystem valuation, identifying the ocean-related services for which economic values have most often been estimated.

Other researchers have examined the full range of ecological services of ocean biomes. An in-depth survey of the value of coastal ecosystem services, differentiated by biomes, makes clear how much is still unknown and identifies needs for further research; it avoids any summary judgment about total values (Barbier et al., 2011). For a thoughtful review of the issues involved in ecosystem valuation, with illustrations from the UK National Ecosystem Assessment, see Bateman et al. (2010).

In the subsequent sections of this chapter, we examine several of the important ocean-related ecosystem services for which it is possible to estimate the value of avoidable damages caused by climate change. The definition of "ocean-related" impacts is inevitably somewhat imprecise: tropical storms, for example, originate on the ocean, but do almost all of their economic damage on land. Economic value is an anthropocentric concept, and people spend almost all of their time and money on dry land; thus, the analysis presented here is inescapably concerned with the impacts of oceans on humans and our terrestrial habitat. For this reason, some of the numbers developed here may seem small by comparison with other climate damages. Since most of our food comes from agriculture, the economic value of climate impacts is much greater in agriculture than in fisheries. If we were whales, we would have the opposite perspective.

FISHERIES AND CLIMATE CHANGE

Numerous analyses of the expected impacts of climate change on fisheries find that the effects will be felt around the world. One study compared the vulnerability of 132 national economies to climate change impacts on capture fisheries (Allison et al., 2009). Vulnerability was estimated from the impacts of warming, the importance of fisheries to national economies and diets, and the capacity to adapt to change. The greatest vulnerability to climate-caused fishing losses is found in least-developed countries, mainly in Africa.

One of the key climate impacts is the projected redistribution of fish species toward the poles, implying big increases in catch volumes in high latitudes, and big decreases elsewhere (Cheung et al., 2010; Pereira et al., 2010). As oceans warm, many fish species will migrate to higher latitudes to remain in their preferred temperature zones. Pole-ward migration, combined with melting of Arctic ice, could increase opportunities for Arctic and subarctic fishing; a complex economic model of this process projects modest economic gains for Iceland, and somewhat greater percentage gains for Greenland (Arnason, 2008). Such gains for small northern-tier economies, however, will be more than offset by losses elsewhere.

The effects of changes to the El Niño/La Niña-Southern Oscillation (ENSO), an important climate-related phenomenon, include loss of fisheries output (McPhaden et al., 2006). Variability in fishing yields due to ENSO results in problems of adaptation for affected communities, as seen in the case of Peruvian scallop fisheries (Badjeck et al., 2009).

Ocean acidification may be particularly harmful to mollusks, as emphasized in a study of the expected effects on U.S. fishing (Cooley and Doney, 2009). In 2007, U.S. commercial fishing had ex-vessel revenues of $3.8 billion; including activities based on fishing (wholesale, retail, and processing), the total value added in the U.S. economy was $38 billion. Mollusks accounted for 19% of ex-vessel revenues, and crustaceans another 30%. Harm to mollusks from greater acidification under the A1FI climate scenario, versus B1, could cause losses of 5% or more of the net present value of U.S. fishing revenues.

Another study projects even greater losses to mollusk fisheries from acidification (Narita et al., 2011). Under the IS92a scenario, implying a 0.4 unit drop in ocean pH by 2100, mollusk harvests could be reduced by 40%. This would amount to a global loss of $6 billion based on current demand (since mollusk fishing accounted for $15 billion worldwide in 2006), or much more if incomes and demand for mollusks continue to rise.

A potentially important linkage between climate and fisheries is the effect of climate on phytoplankton, the ultimate foundation of ocean food chains. Short- to medium-term fluctuations in phytoplankton growth are correlated with climate events such as ENSO and the North Atlantic Oscillation; the long-term decline in phytoplankton, estimated at about 1% per year, is correlated with rising sea-surface temperatures (Boyce et al., 2010).

Despite the impressive body of research, there are numerous remaining uncertainties in estimates of climate impacts on fisheries (Brander, 2007). A useful paper discusses the implications of uncertainties about climate change for seven fishery case studies around the world (McIlgorm et al., 2010). Climate-related variability is a challenge for those whose livelihoods depend on fishing, creating a need for adaptation but making it difficult to achieve (Badjeck et al., 2010).

The most comprehensive economic analysis is found in a World Bank discussion paper that estimates the cost of climate change for marine capture fisheries worldwide (Sumaila and Cheung, 2010). Comparing A1B climate change to current conditions, it projects annual losses in 2050, at today's prices, of $17-41 billion or up to half of current gross revenues of $80 billion. Developing countries, especially in East Asia and Pacific, would suffer the greatest losses in this analysis.

Even this may be an underestimate of losses, in view of some of the latest research. A new study reexamines the potential fisheries catch in the Northeast Atlantic under the A1B climate scenario, incorporating the effects of ocean acidification, changes in oxygen content, and changes in the phytoplankton community as well as the direct effects of warming (Cheung et al., 2011). For the 120 commercially valuable species in the region, this full suite of climate impacts implied an average shift per decade of 52 km northward, and 5.1 m deeper, greater than the estimates in earlier studies.

As a result, catch potentials will decline even faster than in earlier studies, except in the northernmost regions.

For long-term projection of fisheries impacts, it is also necessary to forecast changes in the demand for fish. A study by the U.S. Department of Agriculture's Economic Research Service examines food expenditures across 144 countries in 2005.[6] Data in that study imply an income elasticity of expenditure on fish of 0.639—that is, a 1% increase in income leads on average to a 0.639% increase in spending on fish.[7] In the long-term projections presented below, we will assume that the same relationship applies over time: every 10% increase in income leads to a 6.39% increase in spending on fish.

TOURISM AND CLIMATE CHANGE

Tourism is a very climate-dependent activity, in which the customers—though not the facilities that serve them—are able to move almost immediately to different parts of the world. It is therefore a leading candidate for disruptive economic effects of climate change. The literature in the field includes several attempts to identify the ideal climate for a tourist destination, including not only temperature but also humidity, precipitation, wind speed, and hours of sunlight. It is not surprising to learn that climate change is expected to shift tourist preferences toward higher latitudes and/or nonsummer seasons (Amelung et al., 2007).

A complete description of climate change impacts on tourism would require a complex analysis of expected losses in current warm-weather destinations versus gains in cooler locations. (Winter tourism, such as skiing trips, is small by comparison with warm-weather tourism and is excluded from this discussion.) Even if tourists can switch destinations at once, there may still be important transitional costs, since the hotels, restaurants, and other tourist facilities cannot immediately follow them. For example, as Europe gets hotter,

[6] http://www.ers.usda.gov/Data/InternationalFoodDemand/. Data analysis described here is based on the table "Distribution of Additional $1 Income Across Food Subgroups, 144 Countries, 2005."

[7] Expenditure on fish was calculated as the product of two data columns in the USDA spreadsheet, per capita spending on food and fish share of food spending.

German tourists can easily decide to spend holidays in Norway instead of Spain, but the Spanish tourist industry will be left behind with substantial losses, while Norway will have the expense of building additional hotels.

There have been several econometric analyses of international tourism, but there is not yet a consensus on even the most basic points. For example, the optimum temperature for a tourist destination has been estimated at a 24-h year-round average of 14 °C (Hamilton et al., 2005) or a 24-h average for the warmest month of the year of about 21 °C (Lise and Tol, 2002). In California, San Francisco would be optimal by the former standard, while San Diego is close to perfect by the latter.[8] While other factors are also important to the choice of tourist destinations, most Caribbean and Pacific islands are already well above both of these standards and are likely to become less attractive as temperatures rise.

A detailed examination of climate impacts on Caribbean islands compared two scenarios, assuming 1.2 and 5.4 °C warming from 2000 to 2100 (Bueno et al., 2008). Building on an earlier World Bank methodology for estimating tourism impacts, the study projected climate-related losses in the two scenarios of 7.0% and 35.3% of Caribbean tourism by 2100.

For long-term forecasts of the growth of tourism, it is essential to estimate how demand will change as incomes rise. This is typically expressed in terms of the income elasticity of demand, defined as the percentage increase in demand that occurs when incomes rise by 1%. Products with income elasticity greater than one represent a bigger share of consumption for the rich than the poor; this is characteristic of luxuries, but not necessities. Since international tourism is unmistakably a luxury, it would be expected to have income elasticity greater than one. Although there have been a few estimates of tourism elasticities below one (e.g., Hamilton et al., 2005), most estimates are much higher. A meta-analysis of early studies reports a mean income elasticity of 1.86 (Crouch, 1995). A more recent study reports 79 estimates published since 2000; the median is greater than 2, and only 9 of the estimates are less than 1 (Song et al., 2010). Using a sophisticated econometric methodology, the same study estimates an income elasticity of 1.36 for tourist visits to Hong Kong.

[8] Temperatures from http://www.weatherbase.com.

Quite a bit of money is at stake in ocean tourism. A global estimate of the value of selected marine recreational activities, encompassing recreational fishing, diving (both snorkeling and scuba diving), and whale-watching, found that worldwide expenditure on these activities was $47 billion in 2003, most of it ($40 billion) for recreational fishing. These activities involved 121 million participants and created more than 1 million jobs. The same study cited an estimate that spending in the United States alone on these three forms of recreation was about $30 billion in 2003, suggesting that the $47 billion global estimate could be too low (Cisneros-Montemayor and Rashid Sumaila, 2010). For recreational fishing, the dominant part of this calculation, the negative impacts from climate change may be parallel to those anticipated for commercial fishing.

Coral reefs, a focal point for ocean tourism, are an irreplaceable ecosystem that will be one of the first to be threatened by global environmental change. Some researchers have found that at 1.7 °C above preindustrial temperatures, all warm-water coral reefs will be bleached, and by 2.5 °C, they will be extinct (IPCC 2007 Working Group II, Ch. 4; Carpenter et al., 2008). These temperatures are much lower than for most other kinds of expected climate damages—and the world is already roughly 0.8 °C above preindustrial temperatures. If climate change continues unabated, it seems likely that the colorful reefs, and the tourism industry that has grown up around them, will be gone well before the end of the century. Temperature increases are only one of several threats to coral reefs, along with acidification and pollution; the synergistic effects of all three occurring simultaneously make the picture even more ominous (see Chapter 6). Thus, the avoidable economic impacts of climate change could include the complete loss of the existing coral reef tourism industry.

The threat is all too real: coral bleaching is not only an ecological problem but a crisis for tourism revenues as well. Tourism revenues drop sharply after coral bleaching, as many tourists choose other destinations. Losses attributable to coral bleaching range from tens of millions of dollars for a single country to billions of dollars on a larger scale; the long-run economic costs of a very serious 1998 coral bleaching event in the Indian Ocean may reach $3.5 billion in lost tourism revenues, in addition to almost $5 billion in other coral reef ecosystem services (Pratchett et al., 2008).

There is an extensive literature estimating recreational values of individual coral reefs, producing widely divergent estimates. A meta-analysis found that reef visitors place a higher value on locations with a larger area of dive sites and fewer other tourists. It also found what appeared to be a lower quality of research than in other meta-analyses of environmental values; both methodology and authorship were significant explanatory variables, suggesting a lack of consistency in approach to the question (Brander et al., 2007). For an additional compilation of studies, see Conservation International (2008).

A widely cited 2003 report—supported by World Wildlife Fund and the International Coral Reef Action Network, and written by leading researchers in the field—estimated the net benefit of the world's coral reefs, if well managed and intact, at U.S. $29.8 billion in 2001. Of that amount, $9.6 billion was from tourism and recreation, $9.0 billion from coastal protection, $5.7 billion from fisheries, and $5.5 billion from the value of biodiversity. Half the tourism and recreation value and more than 40% of the total value came from coral reefs in Southeast Asia (Cesar et al., 2003).

As that study and others have noted, reef-related tourism is growing rapidly; the total today is likely well beyond the $9.6 billion estimate for 2001. For example, that global estimate included $1.1 billion in net benefits of coral reef tourism in Australia. An Australian consultants' study (using different definitions and methods) found that Great Barrier Reef tourism contributed $4.9 billion in value added and $6.0 billion in GDP to the Australian economy[9] in 2005-2006 (Access Economics Pty Limited, 2007). For references to additional studies reaching similar conclusions about the values of the Great Barrier Reef, see Stoeckl et al. (2011).

COSTS OF SLR

There are a number of assessments of the economic impact of SLR. They are difficult to compare with one another for several reasons:

[9] These estimates are in Australian dollars; in recent years, the U.S. and Australian dollars have been close to parity.

- They examine different amounts of SLR: many model 0.5 m, but some have used 0.25-0.30 m; a few have looked at 1 m and up. There is no reason to expect damages to have a linear relationship to the height of SLR.
- They make differing assumptions about coastal protection, which is all-important for cost estimates.
- There is a big difference between modeling with or without storm surges: many omit interactions between SLR and storms; among those including storm effects, there is disagreement about the significance of this interaction.
- Damages are complex and site specific, making them difficult to model above the local level; in contrast, the costs of protection are easier to estimate on a regional or global basis.

An early analysis of global costs of 0.5 m SLR estimated as high as $25-43 billion annually without coastal protection, but only $4-$10 billion annually with optimal levels of coastal protection, with the range of values depending on choices about economic valuation methodology (Darwin and Tol, 2001).

A more recent global analysis, modeling 0.25 m SLR by 2050, finds direct costs of $31 billion with no protection, and much less with protection of coasts; tracing secondary economic effects in detail, it identifies small increases in some countries' and sectors' GDP that may result from increased spending on coastal protection and repair of damages (Bosello et al., 2007).

An analysis focusing specifically on developing countries assumes that 1 m SLR is likely, and up to 3-5 m is eventually possible if ice sheets break up unexpectedly rapidly (Dasgupta et al., 2008). The developing countries most affected by 1 m SLR are Vietnam, Egypt, Mauritania, and Surinam. The study estimates the fraction of population, GDP, agricultural land, and other measures that would be impacted by SLR; Vietnam tops the list of vulnerable countries by most measures. A study of Singapore, assuming up to 0.86 m SLR by 2100, found that coastal protection was economically preferable to allowing loss of land; costs of protection could reach $16.8 million by 2100 (Ng and Mendelsohn, 2005).

Two studies have reached seemingly opposite conclusions about the role of storm surges versus ongoing, gradual SLR. One analysis, modeling a hypothetical beach community in a developed country (e.g., in Florida), finds that storm damages are less than 5% of total SLR damages (West et al., 2001). Another study, based on GIS data for three Chesapeake Bay communities (in the U.S. states of Virginia and Maryland), finds that episodic flooding damages over 100 years will be 9 times greater than the loss from complete inundation at 0.9 m SLR, and 28 times greater at 0.6 m SLR. On this basis, it suggests that estimates based solely on losses from complete inundation may be substantial underestimates of total SLR damages (Michael, 2007).

A comprehensive review of research on SLR estimates that for 4 °C of warming (which is quite possible before 2100 under the A2 climate scenario), SLR in this century is likely to be between 0.5 and 2 m, with low but unknown probability of reaching the upper end of the range (Nicholls et al., 2011; Chapter 5 of this book). This could cause the loss of 0.6-1.2% of global land area and could displace 72-187 million people (0.9-2.4% of the global population). Most of the displaced people would be in Asia; the highest percentage of population displaced would be in small island nations. These impacts are potentially avoidable with widespread upgrade of protection against SLR, but at substantial cost. Increases in adaptation costs would reach $25-270 billion per year by 2100, for 0.5 and 2.0 m SLR, respectively. The wide gap between low- and high-end cost estimates reflects the fact that many high-income areas already are protected against 0.5 m SLR, but not against 2.0 m.

STORMY WEATHER

The intensification of tropical cyclones (hurricanes, in the North Atlantic) is one of the best-studied, though still debated, impacts of climate change. As a review of the subject said, "in a warmer, moister world with higher SSTs [sea surface temperatures], higher sea level, altered atmospheric and oceanic circulations, and increased societal vulnerability, it would be surprising if there were no significant changes in tropical cyclone characteristics and their impacts on society" (Anthes et al., 2006).

There is a growing understanding of the correlation between anthropogenic warming, sea surface temperatures, and cyclone intensity (Emanuel, 2005; Emanuel et al., 2008; Mann and

Emanuel, 2006; Chapter 3 of this book). Debate about the details of this analysis has continued, but has not dislodged its central conclusion (Landsea, 2005). Reconstruction of historical Atlantic hurricane records shows a long-term correlation with La Niña-like climate conditions and with tropical Atlantic temperatures (Mann et al., 2009). An emerging consensus suggests that there could be an increase in the intensity of cyclones, even if there is a decrease in overall frequency (Bender et al., 2010; Knutson et al., 2008). The increase in the most intense Category 4 and 5 storms is expected to be seen worldwide; the decrease in overall frequency of cyclones may be more pronounced outside the North Atlantic (Webster et al., 2005).

A principal challenge to this consensus comes from Pielke, who argues that the observed increase in the destructiveness of hurricanes is entirely explained by the growing numbers of people and value of property located in high-risk areas (Pielke, 2005; Pielke et al., 2003). An independent evaluation of this claim finds that "In the period 1971–2005, since the beginning of a trend towards increased intense cyclone activity, losses excluding socioeconomic effects show an annual increase of 4% per annum," which is "more likely than not" due in part to anthropogenic warming (Schmidt et al., 2009b). The same researchers find that over the past 50 years, increased U.S. hurricane losses due to socioeconomic changes (Pielke's explanation) have been three times as large as the increases due to climate-induced changes—implying that while Pielke is partially correct about the increase in losses, there is also a significant contribution from anthropogenic climate change (Schmidt et al., 2009a).

There are two related obstacles to analysis of climate-induced damages from tropical cyclones. First, as suggested by the Pielke debate, the largest cause of increased damages is growth in the amount of property and population at risk. Small changes in the methods of correction for economic growth can cause large differences in the estimated effect of climate change. Second, the functional form of the relationship between cyclone intensity and economic damages is not yet clear. Many analysts have argued that, based on physical principles, the power exerted by a storm, and therefore its economic damages, should be proportional to the cube of maximum wind speed. Nordhaus (2010) has estimated a much larger exponent, finding economic damages proportional to the eighth or ninth power of wind speed; he argues that damages are a nonlinear function of the force exerted by storms, since structures typically survive up to a breaking point at which there are abrupt, large losses. Small changes in data choice and model specification have large effects on the estimated exponent of wind speed (Schmidt et al., 2009a; Nordhaus, 2010).

Estimates of climate impacts from tropical cyclones vary by at least an order of magnitude; because they are based on differing economic projections, it is easier to compare them as percentages of GDP. At the low end, an estimate based on the FUND model projects global losses of 0.006% of world GDP by 2100; most dollar losses occur in the United States and China, while small island states have the largest damages as a proportion of GDP (Narita et al., 2009). At the high end, Nordhaus (2010) estimates that a doubling of the atmospheric concentration of greenhouse gases would slightly more than double hurricane damages; climate-induced damages would amount to 0.08% of U.S. GDP, excluding the effects of either SLR (which would increase damages) or adaptation (which would reduce damages).

A World Bank analysis argues that damages should be modeled as a function of a storm's minimum atmospheric pressure, rather than maximum wind speed (Mendelsohn et al., 2011). It estimates that the A1B climate scenario would cause an increase in climate damages of 0.01% of world GDP by 2100.[10] That estimate reflects the assumption that damages, as a percent of GDP, decline rapidly with rising incomes—presumably because higher incomes allow more robust and adequate adaptation to storms. The same analysis, however, estimates that 73% of global damages will occur in the United States and Japan, countries that are already sufficiently affluent to prepare for storms; it is difficult to imagine that another century of economic growth will increase the resilience of these high-income nations in the face of cyclones. In a sensitivity analysis, assuming that damages grow proportionally with income, the World Bank study found that climate-induced cyclone damages would amount to 0.032% of world GDP by 2100.

[10] For comparison to the Nordhaus estimate, the World Bank study estimated climate-induced cyclone damages in North America of 0.03% of GDP.

Most of the damages come from the handful of most intense storms; by 2100, the worst 10% of storms are expected to cause 93% of damages, while the worst 1% of storms will cause 64% of total damages (Mendelsohn et al., 2011). Thus, the most important damages from cyclones are essentially high-cost, low-probability disasters, which pose a fundamental challenge to the insurance industry; one of the few available strategies for insurers is to provide premium reductions and credits for policyholders who invest in risk-reducing measures (Kunreuther and Michel-Kerjan, 2007).

A small literature on ecosystem services has identified the value of natural barriers to cyclone damages. In a massive cyclone that struck India in 1999, villages with wider mangroves between them and the coast experienced significantly fewer deaths than ones with narrower or no mangroves; although saving fewer lives than an early warning issued by the government, the mangrove effect is statistically robust under many variations of the model (Das and Vincent, 2009). An analysis of damages from major U.S. hurricanes since 1980 found that wind speed and the area of wetlands in the affected region explained most of the variation in relative damages; coastal wetlands in the United States currently provide an estimated $23 billion per year in storm protection services (Costanza et al., 2008).

SHRINKING THE OCEAN CARBON SINK

The oceans currently absorb one-third of the anthropogenic carbon emissions, with the physical potential to absorb much more. Yet, climate conditions affect the pace of this absorption; most analysts have concluded that climate change will reduce the oceans' uptake of atmospheric carbon. As the world becomes warmer and the oceans turn more acidic, how fast will the oceanic appetite for carbon shrink?

This is an area of active research at present, with no clear consensus estimates yet available. Several studies have found that the oceans are absorbing a declining fraction of carbon emissions (Canadell et al., 2007; Le Quéré et al., 2009). Climate change is expected to have several effects on oceanic carbon absorption; direct measurements of these effects are only beginning to be available (Doney et al., 2009):

- Ocean warming reduces the solubility of CO_2 in seawater, slowing absorption from the atmosphere.
- Warming and freshening of upper-ocean water reduces vertical mixing, slowing transfer of carbon to the deep oceans (thereby slowing overall absorption).
- Climate-induced changes in winds over the Southern Ocean may bring more CO_2-rich deep water to the surface, also slowing absorption.
- Climate-driven changes in the "biological pump," or biological transfer of carbon to deep oceans, are uncertain in direction; either increases or decreases in absorption are possible.

Moreover, it is difficult to analyze this effect via direct comparison of two climate scenarios. The difficulty is demonstrated by a comparison of the A1B scenario with the "E1" stabilization scenario, which is similar to the RCP2.6 scenario (Vichi et al., 2011; Johns et al., 2011). Because the stabilization scenario implies much lower atmospheric concentrations of CO_2, there is much less ocean absorption in that scenario, returning to mid-twentieth-century levels by 2100; the difference between the two scenarios in ocean absorption by that time is at least 3 Gt C, or 11 Gt CO_2, per year (Johns et al., 2011, Figure 12). This difference does not, however, measure the effects of climate change on the ocean carbon sink. In the A1B scenario, ocean uptake is a decreasing fraction of emissions; after about 2050, annual ocean uptake of carbon is no longer increasing, even though emissions and atmospheric concentrations continue to rise.

To identify the effect of climate change on ocean carbon uptake, therefore, it is necessary to model carbon uptake at the same atmospheric concentration, with and without other climate effects—an approach taken in many studies.

A comparison of 11 coupled climate-carbon cycle models tested the feedback between climate change and the carbon cycle (Friedlingstein et al., 2006). Modeling the A2 emissions scenario through 2100, the study found that the expected effects of climate change will decrease both land and ocean absorption of carbon emissions.[11] The effect was most pronounced on land (e.g., in tropical forests), but also visible in the oceans. Most of the models estimated significant but small decreases in ocean carbon absorption due to the climate change resulting from rising emissions; the decreases ranged as high as 2 Gt C per year by 2100, but were less than 1 Gt C per year in most models. Increasing atmospheric concentrations of CO_2 alone lead to greater ocean uptake; the models estimated cumulative ocean carbon absorption of 0.8-1.6 Gt C per ppm CO_2 in the atmosphere (with differences among models based on the rate of transfer of carbon from the surface to the deep ocean). Increasing temperatures alone lead to reduced ocean uptake of carbon; a 1 °C global surface warming reduces ocean uptake by about 20 Gt C.

An in-depth look at one of the 11 models in that comparison provides additional insight into the mechanisms affecting ocean carbon absorption (Crueger et al., 2007). The climate-driven decrease in ocean uptake is partially (not fully) offset by melting of sea ice and increased wind speeds, both of which increase absorption of CO_2 into the ocean. That model estimated that A2 climate change would cause a decrease in ocean carbon uptake of about 0.5 Gt C per year by the end of the century and highlighted the crucial role of the Southern Ocean, where data are least available, in ocean carbon dynamics.

Others have remained more agnostic on the overall effect. The absorption of carbon emissions by the oceans is the combined result of physical and biological processes. The physical processes are better understood—but because the two are interacting, nonlinear systems, their operations cannot always be analyzed in isolation from each other. Some researchers have concluded that both the sign and the magnitude of the effect of climate change on ocean carbon absorption remain uncertain (Riebesell et al., 2009).

Another study identified separate physical and biological effects of warming on ocean carbon uptake, concluding that the physical effects were more important, and outweighed a small increase in carbon absorption due to biological processes (Matsumoto et al.,

[11] The models compared the effects of the A2 emissions scenario with and without the expected effects of those emissions on climate.

2010). Using an emissions scenario that leads to stabilization at 650 ppm by 2200, the study found that the effects of climate change caused a net decrease of 8%, or 35 Gt C, in cumulative ocean uptake by 2100. Other researchers have found that climate change will cause a net decrease in ocean carbon absorption due to biological processes (Steinacher et al., 2009).

VALUING THE DAMAGES

Our best estimates of the value of the five categories of damages discussed here are presented in Table 10.3. Under low emissions, low impact scenario, damages rise to $612 billion a year, or 0.11% of world GDP, by 2100. Under high emissions, high impact scenario, the comparable total is $1980 billion, or 0.37% of GDP. The difference between the two—the amount that can be saved by achieving the low emissions rather than the high emission scenario—is $1367 billion, well over a trillion dollars, per year by 2100, or 0.25% of world GDP.

The following sections explain the calculations that are the basis for Table 10.3.

TABLE 10.3 Valuation of Selected Climate Impacts on Oceans

	(Billions of 2010 U.S. dollars)					
	Low Climate Impacts		High Climate Impacts		Difference	
	2050	2100	2050	2100	2050	2100
Fisheries	67.5	262.1	88.4	343.3	20.9	81.2
Sea-level rise	10.3	34.0	111.6	367.2	101.3	333.2
Storms	0.6	14.5	7.0	171.9	6.4	157.4
Tourism	27.3	301.6	58.3	639.4	31.1	337.7
Ocean carbon sink	0.0	0.0	162.8	457.8	162.8	457.8
Total	105.7	612.2	428.1	1979.6	322.5	1367.4
Percent of GDP	0.06	0.11	0.25	0.37	0.18	0.25

SCENARIO DEFINITIONS

Our emission projections are based on the new RCP6 scenario, for high emissions, and RCP2.6, for low emissions (van Vuuren et al., 2011a). To simplify comparisons, it seems desirable to use a single set of population and GDP projections for both scenarios. Population is based on the medium variant of the UN's 2010 projections,[12] reaching 9.3 billion in 2050 and 10.1 billion in 2100; these figures are close to the values for RCP6 and only slightly above the values for RCP2.6 (van Vuuren et al., 2011a).

GDP projections to 2100 are inescapably highly uncertain. Our calculation assumes the same world GDP in 2100 as the average of the four major IPCC SRES scenarios; this is equivalent to assuming 2.1% average annual growth in real GDP per capita (at market prices, not PPP) from 2000 to 2100.

All dollar amounts in Table 10.3 and the following discussion are expressed in constant 2010 dollars; amounts in other years' dollars were converted to 2010 dollars using the U.S. price deflator for GDP (an index of inflation that is more appropriate for this purpose than the consumer price index).

Temperature outcomes for the two scenarios, used in some of our calculations, are based on the comparison of the RCP scenarios to the IPCC's older categories of emission pathways (Bernie, 2010; Masui et al., 2011). RCP2.6 appears typical of Category I (rapid emission reduction) pathways and can be roughly estimated to reach temperatures of 1.4 °C above preindustrial levels by 2050 and 2.2 °C by 2100. RCP6 is at the high end of Category V, peaking at 4.9 °C above preindustrial in the next century; it can be assumed to reach 2.2 °C in 2050 and 4.0 °C in 2100.

Carbon prices, defined as marginal abatement costs, reach \$178/tCO_2 in 2050 and \$250 in 2100 under RCP2.6 (van Vuuren et al., 2011b). Under RCP6, the more leisurely pace of abatement leads to near-zero abatement costs by 2050 and only \$60/tCO_2 in 2100. In the interests of consistency and simplicity of comparison, the RCP2.6 carbon prices were used throughout our calculations.

[12] http://esa.un.org/unpd/wpp/index.htm.

FISHERIES

Calculations are based on the global estimates in Sumaila and Cheung (2010). They compare losses to marine capture fisheries in 2050 under A1B climate change versus continuation of year 2000 conditions, assuming constant year 2000 prices. They project losses in their "more intensive" fisheries exploitation case of $34.7 billion (converted to 2010 dollars) under year 2000 conditions versus $45.5 billion under A1B. Thus, in their analysis, most of the damages are done by the initial stages of climate change, with only moderate incremental damages from much higher emissions.

Our scenarios involve more severe climate change than theirs, both for high and low emissions. It seems likely that fisheries damages due to climate change should be somewhat higher, in both cases, for our scenarios. Lacking any basis for extrapolation, however, we have adopted their estimates; the difference between the two scenarios may be appropriate, even though the absolute amounts are too low. To extrapolate to 2100, we double the 2050 damage estimates.

Sumaila and Cheung argue that constant real prices for fish through 2050 may be appropriate, if factors such as the growth of aquaculture lead to a sufficient increase in supply. Such price stability seems unlikely to be sustained, in the context of a projected century of steady economic growth. As an alternative, we assume that the supply of fish per capita remains constant (which still requires significant growth in supply through aquaculture to feed a growing population). Under this assumption, the real price will rise at a rate determined by the income elasticity of demand for fish; as noted above, this elasticity is estimated at 0.639. Thus, we multiply the Sumaila and Cheung estimates for 2050 by

$$(\text{per capita income}_{2050}/\text{per capita income}_{2000})^{0.639}$$

and similarly for 2100.

SEA-LEVEL RISE

We use the global estimates from Nicholls et al. (2011) for the costs of protection in 2100; converted to 2010 dollars, these are $34 billion for 0.5 m or $367 billion for 2.0 m of SLR. The high estimate is a worst-case possibility under a 4 °C warming scenario, similar to RCP6. The low estimate is so small that there is little need to worry about fine-tuning; it can be used as an estimate for RCP2.6. Costs in 2050 are interpolated by assuming a linear rate of SLR, while costs rise faster than sea levels, consistent with the Nicholls et al. estimates; this implies mid-century costs of about 30% of the 2100 level.

STORMS

We use the global estimate from Mendelsohn et al. (2011), assuming that damages rise in proportion to future incomes (an income elasticity of 1). That study estimates damages under A1B, which we use for damages under RCP6; this is a conservative assumption, as RCP6 implies higher emissions and damages than A1B. Damages in 2100, under this scenario, are 0.032% of world GDP (the Mendelsohn et al. study uses a different GDP projection, so dollar figures are not directly comparable).

To obtain corresponding figures for RCP6 in 2050 and for RCP2.6 in both 2050 and 2100, we assume that wind speeds are proportional to temperature increases (Nordhaus, 2010), and that damages, as a percent of GDP, are proportional to the cube of wind speed. Within each scenario, temperature increases are roughly half as great in 2050 as in 2100, so wind speeds are half as strong, and damages are one-eighth as large in 2050 as in 2100.

TOURISM

We project losses in two categories of ocean-related tourism. First, coral reef-based tourism is assumed to be reduced by 100% (eliminated) under RCP6 and reduced by 50% under RCP2.6, by 2100; percentage losses in 2050 are half of those amounts. This reflects the extreme sensitivity to temperature increases of coral reefs and the likelihood that they will be entirely eliminated under RCP6. Second, tourism in Caribbean and Pacific islands is assumed to be reduced by more moderate amounts. A detailed calculation of Caribbean tourism losses at different temperatures was developed by Bueno et al. (2008), based on the methodology of an earlier World Bank study. Interpolating from the relationship between temperature and tourism losses developed there, we project losses of island tourism of 3.8% by 2050 and 9.1% by 2100 under RCP2.6, and 9.1% by 2050 and 21.2% by 2100 under RCP6.

Data on the value of coral reef tourism are taken from Cesar et al. (2003), excluding Pacific and Caribbean coral reef tourism to avoid double-counting. Data on Pacific and Caribbean island tourism are taken from three sources: for Hawai'i, from the Hawai'i Tourism Authority's 2010 report[13]; for other Pacific islands, from Seidel and Lal (2010)[14]; and for the Caribbean, from Bueno et al. (2008).

Based on the discussion of income elasticity, above, we assume a conservative value of 1.5 for the income elasticity of tourism. Global demand for each category of tourism in 2050, in the absence of climate losses, is calculated from the historical value in a given base year (shown as 20*xx* because the years of available data vary by category of tourism) as:

$$\text{Demand}_{2050} = \text{Demand}_{20xx} * \frac{\text{population}_{2050}}{\text{population}_{20xx}} * \left(\frac{\text{income per capita}_{2050}}{\text{income per capita}_{20xx}}\right)^{1.5}$$

The corresponding calculation is also done for 2100. Each of these demand figures is then multiplied by the loss percentages discussed above to calculate tourism losses due to climate change.

OCEAN CARBON SINK

Under RCP6, impacts were assumed to follow the model of Crueger et al. (2007), implying a reduction in ocean carbon absorption of 0.5 Gt C = 1.83 Gt CO_2 per year by 2100; the impact in 2050 was assumed to be half that amount. Under RCP2.6, which

[13] 2010 Annual Visitor Research Report, http://www.hawaiitourismauthority.org/research.

[14] The large estimate in Seidel and Lal (2010) for Guam, which combines tourism and U.S. military travel, is excluded.

resembles the E1 reduction scenario (Vichi et al., 2011; Johns et al., 2011), losses were assumed to be negligible.

Impacts under both emission scenarios were priced at the carbon prices cited above.

References

Ackerman, F., and E. A. Stanton (2013), *Climate Economics: The State of the Art*, Routledge, London.

Access Economics Pty Limited (2007), *Measuring the Economic & Financial Value of the Great Barrier Reef Marine Park, 2005-06*, Commonwealth of Australia, Townsville, Queensland, http://www.gbrmpa.gov.au/corp_site/info_services/publications/research_publications/rp088/access_economics_report_0607.

Ackerman, F., and L. Heinzerling (2004), *Priceless: On Knowing the Price of Everything and the Value of Nothing*, The New Press, New York.

Allison, E. H., A. L. Perry, M. -C. Badjeck, W. Neil Adger, K. Brown, D. Conway, and A. S. Halls, et al. (2009), Vulnerability of National Economies to the Impacts of Climate Change on Fisheries, *Fish and Fisheries*, **10**(2) (June), 173–196, http://dx.doi.org/10.1111/j.1467-2979.2008.00310.x.

Amelung, B., S. Nicholls, and D. Viner (2007), Implications of Global Climate Change for Tourism Flows and Seasonality, *Journal of Travel Research*, **45**(3) (February), 285–296, http://dx.doi.org/10.1177/0047287506295937.

Anthes, R. A., R. W. Corell, G. Holland, J. W. Hurrell, M. C. MacCracken, and K. E. Trenberth (2006), Hurricanes and Global Warming—Potential Linkages and Consequences, *B. Am. Meteorol. Soc.*, **87**(5), 623–628.

Arnason, R. (2008), Climate Change and Fisheries: Assessing the Economic Impact in Iceland and Greenland, *Natural Resource Modeling*, **20**(2) (June), 163–197, http://dx.doi.org/10.1111/j.1939-7445.2007.tb00205.x.

Badjeck, M. -C., E. H. Allison, A. S. Halls, and N. K. Dulvy (2010), Impacts of Climate Variability and Change on Fishery-based Livelihoods, *Mar. Policy*, **34**(3) (May), 375–383, http://dx.doi.org/10.1016/j.marpol.2009.08.007.

Badjeck, M. -C., J. Mendo, M. Wolff, and H. Lange (2009), Climate Variability and the Peruvian Scallop Fishery: The Role of Formal Institutions in Resilience Building, *Climatic Change*, **94**(1–2) (April), 211–232, http://dx.doi.org/10.1007/s10584-009-9545-y.

Barbier, E. B., S. D. Hacker, C. Kennedy, E. W. Koch, A. C. Stier, and B. R. Silliman (2011), The Value of Estuarine and Coastal Ecosystem Services, *Ecol. Monogr.*, **81**(2) (May), 169–193, http://dx.doi.org/10.1890/10-1510.1.

Bateman, I. J., G. M. Mace, C. Fezzi, G. Atkinson, and K. Turner (2010), Economic Analysis for Ecosystem Service Assessments, *Environ. Resour. Econ.*, **48**(2) (October 13), 177–218, http://dx.doi.org/10.1007/ s10640-010-9418-x.

Bender, M. A., T. R. Knutson, R. E. Tuleya, J. J. Sirutis, G. A. Vecchi, S. T. Garner, and I. M. Held (2010), Modeled Impact of Anthropogenic Warming on the Frequency of Intense Atlantic Hurricanes, *Science*, **327**(5964) (January 22), 454–458, http://dx.doi.org/10.1126/science.1180568.

Bernie, D. (2010), *Temperature Implications from the IPCC 5th Assessment Representative Concentration Pathways (RCP)*, Met Office Hadley Centre, London, http://www.metoffice.gov.uk/avoid/files/resources.researchers/AVOID_WS2_D1_11_20100422.pdf.

Bosello, F., R. Roson, and R. S. J. Tol (2007), Economy-wide Estimates of the Implications of Climate Change: Sea Level Rise, *Environ. Resour. Econ.*, **37**(3) (January 13), 549–571, http://dx.doi.org/10.1007/ s10640-006-9048-5.

Boyce, D. G., M. R. Lewis, and B. Worm (2010), Global Phytoplankton Decline over the Past Century, *Nature*, **466**(7306) (July), 591–596, http://dx.doi.org/10.1038/nature09268.

Brander, K. M. (2007), Climate Change and Food Security Special Feature: Global Fish Production and Climate Change, *Proceedings of the National Academy of Sciences*, **104**(50) (December), 19709–19714, http://dx.doi.org/10.1073/ pnas.0702059104.

Brander, L., P. Vanbeukering, and H. Cesar (2007), The Recreational Value of Coral Reefs: A Meta-analysis, *Ecol. Econ.*, **63**(1) (June), 209–218, http://dx.doi.org/10.1016/j.ecolecon.2006.11.002.

Bueno, R., C. Herzfeld, E. A. Stanton, and F. Ackerman (2008), *The Caribbean and Climate Change: The Costs of Inaction*, Stockholm Environment Institute-U.S. Center, Somerville, MA, http://sei-us.org/publications/id/86.

Canadell, J. G., C. Le Quere, M. R. Raupach, C. B. Field, E. T. Buitenhuis, P. Ciais, T. J. Conway, N. P. Gillett, R. A. Houghton, and G. Marland (2007), Contributions to Accelerating Atmospheric CO2 Growth from Economic Activity, Carbon Intensity, and Efficiency of Natural Sinks, *Proceedings of the National Academy of Sciences*, **104**(November 20), 18866–18870, http://dx.doi.org/10.1073/pnas.0702737104.

Carpenter, K. E., M. Abrar, G. Aeby, R. B. Aronson, S. Banks, A. Bruckner, and A. Chiriboga, et al. (2008), One-Third of Reef-Building Corals Face Elevated Extinction Risk from Climate Change and Local Impacts, *Science*, **321**(5888) (July 25), 560–563, http://dx.doi.org/10.1126/science.1159196.

Cesar, H., L. Burke, and L. Pet-Soede (2003), *The Economics of Worldwide Coral Reef Degradation*, Cesar Environmental Economics Consulting, Arnhem, Netherlands, http://pdf.wri.org/cesardegradationreport100203.pdf.

Cheung, W. W. L., J. Dunne, J. L. Sarmiento, and D. Pauly (2011), Integrating Ecophysiology and Plankton Dynamics into Projected Maximum Fisheries Catch Potential Under Climate Change in the Northeast Atlantic, *ICES J. Mar. Sci.*, **68**(6) (April), 1008–1018, http://dx.doi.org/10.1093/icesjms/ fsr012.

Cheung, W. W. L., V. W. Y. Lam, J. L. Sarmiento, K. Kearney, R. Watson, D. Zeller, and D. Pauly (2010), Large-scale Redistribution of Maximum Fisheries Catch Potential in the Global Ocean Under Climate Change, *Glob. Change Biol.*, **16**(1) (January), 24–35, http://dx.doi.org/10.1111/j.1365-2486.2009.01995.x.

Cisneros-Montemayor, A. M., and U. Rashid Sumaila (2010), A Global Estimate of Benefits from Ecosystem-based Marine Recreation: Potential Impacts and Implications for Management, *Journal of Bioeconomics*, **12**(3) (August), 245–268, http://dx.doi.org/10.1007/s10818-010-9092-7.

Conservation International (2008), *Economic Values of Coral Reefs, Mangroves, and Seagrasses: A Global Compilation*, Center for Applied Biodiversity Science,

Conservation International, Arlington, VA, http://02cbb49.netsolhost.com/library/Economic_values_globalcompilation.pdf.

Cooley, S. R., and S. C. Doney (2009), Anticipating Ocean Acidification's Economic Consequences for Commercial Fisheries, *Environmental Research Letters*, **4**(2) (June), 024007, http://dx.doi.org/10.1088/1748-9326/4/2/024007.

Costanza, R., R. d' Arge, R. de Groot, S. Farber, M. Grasso, B. Hannon, K. LImburg, et al. (1997), The Value of the World's Ecosystem Services and Natural Capital, *Nature*, **387**, 253–260, http://dx.doi.org/10.1038/387253a0.

Costanza, R., O. Pérez-Maqueo, M. Luisa Martinez, P. Sutton, S. J. Anderson, and K. Mulder (2008), The Value of Coastal Wetlands for Hurricane Protection, *AMBIO*, **37**(4) (June), 241–248, http://dx.doi.org/10.1579/0044-7447(2008)37[241: TVOCWF]2.0.CO;2.

Crouch, G. (1995), A Meta-analysis of Tourism Demand, *Ann. Tourism Res.*, **22**, 103–118, http://dx.doi.org/10.1016/0160-7383(94)00054-V.

Crueger, T., E. Roeckner, T. Raddatz, R. Schnur, and P. Wetzel (2007), Ocean Dynamics Determine the Response of Oceanic CO2 Uptake to Climate Change, *Clim. Dynam.*, **31**(2–3) (December), 151–168, http://dx.doi.org/10.1007/s00382-007-0342-x.

Darwin, R., and R. S. J. Tol (2001), Estimates of the Economic Effects of Sea Level Rise, *Environ. Resour. Econ.*, **19**, 113–129.

Das, S., and J. R. Vincent (2009), Mangroves Protected Villages and Reduced Death Toll During Indian Super Cyclone, *Proc. Natl. Acad. Sci. U. S. A.*, **106**(18) (April), 7357–7360, http://dx.doi.org/10.1073/pnas.0810440106.

Dasgupta, S., B. Laplante, C. Meisner, D. Wheeler, and J. Yan (2008), The Impact of Sea Level Rise on Developing Countries: a Comparative Analysis, *Climatic Change*, **93**(3–4) 379–388, http://dx.doi.org/10.1007/s10584-008-9499-5.

Doney, S. C., B. Tilbrook, S. Roy, N. Metzl, C. Le Quéré, M. Hood, R. A. Feely, and D. Bakker (2009), Surface-ocean CO2 Variability and Vulnerability, *Deep-Sea Res. Pt. II*, **56** (April), 504–511, http://dx.doi.org/10.1016/j. dsr2.2008.12.016.

Emanuel, K. (2005), Increasing Destructiveness of Tropical Cyclones over the Past 30 Years, *Nature*, **436**(7051), 686–688, http://dx.doi.org/10.1038/ nature03906.

Emanuel, K., R. Sundararajan, and J. Williams (2008), Hurricanes and Global Warming: Results from Downscaling IPCC AR4 Simulations, *B. Am. Meteorol. Soc.*, **89**(3) (March), 347–367, http://dx.doi.org/10.1175/BAMS-89-3-347.

Friedlingstein, P., P. Cox, R. Betts, L. Bopp, W. von Bloh, V. Brovkin, P. Cadule, et al. (2006), Climate–Carbon Cycle Feedback Analysis: Results from the C4MIP *Model Intercomparison, J. Climate*, **19**(14) 3337–3353, http://dx.doi.org/10.1175/JCLI3800.1.

de Groot, R. S., P. Kumar, S. van der Ploog, and P. Sukhdev (2010), *Estimates of Monetary Values of Ecosystem Services*, The Economics of Ecosystems and Biodiversity: Ecological and Economic Foundations. Wageningen and Brussels, http://www.teebweb.org/LinkClick.aspx?fileticket=tgNd9iW2zH4%3d&tabid=1018&language=en-US.

Hamilton, J. M., D. J. Maddison, and R. S. J. Tol (2005), Climate Change and International Tourism: A Simulation Study, *Global Environ. Chang.*, **15** (October), 253–266, http://dx.doi.org/10.1016/j. gloenvcha.2004.12.009.

Intergovernmental Panel on Climate Change (2007), *Climate Change 2007—IPCC Fourth Assessment Report*, Cambridge, UK. Cambridge University Press.

Johns, T. C., J. -F. Royer, I. Höschel, H. Huebener, E. Roeckner, E. Manzini, W. May, et al. (2011), Climate Change Under Aggressive Mitigation: The ENSEMBLES Multi-model Experiment, *Clim. Dynam.*, **37** (February 11), 1975–2003, http://dx.doi.org/10.1007/s00382-011-1005-5.

Knutson, T., J. Sirutis, S. Garner, G. Vecchi, and I. Held (2008), Simulated Reduction in Atlantic Hurricane Frequency Under Twenty- first-century Warming Conditions, *Nature Geoscience*, **1**(7) (June), 359–364, http://dx.doi.org/10.1038/ngeo229.

Kunreuther, H., and E. O. Michel-Kerjan (2007), *Climate Change, Insurability of Large-scale Disasters and the Emerging Liability Challenge,* Working Paper. National Bureau of Economic Research, Cambridge, MA.

Landsea, C. W. (2005), Meteorology: Hurricanes and Global Warming, *Nature*, **438**(7071) (December 22), E11–E12, http://dx.doi.org/10.1038/ nature04477.

Lise, W., and R. S. J. Tol (2002), Impact of Climate on Tourist Demand, *Climatic Change*, **55**(4), 429–449, http://dx.doi.org/10.1023/A:1020728021446.

Mann, M. E., and K. A. Emanuel (2006), Atlantic Hurricane Trends Linked to Climate Change, *Eos T. Am. Geophys. Un.*, **87**(24) (June), http://dx.doi.org/10.1029/2006EO240001, http://www.crossref.org/guestquery/.

Mann, M. E., J. D. Woodruff, J. P. Donnelly, and Z. Zhang (2009), Atlantic Hurricanes and Climate over the Past 1,500 Years, *Nature*, **460**(7257), 880–883, http://dx.doi.org/10.1038/nature08219.

Masui, T., K. Matsumoto, Y. Hijioka, T. Kinoshita, T. Nozawa, S. Ishiwatari, E. Kato, P. R. Shukla, Y. Yamagata, and M. Kainuma (2011), An Emission Pathway for Stabilization at 6Wm^2 Radiative Forcing, *Climatic Change*, **109**(1–2), (August 13), 59–76, http://dx.doi.org/10.1007/s10584-011-0150-5.

Matsumoto, K., K. S. Tokos, M. O. Chikamoto, and A. Ridgwell (2010), Characterizing Post-industrial Changes in the Ocean Carbon Cycle in an Earth System Model, *Tellus B*, **62**(4) (September), 296–313, http://dx.doi.org/10.1111/j.1600-0889.2010.00461.x.

McIlgorm, A., S. Hanna, G. Knapp, P. Le Floc'H, F. Millerd, and M. Pan (2010), How Will Climate Change Alter Fishery *Governance? Insights from Seven International Case Studies, Mar. Policy*, **34**(1) (January):, 170–177, http://dx.doi.org/10.1016/j.marpol.2009.06.004.

McPhaden, M. J., S. E. Zebiak, and M. H. Glantz (2006), ENSO as an Integrating Concept in Earth Science, *Science*, **314**(5806) (December), 1740–1745, http://dx.doi.org/10.1126/science.1132588.

Mendelsohn, R., K. Emanuel, and S. Chonabayashi (2011), *The Impact of Climate Change on Global Tropical Storm Damages,* World Bank, http://papers.ssrn.com/sol3/papers.cfm?abstract_id=1955106.

Michael, J. (2007), Episodic Flooding and the Cost of Sea-level Rise, *Ecol. Econ.*, **63** (1) (June), 149–159, http://dx.doi.org/10.1016/j.ecolecon.2006.10.009.

Narita, D., K. Rehdanz, and R. S. J. Tol (2011), *Economic Costs of Ocean Acidification: A Look into the Impacts on Shellfish Production,* Kiel Institute for the World Economy, http://ideas.repec.org/p/kie/kieliw/1710.html.

Narita, D., R. Tol, and D. Anthoff (2009), Damage Costs of Climate Change Through Intensification of Tropical Cyclone Activities: An Application of FUND, *Climate Res.*, **39** (June), 87–97, http://dx.doi.org/10.3354/cr00799.

Ng, W. -S., and R. Mendelsohn (2005), The Impact of Sea Level Rise on Singapore, *Environment and Development Economics*, **10**(2) (April), 201–215, http://dx.doi.org/10.1017/S1355770X04001706.

Nicholls, R. J., N. Marinova, J. A. Lowe, S. Brown, P. Vellinga, D. de Gusmao, J. Hinkel, and R. S. J. Tol (2011), Sea-level Rise and Its Possible Impacts Given a 'Beyond 4 C World' in the Twenty-first Century, *Philosophical Transactions of the Royal Society A: Mathematical, Physical and Engineering Sciences*, **369**(1934) (January 13), 161–181, http://dx.doi.org/10.1098/rsta.2010.0291.

Nordhaus, W. D. (2010), The Economics of Hurricanes and Implications of Global Warming, *Climate Change Economics*, **01**(01), 1, http://dx.doi.org/10.1142/S2010007810000054.

Pearce, D. (1998), Auditing the Earth: The Value of the World's Ecosystem Services and Natural Capital, *Environment*, **40**(2), 23–28.

Pereira, H. M., P. W. Leadley, V. Proenca, R. Alkemade, J. P. W. Scharlemann, J. F. Fernandez-Manjarres, M. B. Araujo, et al. (2010), Scenarios for Global Biodiversity in the 21st Century, *Science*, **330**(6010) (October 26), 1496–1501, http://dx.doi.org/10.1126/science.1196624.

Pielke, R. A. (2005), Meteorology: Are There Trends in Hurricane Destruction? *Nature*, **438**(7071), (December 22), E13, http://dx.doi.org/10.1038/ nature04426.

Pielke, R. A., J. Rubiera, C. Landsea, M. L. Fernández, and R. Klein (2003), Hurricane Vulnerability in Latin America and The Caribbean: Normalized Damage and Loss Potentials, *Natural Hazards Review*, **4**(3) (August), 101–114, http://dx.doi.org/10.1061/(ASCE)1527-6988(2003) 4:3(101).

Pratchett, M. S., P. L. Munday, S. K. Wilson, N. A. J. Graham, J. E. Cinner, D. R. Bellwood, G. P. Jones, N. V. C. Polunin, and T. R. McClanahan (2008); Effects of Climate-induced Coral Bleaching on Coral-reef Fishes–Ecological and Economic Consequences. In: *Oceanography and Marine Biology: An Annual Review* [Gibson R. N., R. J. A. Atkinson and J. D. M. Gordon (eds.)]. 46. Taylor & Francis, 251–296, http://eprints.jcu.edu.au/4901/.

Le Quéré, C., M. R. Raupach, J. G. Canadell, G. Marland, et al. (2009), Trends in the Sources and Sinks of Carbon Dioxide, *Nature Geoscience*, **2**(12) (December), 831–836, http://dx.doi.org/10.1038/ngeo689.

Riebesell, U., A. Kortzinger, and A. Oschlies (2009), Sensitivities of Marine Carbon Fluxes to Ocean Change, *Proceedings of the National Academy of Sciences*, **106**(49) (December), 20602–20609, http://dx.doi.org/10.1073/pnas.0813291106.

Schmidt, S., C. Kemfert, and P. Höppe (2009a), The Impact of Socio- economics and Climate Change on Tropical Cyclone Losses in the USA, *Regional Environmental Change*, **10**(1) (January), 13–26, http://dx.doi.org/10.1007/s10113-008-0082-4.

Schmidt, S., C. Kemfert, and P. Höppe (2009b), Tropical Cyclone Losses in the USA and the Impact of Climate Change—A Trend Analysis Based on Data from a New Approach to Adjusting Storm Losses, *Environ. Impact Asses.*, **29** (6) (November), 359–369, http://dx.doi.org/10.1016/j.eiar.2009.03.003.

Seidel, H., and P. N. Lal (2010), *Economic Value of the Pacific Ocean to the Pacific Island Countries and Territories*, IUCN, Gland, Switzerland.

Serafy, S. El (1998), Pricing the Invaluable: The Value of the World's Ecosystem Services and Natural Capital, *Ecol. Econ.*, **25**(1) (April), 25–27, http://dx.doi.org/10.1016/S0921-8009(98)00009-3.

Song, H., J. H. Kim, and S. Yang (2010), Confidence Intervals for Tourism Demand Elasticity, *Ann. Tourism Res.*, **37** (April), 377–396, http://dx.doi.org/10.1016/j.annals.2009.10.002.

Steinacher, M., F. Joos, T. L. Frölicher, L. Bopp, P. Cadule, S. C. Doney, M. Gehlen, B. Schneider, and J. Segschneider (2009), Projected 21st Century Decrease in Marine Productivity: a Multi-model Analysis, *Biogeosciences Discussions*, **6**(4) (August), 7933–7981, http://dx.doi.org/10.5194/bgd-6-7933-2009.

Stoeckl, N., C. C. Hicks, M. Mills, K. Fabricius, M. Esparon, F. Kroon, K. Kaur, and R. Costanza (2011), The Economic Value of Ecosystem Services in the Great Barrier Reef: Our State of Knowledge, *Ann. N. Y. Acad. Sci.*, **1219**(1) (February), 113–133, http://dx.doi.org/10.1111/j.1749-6632.2010.05892.x.

Sumaila, U. R., and W. W. L. Cheung (2010), *Cost of Adapting Fisheries to Climate Change*, Discussion Paper. World Bank. http://water.worldbank.org/water/publications/cost-adapting-fisheries-climate-change.

Toman, M. (1998), Why Not to Calculate the Value of the World's Ecosystem Services and Natural Capital, *Ecol Econ.*, **25**(1) (April), 57–60, http://dx.doi.org/10.1016/S0921-8009(98)00017-2.

Vichi, M., E. Manzini, P. G. Fogli, A. Alessandri, L. Patara, E. Scoccimarro, S. Masina, and A. Navarra (2011), Global and Regional Ocean Carbon Uptake and Climate Change: Sensitivity to a Substantial Mitigation Scenario, *Clim. Dynam.* (published online May 3) (May), http://dx.doi.org/10.1007/s00382-011-1079-0 http://www.springerlink.com/index/10.1007/s00382-011-1079-0.

van Vuuren, D. P., J. Edmonds, M. Kainuma, K. Riahi, A. Thomson, K. Hibbard, and G. C. Hurtt, et al. (2011), The *Representative Concentration Pathways: An Overview, Climatic Change*, **109**(1–2) (August 5), 5–31, http://dx.doi.org/10.1007/s10584-011-0148-z.

van Vuuren, D. P., E. Stehfest, M. G. J. Elzen, T. Kram, J. Vliet, S. Deetman, M. Isaac, et al. (2011), RCP2.6: Exploring the Possibility to Keep Global Mean Temperature Increase Below 2C, *Climatic Change*, **109**(1–2) (August 5), 95–116, http://dx.doi.org/10.1007/s10584-011-0152-3.

Webster, P. J., G. J. Holland, J. A. Curry, and H. R. Chang (2005), Changes in Tropical Cyclone Number, Duration, and Intensity in a Warming Environment, *Science*, **309**(5742) 1844–1846, http://dx.doi.org/10.1126/science.1116448.

West, J. J., M. J. Small, and H. Dowlatabadi (2001), Storms, Investor Decisions, and the Economic Impacts of Sea Level Rise, *Climatic Change*, **48**(2–3) 317–342, http://dx.doi.org/10.1023/A:1010772132755.

CHAPTER

11

Managing Multiple Human Stressors in the Ocean: A Case Study in the Pacific Ocean

William W.L. Cheung, Ussif Rashid Sumaila*[†]

*Changing Ocean Research Unit, Fisheries Centre, The University of British Columbia, Vancouver, British Columbia, Canada

[†]Fisheries Economics Research Unit, Fisheries Centre, The University of British Columbia, Vancouver, British Columbia, Vancouver, Canada

INTRODUCTION

The Pacific Ocean is the largest ocean in the world, providing a wide variety of goods and services to human society. It represents almost half of the world's ocean area, bordering the coastline of 50 countries or territories (Figure 11.1). Supported by the existence of a wide range of habitats such as coral reefs, mangroves, seagrass, and seamounts, the Pacific Ocean hosts much of the world's marine biodiversity (Cheung et al., 2005). The Pacific Ocean also plays an important role in the regulation of global climate and biogeochemical cycles. However, the Pacific Ocean is being changed by a variety of human-induced drivers, including overexploitation, climate change, pollution, habitat destruction, and invasive species. These factors threaten its diversity as well as many goods and services that the marine ecosystems provide.

The scope of this chapter is to provide a summary of the status of marine ecosystems in the Pacific Ocean. Specifically, we discuss the

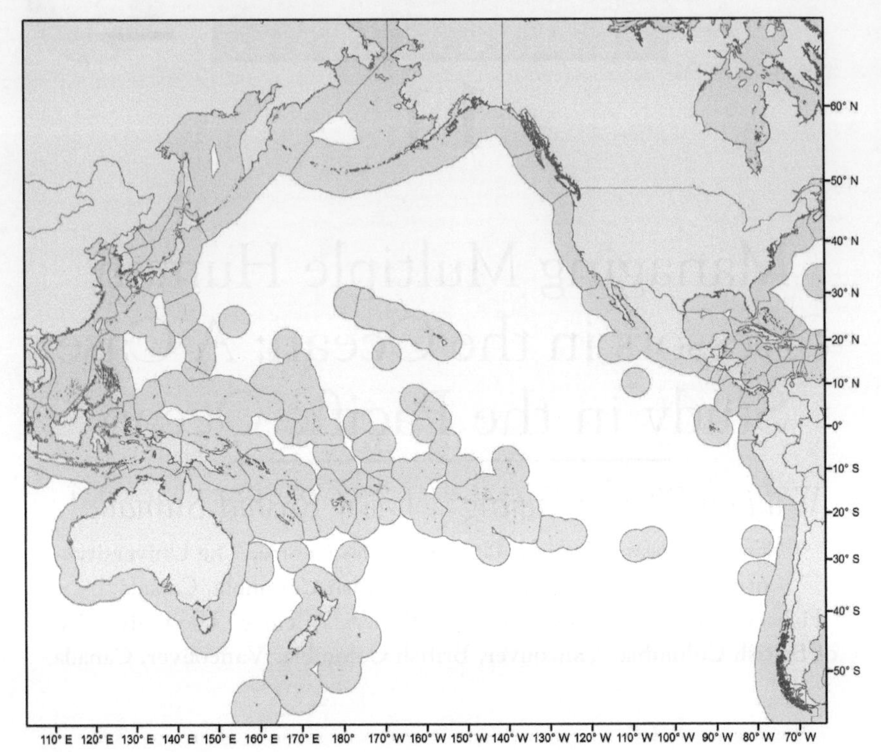

FIGURE 11.1 Pacific Ocean with boundaries of the Exclusive Economic Zones of countries and territories in the region (blue).

major human drivers of changes and how those affect the current and, potentially, future status of the marine ecosystems. The drivers that we focus on in this chapter are overexploitation, climate change, pollution, and habitat destruction. A number of case studies, representing different ecological and socioeconomic context in the Pacific Ocean, are discussed. Finally, we review proposed policy options to improve the management and conservation of marine ecosystems and biodiversity in the region.

PATTERN OF BIODIVERSITY AND MARINE LIVING RESOURCES

The Pacific Ocean consists of diverse habitat types, ranging from shallow coasts, reefs, and islands to the abyssal zone that reaches thousands of meters in depth (Figure 11.2a). It consists of all major marine habitats, including coral reefs, seagrass, mangroves,

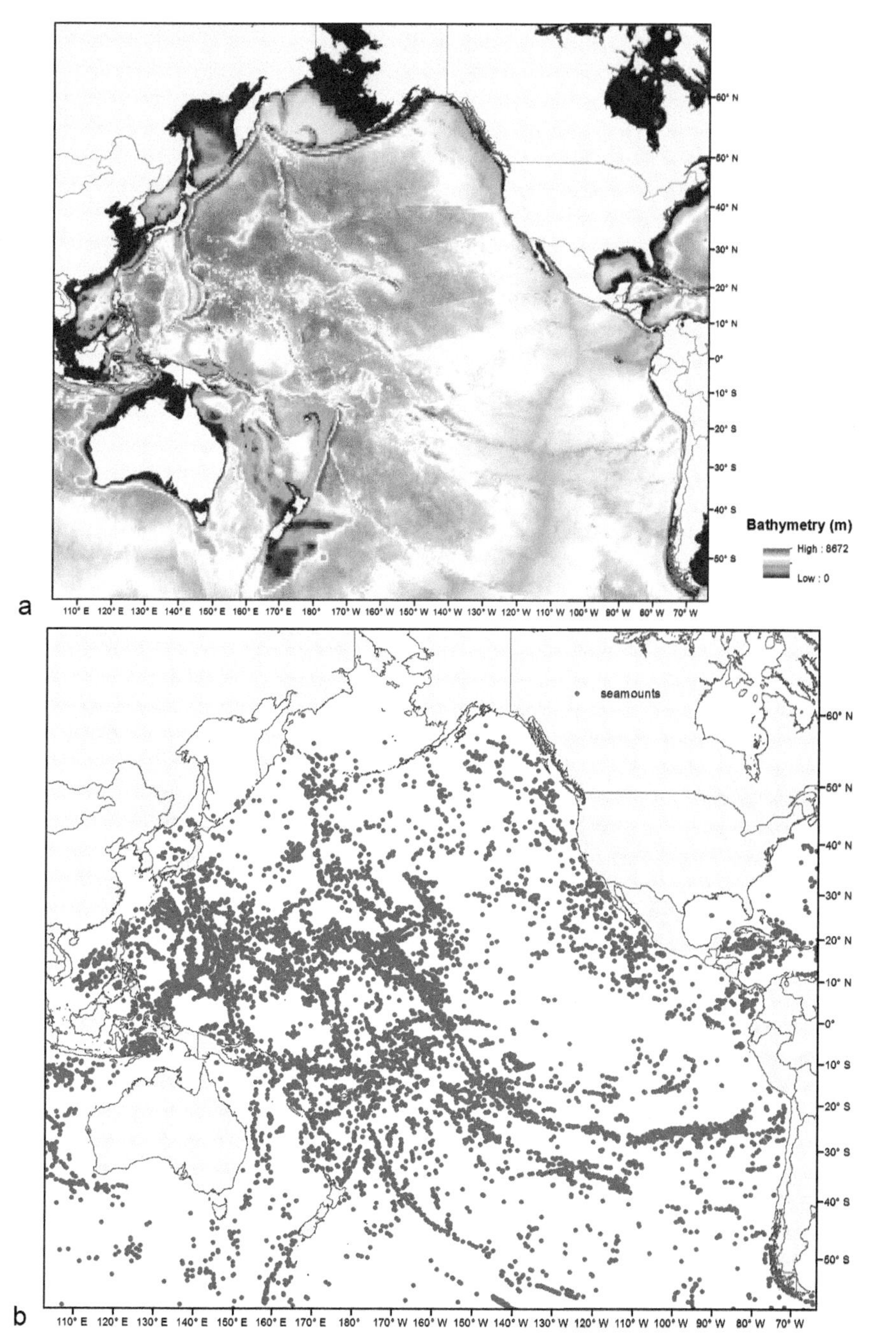

FIGURE 11.2 The diversity of topography and habitat types houses high diversity of marine organisms in the Pacific Ocean (a) bathymetry, (b) large seamounts identified from a mid-resolution bathymetric map, using methods outlined in Kitchingman et al. (2007)

Continued

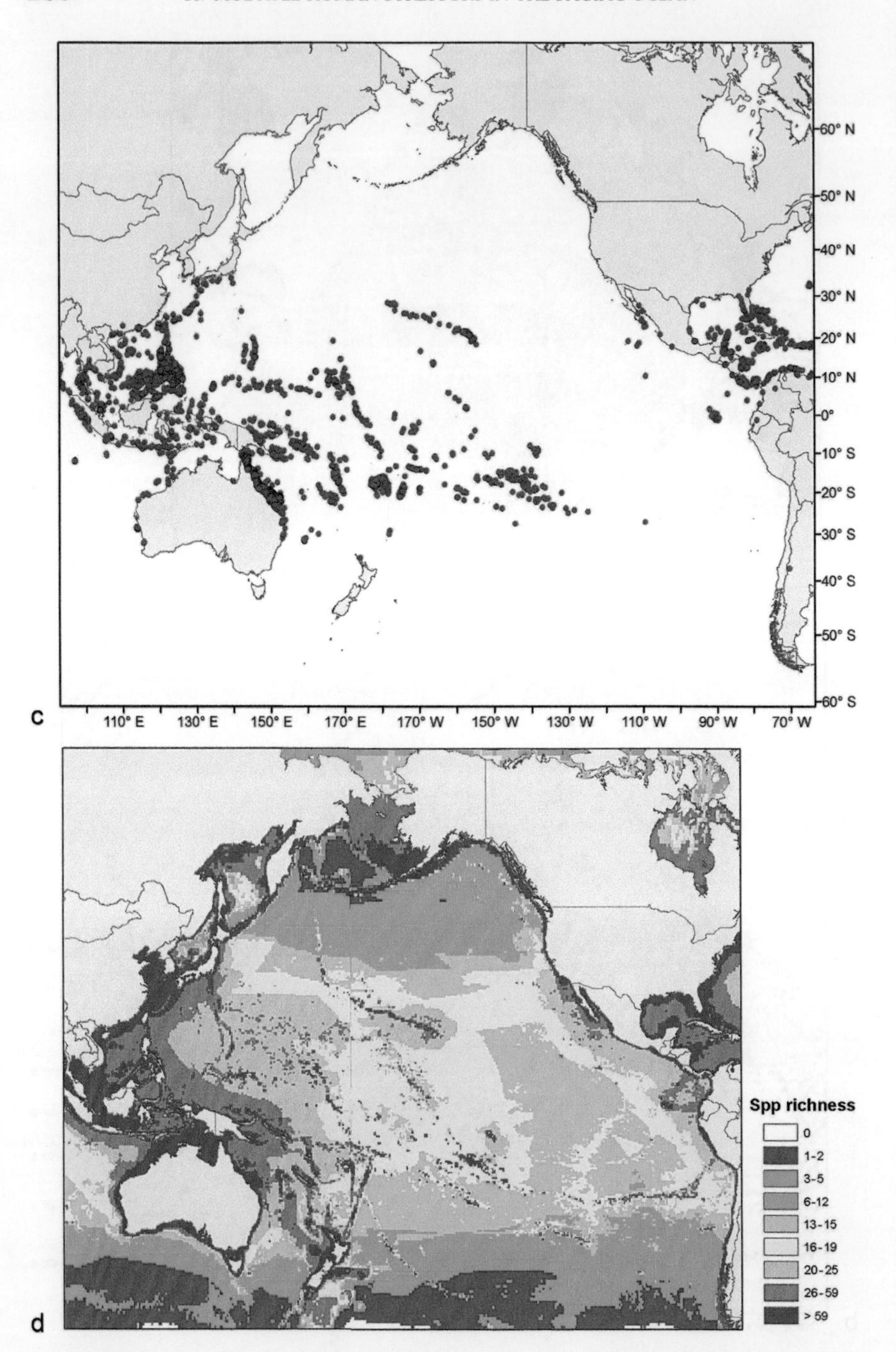

FIGURE 11.2—cont'd (c) coral reefs (www.reefbase.org); (d) predicted species richness of exploited marine fishes and invertebrates (Cheung et al., 2009).

seamounts, and estuaries (Figure 11.2b,c). These habitats house a high diversity of marine organisms. For example, the coral triangle in the Indo-Pacific region is recognized as a center of marine biodiversity. Using predicted distributions of exploited marine fishes and invertebrates ($N=1066$), it is suggested that the Pacific Ocean has the highest species richness of exploited species (Figure 11.2d) (Cheung et al., 2005, 2009).

The Pacific Ocean supports rich fisheries resources, providing us with animal protein and a source of revenue particularly for coastal communities. In the 2000s, global reported fisheries landings were around 80 million tons (FAO, 2010), more than half of which was estimated to be caught from the Pacific Ocean (www.seaaroundus.org) (Figure 11.3). Globally, the ocean supports capture fisheries with gross revenues between US$80-85 billion annually (Sumaila et al., 2012), while gross revenues generated in the Pacific Ocean are close to 60% of the global total. The majority of the catch is from the continental shelf region, while the pelagic resources such as tunas in the Pacific also support significant fisheries.

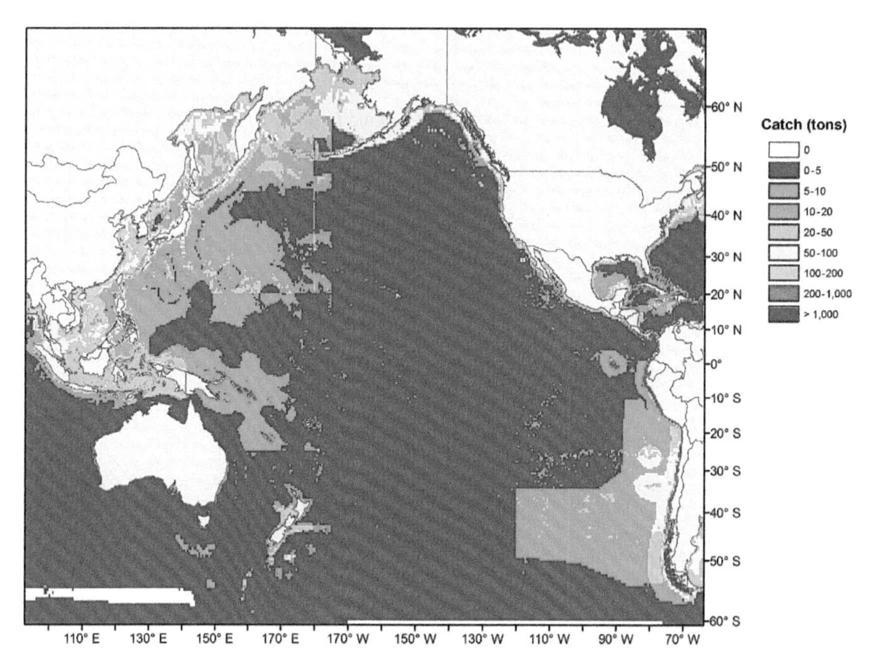

FIGURE 11.3 Estimated annual average fisheries catch from the Pacific Ocean in the 2000s from an algorithm that allocation of fisheries records onto a 30′ latitude × 30′ longitude grid (see Watson et al., 2004 for details. Data source: Sea Around Us Project).

KEY HUMAN PRESSURES IN THE PACIFIC OCEAN

Overfishing

Based on available data on fisheries catch, it appears that the fisheries resources in the Pacific Ocean are generally fully exploited. Available data show that the total amount of fish catches from the Pacific Ocean increased from around 10 million tons in the 1950s, peaked at around 50 million tons in the early 1990s, and decreased to around 45 million tons in the 2000s (Figure 11.4). If catches from China, previously suggested as largely overreported in the FAO Fisheries Statistics (Watson and Pauly, 2001), were excluded, the decline in total catch is even steeper (Figure 11.4a). The landed values from the Pacific Ocean increased from around

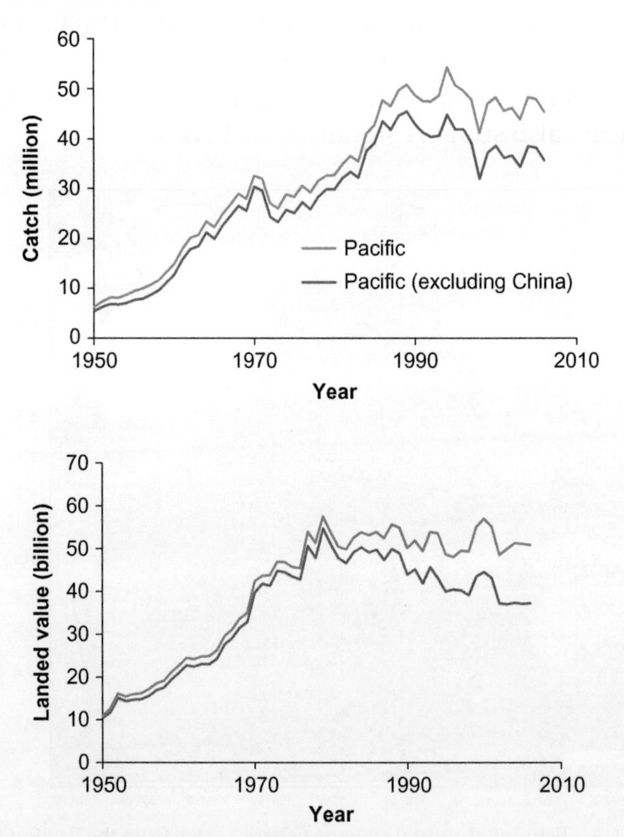

FIGURE 11.4 Estimated fisheries catch (a) and landed value (b) from the Pacific Ocean (data source: Sea Around Us Project).

US$10 billion in the 1950s to around US$50 billion by the 1990s and remained relatively stable. However, if landings from China are excluded, landed value declined gradually to around US$40 billion in the 2000s (Figure 11.4b).

Evidence suggests that fisheries in many regions in the western Pacific and the small islands developing countries (SIDC) in Oceania and Asia are overfished, while fisheries resources in the Northeast Pacific are better managed. For example, in the northern South China Sea, fisheries are underperforming in both ecological and economic objectives (Cheung and Sumaila, 2008). Abundance of 12 out of 17 studied taxa declined by over 70% in 15 years over the past four decades (Cheung and Pitcher, 2008). Vulnerable fish species such as skates, rays, and large-bodied croakers have declined by over 90% during this period (Sadovy and Cheung, 2003; Cheung and Sadovy, 2004; Cheung and Pitcher, 2008). Moreover, fisheries in many SIDC in the Pacific are also overexploited. Analysis of catch time-series trends suggests that over 50% of the fish stocks in the SIDC in Oceania and Asia are likely to be overfished, leading to an estimated loss of fisheries production of 55-70% (Srinivasan et al., 2010). In contrast, the status of the fisheries resources in the Northeast Pacific appears to be better. Based on a study analyzing fish stocks with assessment or survey data, the waters around the United States and Canada in the Pacific are considered "not overfished" and "low exploitation rate, with biomass rebuilding from overfishing" (Worm et al., 2009).

CHALLENGES TO SUSTAINABLE MANAGEMENT OF FISHERIES RESOURCES IN THE PACIFIC OCEAN

Open-access fisheries are those in which fishers may use the marine ecosystem, either totally uncontrolled or where no well-defined access rights—whether individual, communal, or state—exist or are enforced. It has been convincingly shown that under open access, the tendency is to overcapitalize fisheries, resulting in overexploitation of the resource. Since the Law of the Sea turned what used to be global commons into the property of coastal nations, in theory, there is no open-access fishery within exclusive economic zones (EEZs) in the Pacific but given the difficulty in putting in place effective management by many countries in the region, it can be argued that in this regard alone, *de facto* open access exists in the region. This is one of the fundamental causes of overexploitation in many of these regions in the Pacific.

Fishing gear and vessel technology have achieved the capacity to impact radically the Earth's marine ecosystems. Fleets have become powerful enough to overexploit essentially all stocks in the world. In fact, technological progress in fishing gear has virtually removed the natural protection afforded fish in earlier times when fishing was conducted by less powerful technology. Also, improvement in fish product preservation and transportation technology has significantly increased the scope of international trade in fish products, thereby removing market barriers to fishing. Fishers who were able to do business in a domestic market now have access to a global market with significantly higher demand for fish.

Another root cause of overexploitation is subsidies to the fisheries sector. Fisheries subsidies are the practice by governments of providing financial support, whether direct or indirect, to the fishing sector. Globally, studies put the amount of subsidies paid by governments around the world at between US$15-27 billion (Milazzo, 1998; Sumaila et al., 2010). This is a substantial amount given that the total gross revenue from the world's fisheries is estimated at between US$80-85 billion (FAO, 2010). Fisheries subsidies that are variously labeled "bad," "overfishing," "harmful," or "capacity enhancing" are recognized worldwide as serious threats to sustainable fisheries management, as they intensify overcapitalization and overfishing (Munro and Sumaila, 2002). Total subsides to fisheries in Asia are almost US$14 million (Figure 11.5a). In North America, the estimated fisheries subsides are around US$3.5 million.

Illegal, unregulated, and unreported (IUU) fishing occurs in many places—not only in the high seas but also within EEZs that are not properly regulated (Sumaila et al., 2006) (Figure 11.5b). Particularly, illegal fishing is estimated to be substantial in some regions of the Pacific. Overall, illegal catch in the Pacific Ocean in 2000-2003 is estimated to be around 3.5 to 8.1 million tons, with a value of US$3102-7312 million (Agnew et al., 2009). Illegal fishing appears to be particularly serious in the northwest and west central Pacific, amounting up to 48% of the reported catch. Illegal fishing in the Northeast and Southwest Pacific is estimated to be lowest at

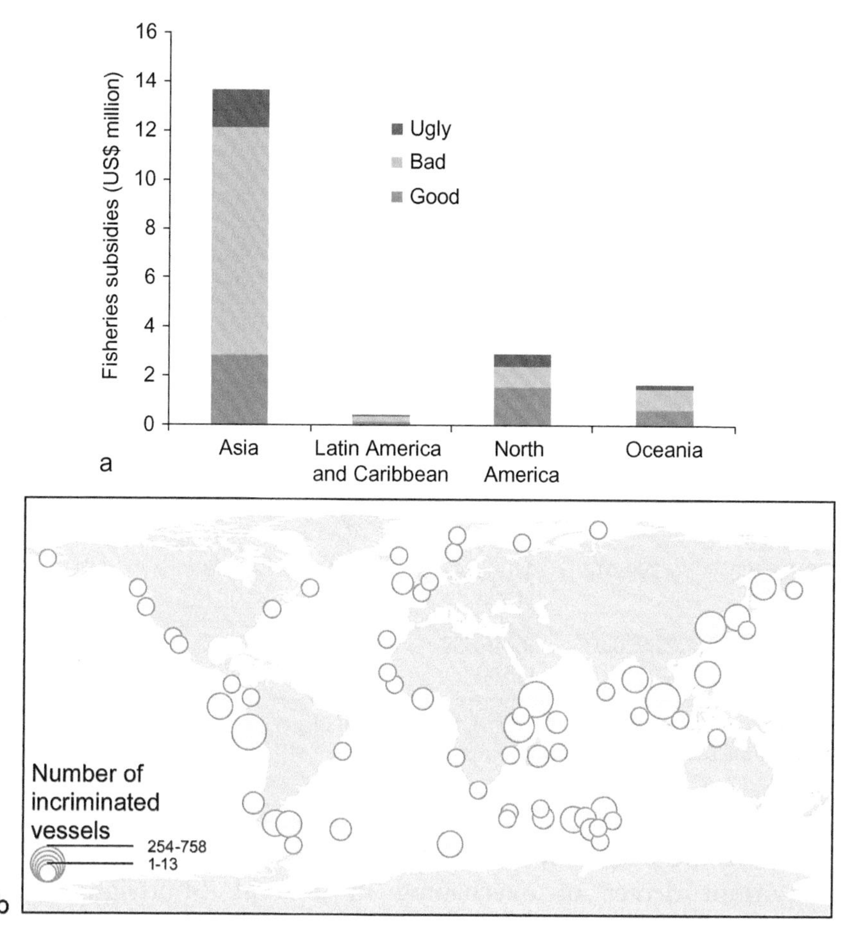

FIGURE 11.5 Estimated fisheries subsidies (a) (Sumaila et al., 2010) and the number of incriminated vessels because of illegal fishing (b) (Sumaila et al., 2006). Good subsidies consist mainly of research and management that contribute to the sustainability of the fisheries; bad subsidies are those that fund fleet capacity growth; ugly subsidies are government transfers whose impact on sustainability depends on the context (Sumaila et al. 2010). *Panel B: From Sumaila et al. (2006), figure 1. © Elsevier.*

1-7% of the total catch. IUU fishing can lead to a failure to achieve both management goals and sustainability of fisheries. When stock assessments are performed on fish stocks, reported catch and effort data are used. The underreporting of illegal catches, however, results in the absence of a significant part of the annual catch and therefore makes the stock assessments inaccurate (Pauly et al., 2002). In addition, IUU fishing distorts the market to the disadvantage of legal fishers (Table 11.1).

TABLE 11.1 Estimated Illegal Fisheries Catch by Weight and Value in the Pacific Region, Averaged over 2000-2003

		Illegal Catch		Value (US$m)	
Region	**Reported Catch**	**Lower Estimate**	**Upper Estimate**	**Lower Estimate**	**Upper Estimate**
NW Pacific	7,358,470	1,325,763 (18%)	3,505,600 (48%)	1193	3155
NE Pacific	196,587	2326 (1%)	8449 (4%)	2	8
W Central Pacific	3,740,192	785,897 (21%)	1,729,588 (46%)	707	1557
E Central Pacific	1,374,062	129,772 (9%)	278,450 (20%)	117	251
SW Pacific	451,677	5227 (1%)	32,848 (7%)	5	30
SE Pacific	9,799,047	1,197,547 (12%)	2,567,890 (26%)	1078	2311
Total	22,920,035	3,446,532 (15%)	8,122,825 (35%)	3102	7312

Estimates are from Agnew et al. (2009). Percentage of illegal catch over the reported catch of the sampled species is shown in parenthesis.

An inherent tendency to focus on short-term benefits while ignoring the long-term costs from fishing is suggested to be an important driver of overfishing in general. Shortsightedness stems from the general human perception that what is closest to us appears to be large and weighty, while size and weight decrease with our distance from things, both temporally and spatially. This human tendency to be shortsighted is captured by the economic concept of discounting—that is, the approach by which values to be received in the future are reduced to their present value equivalent using a discount rate (Koopmans, 1960). Discount rate assumptions, as used to reduce a stream of net benefits into net present value, can have a big impact on the apparent best policy or project (Nijkamp and Rouwendal, 1988). In particular, high discount rates favor myopic fisheries policies that can result in global overfishing (Clark, 1973). Essentially, this tendency drives fishers to frontload fisheries benefits resulting in overfishing and unsustainability (Sumaila, 2004).

CLIMATE CHANGE

Climate change is considered as one of the main human drivers threatening the global marine ecosystems. Anthropogenic climate change is already causing long-term changes in atmospheric and oceanographic conditions that affect marine ecosystems (Brierley and Kingsford, 2009; Sumaila et al., 2011; Doney et al., 2012). Overall, there is compelling evidence that the ocean has become warmer (with an estimated increase in average temperature of 0.2 °C within the top 300 m depth of the ocean, between the mid-1950s and the mid-1990s), with less sea ice (summer Arctic sea ice extent is decreasing at about 7.4% per decade and the Arctic Ocean may become ice-free in summer by 2030), more stratified and more acidic in the twentieth century (pH of surface ocean waters has dropped by an average of about 0.1 units from preindustrial levels, particularly in high-latitude regions), and that these trends are expected to continue in the next century under the climate change scenarios considered by the IPCC (Sumaila et al., 2011). Many regions in the Pacific Ocean have warmed considerably over the past five decades. For example, in the Northwest and Southwest Pacific, average annual sea surface temperature increased by around 0.7 and 0.3 °C, respectively, by the 2000s relative to the 1960s (Figure 11.6a). Also, available evidence indicates that climate change is expected to result in expansion of oxygen minimum zones, changes in primary productivity, changes in ocean circulation patterns, sea level rises, and increase in extreme weather events.

Ocean warming causes shifts in geographic ranges of marine organisms, affecting the distribution of marine biodiversity. For example, community-wide northward distribution shifts were recently observed on the Bering Sea continental shelf where the centers of distribution of 40 taxa of fishes and invertebrates shifted northward by an average of 34 km from 1982 to 2006, thereby causing invasion of subarctic fauna and an increase in species richness and average trophic level in an area that was formerly covered by seasonal sea ice (Mueter and Litzow, 2008). In the Northeast Pacific (including the Gulf of Alaska and California Current large marine ecosystems), sea surface temperature increased by an average of 0.3-0.6 °C from 1982 to 2006 (Belkin, 2009) and is projected to increase by 1.0-1.5 °C by 2050 relative to 2000

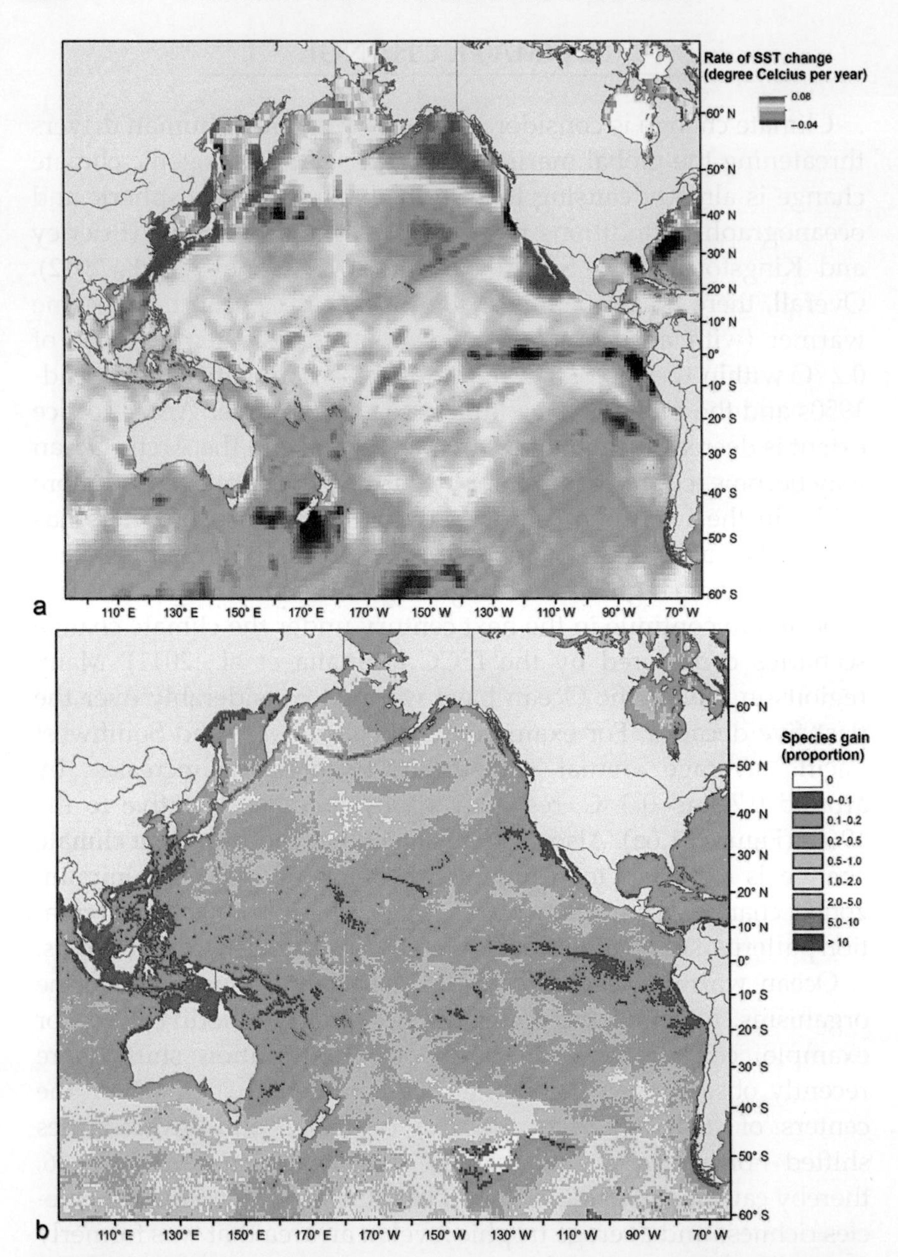

FIGURE 11.6 Rate of changes in Sea Surface Temperature (SST) in the Pacific from 1950 to 2007 (20-year average) based on data available from the Hadley Centre (a) and projected intensity of species gain (b)

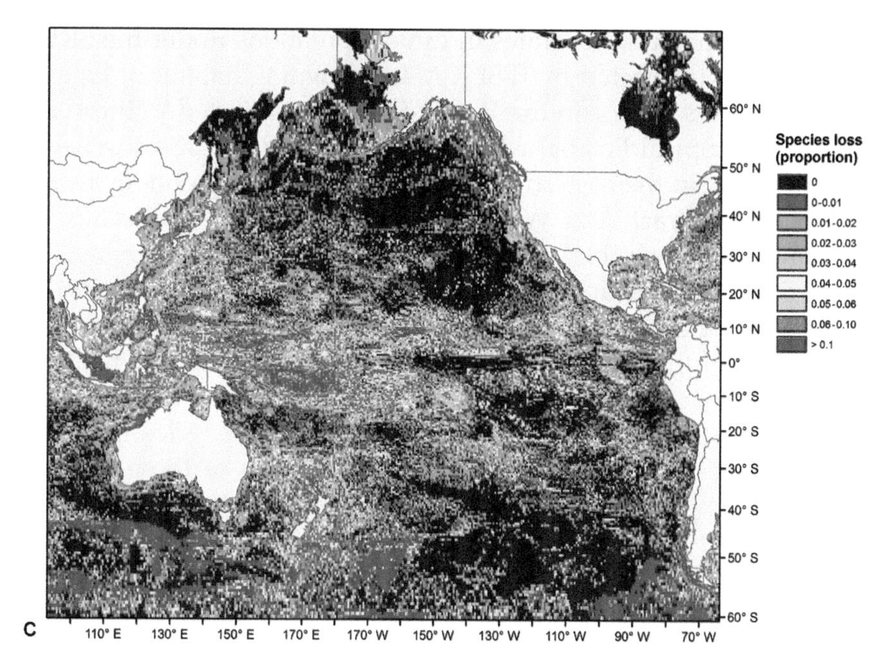

FIGURE 11.6—cont'd and loss (c) by 2050 relative to 2005 (10-year average) under the SRES A1B scenario (redrawn from Cheung et al. 2009). Intensity is expressed as proportion (prop.) of species relative to the initial species richness of the cell.

(Overland and Wang, 2007). Such changes are likely to have large, yet poorly understood, implications for species distributions and the overall biological community of this region (King et al., 2011). The decline of eulachon (*Thaleichthys pacificus*) in the southern part of its range (Moody and Pitcher, 2010) and the recent invasion of the Humboldt squid (*Dosidicus gigas*) along the west coast of North America from Central and South America may be examples of some of the expected linkages between changes in climate and the resulting oceanographic conditions (Brodeur et al., 2006; Zeidberg and Robison, 2007).

Distributions of marine species in the Pacific Ocean are projected to continue to shift in the coming decades. It is projected that distributions of all exploited marine fishes and invertebrates in the world, including those in the Pacific Ocean, will shift by 2050 under the SRES A1B scenario (Cheung et al., 2009). The distribution shift is generally poleward. Lehodey et al. (2010) use a spatial ecosystem and population dynamic model (SEAPODYM) to simulate potential changes in tuna populations under the SRES A2 scenario (projected by IPSL-CM4). Spawning habitat of bigeye tuna (*Thunnus obesus*) in the Pacific Ocean is projected to improve in the subtropical Pacific, while both spawning and feeding habitats improve in the eastern tropical Pacific regions (Lehodey et al., 2010). In contrast, the Western Central Pacific is projected to become less favorable for spawning of bigeye tuna. Using an empirical relationship developed from catch-per-unit-effort of 14 species of tunas and billfishes from long-line fisheries and SST, Hobday (2010) projects changes in tuna distributions and relative abundance on the east and west coast of Australia. The analysis uses an ensemble of 9 climate models with 25 scenarios for each species. The results suggest that the core ranges of these pelagic fishes are expected to shift toward higher latitudes (southward) and contract by 2100 relative to 1990-2000. The average rate of range shift is projected to be 40 km per decade (Hobday, 2010). Furthermore, changes in future climate are projected to favor poleward invasion of non-native species introduced by human activities, e.g., potential habitat of four marine species that have invaded the Pacific coast of North America is projected to expand northward by 2090-2099 relative to 1980-1999 under the SRES A2 scenario (de Rivera et al., 2011). The poleward shifts in distributions are expected to lead to high rate of species invasion (species moving into a new habitat relative to the number of species in the original habitat) in high-latitude regions such as the North Pacific (Figure 11.6b). In contrast, high rate of local extinction is projected for the tropical Pacific, as well as some parts of the Northeast and Northwest Pacific (Figure 11.6c).

Coral reefs are also expected to be heavily impacted by ocean warming and acidification and increased incidents of coral bleaching from more frequent extreme temperature. For example, assuming a monthly SST threshold for coral bleaching and using projected future SSTs from three different Global Circulation Models (including ECHAM4, ECHAM3, and CSIRO DAR) under the IS92a scenario, Hoegh-Guldberg (1999) projects that coral bleaching in French Polynesia, Jamaica, Rarotonga, Thailand, and at three sites on the Great Barrier Reef would occur biannually within 20-40 years from 2000. It is suggested that some coral reefs may be able to adapt to warming.

Primary production is expected to change under climate change in the Pacific Ocean. For example, using a regional fully coupled ecosystem model (NEMURO), chlorophyll-*a* concentration in the western North Pacific is projected to decrease by 0.1-0.3 mg m^{-3} in the subarctic and transition regions and by less than 0.1 mg m^{-3} in the subtropical region by 2090-2100 relative to the present day under the IS92a scenario (Hashioka and Yamanaka, 2007). Also, NEMURO projects a shift in dominant phytoplankton groups from diatoms to other smaller phytoplankton, with annually averaged percentage of diatoms decrease by 18% at the maximum in the subtropical region and 6% in the subarctic region (Hashioka and Yamanaka, 2007). In addition, an analysis that uses the GFDL ESM2.1 suggests an expansion of the subtropical biome in the North Pacific of ~30% by 2100 relative to 2000 under the SRES A2 scenario, and a contraction of the temperate and equatorial upwelling biomes of 34% and 28%, respectively (Polovina et al., 2011). Simultaneously, primary production is projected to increase by 26% in the subtropical biome and decrease by 38% and 15% in the temperate and equatorial biomes, respectively (Polovina et al., 2011).

The changes in species distribution and primary production are expected to alter potential fisheries catch in the Pacific (Cheung et al., 2010). Cheung et al. (2010) project that the tropical Pacific will suffer from a large reduction in potential catch by 2050 relative to the 2000s under the SRES A1B scenario (Figure 11.7). In contrast, high-latitude regions such as the North Pacific are projected to increase in potential catch. Similar regional analysis using ecosystem models has been carried out in the Northeast Pacific region. Regionally, using five ecosystem models representing the major marine ecosystems along the Northeast Pacific coast, Ainsworth et al. (2011) show that fisheries landings decline in response to cumulative effects of changes in ocean biogeochemistry under the SRES A1B scenario, with possible synergistic effects when multiple factors are considered.

Ocean acidification is expected to affect some marine organisms and biological processes, particularly those that involve calcifying organisms (see Chapter 2). The impacts are potentially strong particularly to organisms that form calcium carbonate exoskeletons or shells, such as molluscs and corals (e.g., Kroeker et al., 2010). Ocean acidification may also affect fishes through behavior changes

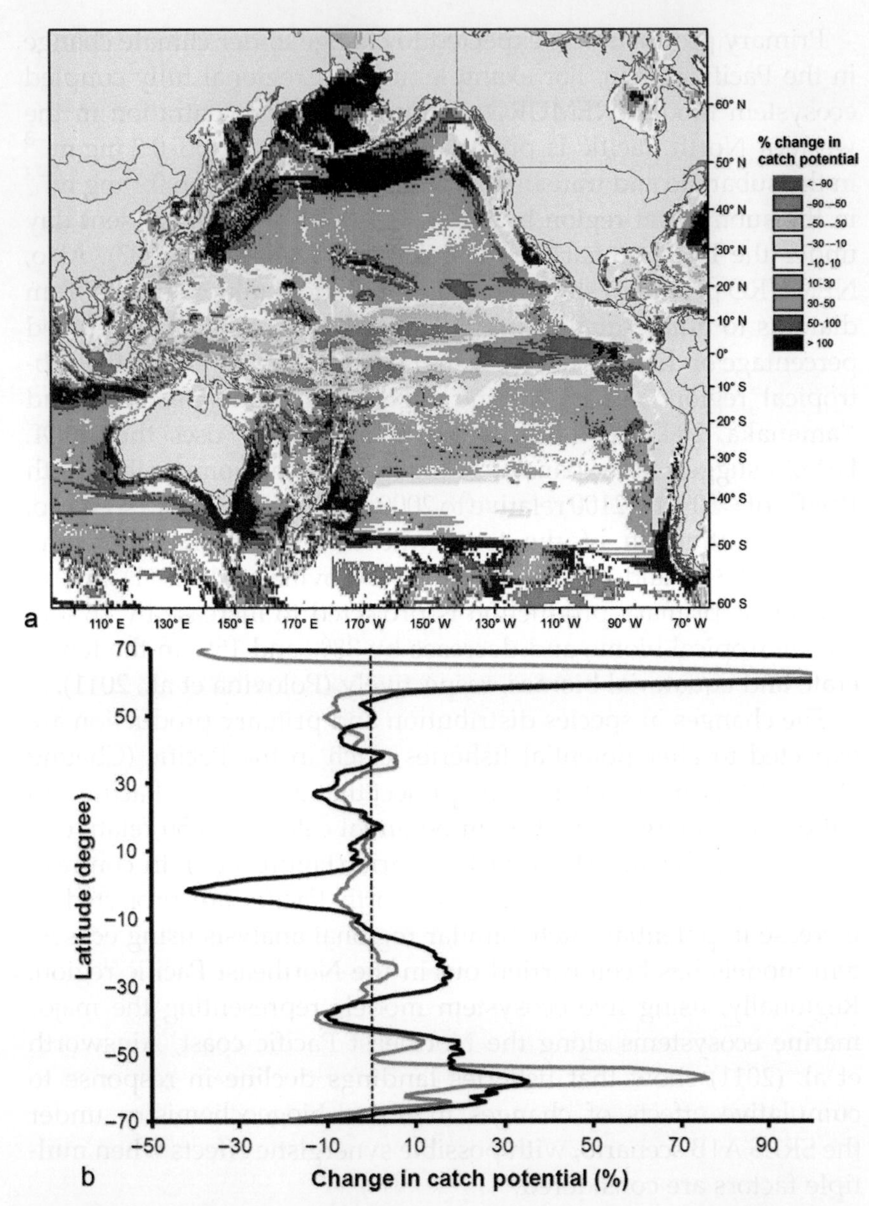

FIGURE 11.7 Projected changes in maximum catch potential by 2055 relative to 2005 (10-year average) in the Pacific Ocean: (a) map of changes in maximum catch potential under the SRES A1B scenario and (b) latitudinal zonal average changes under SRES A1B (black) and B2 (grey) scenarios (redrawn from Cheung et al. 2010).

(e.g., Munday et al., 2010) and negative effects on early life history stages (e.g., Pacific oyster). Thus, the large area of tropical coral reef in the region will become particularly vulnerable to ocean acidification. Moreover, modeling projections for marine fishes and invertebrates suggest that ocean acidification, together with warming and deoxygenation, may lead to up to 30% additional loss of catch potential (Cheung et al., 2011). Although the impacts depend strongly on the assumed sensitivity of marine organisms to ocean acidification, the findings from Cheung et al. (2011) highlight the potential impacts of ocean acidification and the effects from multiple climate stressors on marine fisheries.

Pollution and Other Human Impacts

Because of the large population living along the coast of the Pacific Ocean, pollution is a major threat to the health of the marine ecosystems. One of the most important pathways of pollution impact is through nutrient enrichment from the discharge of sewage, agriculture and industrial waste into the ocean, ultimately leading to oxygen depletion (Diaz and Rosenberg, 2008). Over 10% of the world's known eutrophic/hypoxic zones are found in the Pacific (Figure 11.8). These can directly lead to mortality of marine organisms and degradation of their habitat quality, increasing the chances of disease outbreak, and change in ecosystem structure. Main economic impacts include increased mortality of fishes and shellfishes (particularly aquaculture), reduced fish abundance and productivity, impact on tourism, and ecosystem functions (Diaz and Rosenberg, 2008).

Marine ecosystems in the Pacific are facing a range of other human impacts (Halpern et al., 2008). Rapid coastal development threatens coastal habitats and resources. This is particularly the case in the western Pacific region where the rapid economic growth and usage/modification of coastal habitats threaten the rich biodiversity and marine ecosystems in the region. High shipping volume within the Pacific and other parts of the world poses additional ecological risk on the Pacific marine ecosystems through pollution and invasion species.

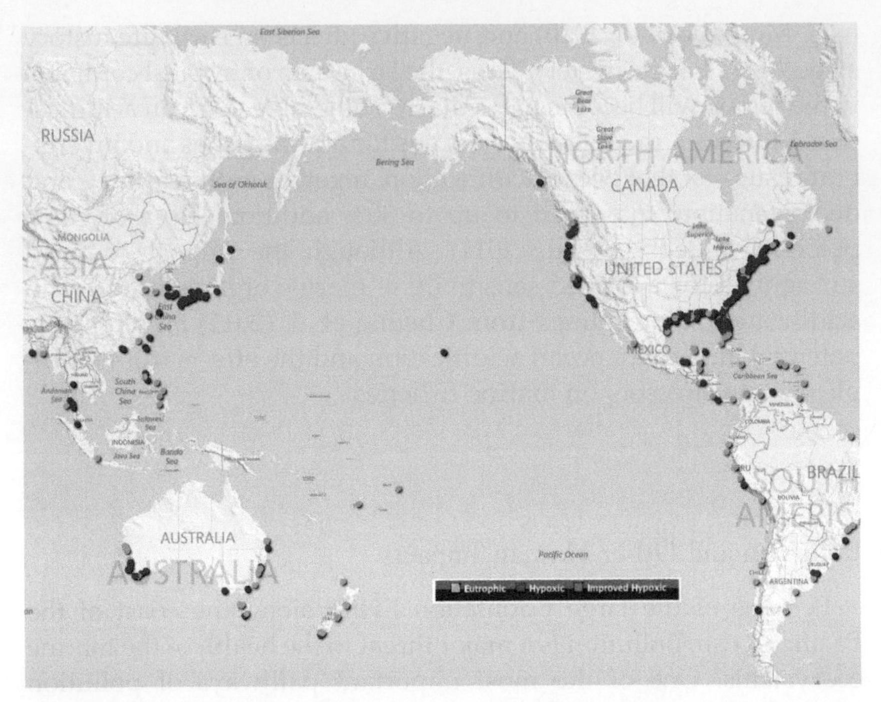

FIGURE 11.8 Eutrophic and hypoxic areas in the Pacific (Diaz and Selman 2010). *From http://www.wri.org/map/world-hypoxic-and-eutrophic-coastal-areas. © World Resources Institute.*

SUSTAINABLE MANAGEMENT OF THE PACIFIC OCEAN

Here, we summarize some of the proposed policy options from the literature that aim to ensure the sustainability of marine ecosystems for the benefits of the current and future generations of our society. First, we need to reestablish the natural protection afforded fish by establishing marine protected areas (MPAs). These would help provide the protection that has been lost as a result of technological progress (Pauly et al., 2002). MPAs would not only enable key stocks to recover but would also provide insurance against assessment errors, which are acknowledged to be a common cause of collapses. MPAs can also act as controls for large-scale monitoring of climate change impacts in the ocean where the effects of climate change can be separated from fishing and other human impacts.

The elimination of harmful subsidies to fisheries provides incentive to reduce overcapacity. The global community mandated the World Trade Organization to discipline subsidies in 2001, but it has failed because negotiators unrealistically aimed for a single all-inclusive undertaking to embrace domestic or international, and small- or large-scale, fisheries. This does not align the incentive to remove subsidies with national interests—which is a prerequisite for successful international action. To succeed, Sumaila (2012) argues that we need to divide the world's fisheries into international stocks and domestic ones (i.e., those that are fished, and spend all their lives, within EEZs). This split is necessary because the incentives for countries to eliminate subsidies differ between domestic and international situations. For the former, the heavy lifting should be at the home front, while for the latter, a coordinated international approach is needed because unilateral action by one country will not eliminate overfishing. Turning harmful subsidies to funding to support sustainable management of fisheries will ultimately result in long-term benefits of the society.

To help overcome the "open-access" problem in fisheries, there is a need for the introduction of more effective ownership structures at different levels, from the local to the national, and beyond, in the case of straddling stocks and high sea fisheries (Norse et al., 2012), or the problem that comes about due to the transboundary (Munro, 1979) or shared, i.e., common property (Munro, 1979) nature of fishery resources. The latter is particularly relevant for the management of tuna fisheries that are economically important for countries in tropical Pacific region. Game theory studies have shown that shared fish stocks need to be managed cooperatively to avoid both biological and economic waste in a fishery (Munro, 1979). Cooperative management of shared stocks takes away the economically rational tendency to overexploit fish stocks by each fishing entity.

To help tackle the problem, economists have developed a framework to examine the motivation and drivers behind IUU fishing. These models rest upon economic models developed by Gary Becker on the economics of crime (Becker, 1968). The basic idea is that a vessel contemplating engaging in IUU fishing will undertake a subjective cost-benefit analysis. The expected net return from engaging in IUU fishing will depend upon (i) the net monetary returns from harvesting and selling the fish; (ii) the expected penalty, which will in turn depend both upon the actual penalty if apprehended and the probability of being apprehended; and (iii) the cost of engaging in avoidance activities. Nonmonetary drivers include (iv) the vessel owner's moral sense and (v) his or her social standing in society.

The issue of IUU fishing has begun to receive the attention of scholars, fisheries managers, and governmental, intergovernmental, and nongovernmental organizations. With support from many countries, FAO has begun the implementation of an International Plan of Action, which encourages all states and regional fisheries organizations to introduce effective and transparent actions to prevent, deter, and eliminate IUU fishing and related activities. The Organization of Economic Cooperation and Development sponsored a Workshop on IUU Fishing Activities in April 2004 as part of their effort to help find solutions to the problem. And the United States Congress is currently working on legislation to tackle IUU fishing.

All in all, it is important to move toward ecosystem-based management of the Pacific Ocean. The Pacific Ocean is facing a multitude of human stressors, from overfishing to climate change, population, and coastal development. These stressors should be addressed and managed concurrently. Solutions to these challenges must be holistic, multisectoral, and cross-scale, putting into effect the concept of optimizing a complex system presented in Chapter 8.

Acknowledgments

W. W. L. C. is grateful for funding support from National Geographic Society and Natural Sciences and Engineering Research Council of Canada. R. S. acknowledges funding support from Social Sciences and Humanities Research of Canada. We thank the two anonymous referees and the editors for suggestions and comments on the manuscript.

References

Agnew, D. J., J. Pearce, G. Pramod, T. Peatman, R. Watson, J. R. Beddington, and T. J. Pitcher (2009), Estimating the Worldwide Extent of Illegal Fishing, *PLoS One*, **4**, e4570.

Ainsworth, C. H., J. F. Samhouri, D. S. Busch, W. W. L. Cheung, J. Dunne, and T. A. Okey (2011), Potential impacts of climate change on Northeast Pacific marine foodwebs and fisheries, *ICES J. Mar. Sci.*, **68**, 1217–1229.

Becker, C. (1968), Punishment: An Economic Approach, 76 *J. Pol. Econ.*, **169**, 183–184.

Belkin, I. M. (2009), Rapid warming of Large Marine Ecosystems, *Prog. Oceanogr.*, **81**, 207–213.

Brierley, A. S., and M. J. Kingsford (2009), Impacts of Climate Change on Marine Organisms and Ecosystems, *Curr. Biol.*, **19**, R602–R614.

Brodeur, R. D., S. Ralston, R. L. Emmett, M. Trudel, T. D. Auth, and A. J. P. (2006), Recent trends and anomalies in pelagic nekton abundance, distribution, and apparent recruitment in the Northeast Pacific Ocean, *Geophys. Res. Lett.*, **33**, L22S08.

Cheung, W. W. L., J. Alder, K. Vasiliki, R. Watson, V. Lam, C. Day, K. Kaschner, et al. (2005), *Patterns of species richness in the high seas*, Technical Series no. 20., Secretariat of the Convention on Biological Diversity Montreal.

Cheung, W. W. L., J. Dunne, J. L. Sarmiento, and D. Pauly (2011), Integrating eco-physiology and plankton dynamics into projected maximum fisheries catch potential under climate change in the Northeast Atlantic, *ICES J. Mar. Sc.*, **68**, 1008–1018.

Cheung, W. W. L., V. W. Y. Lam, J. L. Sarmiento, K. Kearney, R. Watson, and D. Pauly (2009), Projecting global marine biodiversity impacts under climate change scenarios, *Fish and Fisheries*, **10**, 235–251.

Cheung, W. W. L., V. W. Y. Lam, J. L. Sarmiento, K. Kearney, R. E. G. Watson, D. Zeller, and D. Pauly (2010), Large-scale redistribution of maximum fisheries catch potential in the global ocean under climate change, *Glob. Change Biol.*, **16**, 24–35.

Cheung, W. W. L., and T. J. Pitcher (2008), Evaluating the status of exploited taxa in the northern South China Sea using intrinsic vulnerability and spatially explicit catch-per-unit-effort data, *Fish. Res.*, **92**, 28–40.

Cheung, W. W. L., and Y. Sadovy (2004), Retrospective evaluation of data-limited fisheries: a case from Hong Kong, *Rev. Fish Biol. Fisher.*, **14**, 181–206.

Cheung, W. W. L., and U. R. Sumaila (2008), Trade-offs between conservation and socio-economic objectives in managing a tropical marine ecosystem, *Ecol. Econ.*, **66**, 193–210.

Clark, C. W. (1973), The economics of overexploitation, *Science*, **181**, 630–634.

de Rivera, C. E., B. P. Steves, P. W. Fofonoff, A. H. Hines, and G. M. Ruiz (2011), Potential for high-latitude marine invasions along western North America, *Diversity and Distributions*, **17**, 1198–1209.

Diaz, R. J., and R. Rosenberg (2008), Spreading Dead Zones and Consequences for Marine Ecosystems, *Science*, **321**, 926–929.

Doney, S. C., M. Ruckelshaus, J. Emmett Duffy, J. P. Barry, F. Chan, C. A. English, H. M. Galindo, et al. (2012), Climate Change Impacts on Marine Ecosystems, *Annual Review of Marine Science*, **4**, 11–37.

FAO (2010), *The State of World Fisheries and Aquaculture*, 197 pp., FAO, Rome.

Halpern, B. S., S. Walbridge, K. A. Selkoe, C. V. Kappel, F. Micheli, C. D'Agrosa, J. F. Bruno, et al. (2008), A Global Map of Human Impact on Marine Ecosystems, *Science*, **319**, 948–952.

Hashioka, T., and Y. Yamanaka (2007), Ecosystem change in the western North Pacific associated with global warming using 3D-NEMURO, *Ecol. Model.*, **202**, 95–104.

Hobday, A. J. (2010), Ensemble analysis of the future distribution of large pelagic fishes off Australia, *Prog. Oceanogr.*, **86**, 291–301.

Hoegh-Guldberg, O. (1999), Climate change, coral bleaching and the future of the world's coral reefs, *Mar. Freshwater Res.*, **50**, 839–866.

King, J. R., V. N. Agostini, C. J. Harvey, G. A. McFarlane, M. G. Foreman, J. E. Overland, E. Di Lorenzo, et al. (2011), Climate forcing and the California Current ecosystem, *ICES J. Mar. Sci.*, **68**, 1199–1216.

Kitchingman, A., S. Lai, T. Morato, and D. Pauly (2007), *How many seamounts are there and where are they located*, 26–40 pp., Seamounts: Ecology, Fisheries and Conservation.

Koopmans, T. C. (1960), *Stationary ordinal utility and impatience*, 287–309 pp., Econometrica: Journal of the Econometric Society.

Kroeker, K. J., R. L. Kordas, R. N. Crim, and G. G. Singh (2010), Meta-analysis reveals negative yet variable effects of ocean acidification on marine organisms, *Ecol. Lett.*, **13**, 1419–1434.

Lehodey, P., I. Senina, J. Sibert, L. Bopp, B. Calmettes, J. Hampton, and R. Murtugudde (2010), Preliminary forecasts of Pacific bigeye tuna population trends under the A2 IPCC scenario, *Prog. Oceanogr.*, **86**, 302–315.

Milazzo, M. (1998), *Subsidies in world fisheries: a reexamination*, World Bank Publications.

Moody, M. F., and T. J. Pitcher (2010), Eulachon (Thaleichthys pacificus): past and present, *Fisheries Centre Research Reports*, **18**(2).

Mueter, F. J., and M. A. Litzow (2008), Sea Ice Retreat Alters The Biogeography Of The Bering Sea Continental Shelf, *Ecol. Appl.*, **18**, 309–320.

Munday, P. L., D. L. Dixson, M. I. McCormick, M. Meekan, M. C. O. Ferrari, and D. P. Chivers (2010), Replenishment of fish populations is threatened by ocean acidification, *Proceedings of the National Academy of Sciences*, **107**, 12930–12934.

Munro, G., and U. R. Sumaila (2002), The impact of subsidies upon fisheries management and sustainability: the case of the North Atlantic, *Fish and Fisheries*, **3**, 233–250.

Munro, G. R. (1979), The optimal management of transboundary renewable resources, *Can. J. Economics*, 355–376.

Nijkamp, P., and J. Rouwendal (1988), 10 Time, discount rate and public decision making. The Formulation of time preferences in a multidisciplinary perspective: their consequences for individual behaviour and collectivedecision-making: 131.

Norse, E. A., S. Brooke, W. W. L. Cheung, M. R. Clark, I. Ekeland, R. Froese, K. M. Gjerde, et al. (2012), Sustainability of deep-sea fisheries, *Mar. Policy*, **36**, 307–320.

Overland, J. E., and M. Wang (2007), Future climate of the north Pacific Ocean, *Eos T. Am. Geophys. Un.*, **88**, 178–182.

Pauly, D., V. Christensen, S. Guenette, T. J. Pitcher, U. R. Sumaila, C. J. Walters, R. Watson, et al. (2002), Towards sustainability in world fisheries, *Nature*, **418**, 689–695.

Polovina, J. J., J. P. Dunne, P. A. Woodworth, and E. A. Howell (2011), Projected expansion of the subtropical biome and contraction of the temperate and equatorial upwelling biomes in the North Pacific under global warming, *ICES J. Mar. Sci.*, **68**, 986–998.

Sadovy, Y., and W. L. Cheung (2003), Near extinction of a highly fecund fish: the one that nearly got away, *Fish and Fisheries*, **4**, 86–99.

Srinivasan, U., W. Cheung, R. Watson, and U. Sumaila (2010), Food security implications of global marine catch losses due to overfishing, *Journal of Bioeconomics*, **12**, 183–200.

Sumaila, U., A. Khan, A. Dyck, R. Watson, G. Munro, P. Tydemers, and D. Pauly (2010), A bottom-up re-estimation of global fisheries subsidies, *Journal of Bioeconomics*, **12**, 201–225.

Sumaila, U. R. (2004), Intergenerational cost-benefit analysis and marine ecosystem restoration, *Fish and Fisheries*, **5**, 329–343.

Sumaila, U. R. (2012), Overfishing: call to split fisheries at home and abroad, *Nature*, **481**, 265.

Sumaila, U. R., J. Alder, and H. Keith (2006), Global scope and economics of illegal fishing, *Mar. Policy*, **30**(6), 696–703.

Sumaila, U. R., W. Cheung, A. Dyck, K. Gueye, L. Huang, V. Lam, D. Pauly, et al. (2012), Benefits of Rebuilding Global Marine Fisheries Outweigh Costs, *PLoS ONE*, **7**, e40542.

Sumaila, U. R., W. W. L. Cheung, V. W. Y. Lam, D. Pauly, and S. Herrick (2011), Climate change impacts on the biophysics and economics of world fisheries, *Nature Climate Change*, **1**, 449–456.

Watson, R., A. Kitchingman, A. Gelchu, and D. Pauly (2004), Mapping global fisheries: sharpening our focus, *Fish and Fisheries*, **5**, 168–177.

Watson, R., and D. Pauly (2001), Systematic distortions in world fisheries catch trends, *Nature*, **414**, 534–536.

Worm, B., R. Hilborn, J. K. Baum, T. A. Branch, J. S. Collie, C. Costello, M. J. Fogarty, et al. (2009), Rebuilding Global Fisheries, *Science*, **325**, 578–585.

Zeidberg, L. D., and B. H. Robison (2007), Invasive range expansion by the Humboldt squid, Dosidicus gigas, in the eastern North Pacific, *Proceedings of the National Academy of Sciences*, **104**, 12948–12950.

CHAPTER 12

Paths to Sustainable Ocean Resources*

Kateryna M. Wowk
U.S. National Oceanic and Atmospheric Administration, Washington, District of Columbia, USA

INTRODUCTION

The benefits provided to humanity by marine ecosystems and resources are invaluable. The ocean is the primary regulator of climate, generates half the oxygen we breathe, and mitigates global climate change. The ocean has absorbed about one-third of all anthropogenic CO_2 released to the atmosphere, as well as 80% of all heat added to the global system (IPCC, 2007). Coastal habitats provide further benefits by filtering pollutants, such as herbicides, pesticides, and heavy metals out of the water flowing through them, and by providing coastal protection and the retention of shorelines. Ocean and coastal areas are also troves of rich biodiversity and provide food for billions of people, and many coastal and ocean ecosystems exhibit unique genetic characteristics and resources with untold values for medicinal, pharmaceutical, cosmetic, and biotechnological fields.

* The author writes in her personal capacity—any views expressed do not necessarily reflect those of the U.S. National Oceanic and Atmospheric Administration.

Yet, as shown in this book, viewing the ocean and its resources as invaluable does little to provide clarity on the merits of securing continued marine ecosystem services in the face of climate and other environmental change, including ocean acidification, and whether the costs of action outweigh the costs of inaction. Throughout this study, we have seen that human-caused environmental change is already impacting the ability of marine ecosystems and coastal habitats to provide food security, secure livelihoods, and provide protection from increasingly intense weather events and storm surges, particularly in vulnerable areas and small island developing States (SIDS), and further more recent observations show that myriad impacts on the ocean far exceed the findings of the 2007 Intergovernmental Panel on Climate Change (IPCC) report.

In today's world of increasingly complex challenges and limited funds to address them, it is essential that States, regions, and the international community prioritize and concert efforts toward addressing the most pressing issues that cannot wait. Though specific efforts may vary spatially and temporally, the research clearly shows that long-standing stresses to the marine environment lessen its ability to resist and recover, and thus the time for urgent and concerted action is now. It is not only a matter of ecological health but also a matter of economic and social well-being for coastal nations, and in particular, for SIDS and vulnerable coastal communities, for which emerging and anticipated disruptions in the provision of marine ecosystem services would only impede economic development, exacerbate displacement and the exodus of refugees, and contribute to national insecurity.

This final chapter presents specific policy recommendations on the way forward to address some of the challenges presented in this book, focused at the global level but with implementation centered at the national and local levels. Though consideration of individual impacts is important, it emphasizes the need to consider multiple stressors in the marine environment and the often negatively synergistic impacts that result. Throughout the book, the authors have focused on the need for a holistic approach to understanding and dealing with the different—but closely interrelated—threats to the global ocean. This emphasis on a holistic approach is also true of our recommendations. Making well-informed decisions about marine issues at local scales that collectively move us forward on the global scale requires us to embrace the complexity of the threats and challenges described in the previous chapters. In doing so, we may find opportunities and solutions that allow us to "manage the unavoidable and avoid the unmanageable."[1] The following presents recommendations that are specific to each threat, followed by higher-level recommendations to enable and promote management strategies that move us in the right direction on the global scale.

IMPLICATIONS OF MAJOR THREATS AND POLICY RECOMMENDATIONS

The following summarizes some of the global, regional, and national implications of each threat investigated throughout the research effort, as well as high-priority policy recommendations specific to each threat. However, while it is important to consider the specific implications underlying each threat, it should be underscored that although the severity of individual threats may differ, and therefore various responses must be pursued, an overarching point remains that, in this era of multiple threats, responses must be embedded in the larger picture of change at the ecosystem and global scale. If we are to secure the sustainability of ocean resources and services, we must begin to address these complex challenges with multidimensional management schemes that can simultaneously account for optimization against more than one goal. This point is further addressed in the final section.

Ocean Acidification

Increased acidification is a growing and severe threat to the global ocean. A direct result of increased CO_2 emissions to the atmosphere over the past 200 years, about a third of which has been

[1] Scientific Expert Group on Climate Change (SEG) (2007). Confronting Climate Change: Avoiding the Unmanageable and Managing the Unavoidable [Rosina, M. B., John, P. H., Michael C. M., Richard H. M., and Peter H. R. (eds.)]. Report prepared for the United Nations Commission on Sustainable Development. Sigma Xi, Research Triangle Park, NC, and the United Nations Foundation, Washington, D.C., 144 pp.

absorbed by the ocean, ocean acidification is causing and will continue to cause damaging, rapid, and extremely long-term changes to ocean carbonate chemistry, which threaten the survival of many shelled animals and corals, and may impact fish populations (Sigler et al., 2008). Ocean acidification stands to significantly change biodiversity and ecosystems as we know them today, altering our ability to depend on the goods and services they currently provide.

Perhaps of greater concern, as emissions of CO_2 continue largely unabated, is whether the ocean's buffering capacity may also be compromised. Overall, the ocean absorbs about 2 gigatons of carbon (Gt C; 1 gigaton $= 10^{15}$ g) per annum, representing a substantial proportion of the approximate 6 Gt C released by anthropogenic activities each year (The Royal Society, 2005). Were the ocean not absorbing such a large amount of anthropogenic CO_2 released into the atmosphere each year, the effects of climate change would be felt much more rapidly. Recent evidence indicates that the greater amount of CO_2 distributed throughout the atmosphere, the lesser capacity the ocean has to buffer additional CO_2, resulting in a greater proportion of CO_2 staying in the atmosphere (Canadell et al., 2007). There is concern that the buffering capacity of this massive sink is slowing down, due to both overall higher atmospheric CO_2 levels and additional climate-related issues (e.g., ocean warming, which reduces the solubility of CO_2, thus reducing uptake by the ocean).

It is also essential to recognize that ocean and coastal areas not only absorb but also store carbon. The chemical-physical process of air-sea gas transfer of CO_2 from the atmosphere into the depths of the ocean and the activity of the biological pump carried out by plankton in the ocean are incredibly effective in absorbing, sinking, storing, and cycling carbon, with a significant total ocean reservoir of about 38,000 Gt C, or, approximately 95% of total carbon on earth (The Royal Society, 2005). Recent estimates also indicate that the ocean's vegetated habitats, especially mangroves, salt marshes, and seagrasses, are among the most effective carbon sinks on the planet, capturing and storing between 0.24 and 0.45 Gt C per annum. This is comparable to up to half of emissions from the global transport sector, estimated at about 1 Gt C per annum (Nellemann et al., 2009). Yet once degraded or destroyed, which in some areas is occurring at alarming rates, these same ecosystems can become major sources of carbon. For example, it is estimated that 13.5 Gt CO_2 will be released over the next 50 years due to the clearance of 35,000 km^2 of mangrove forests, equivalent to all transport-related emissions in the 27 European Union countries from 1997 to 2005 (World Bank et al., 2010). To assist in mitigation, action is needed to better value key "blue carbon"[2] ecosystems, halt their degradation, and catalyze restoration for those habitats that have been lost. With concerted action, marine ecosystems could be an integral component of carbon financing (Murray et al., 2010).

However, while marine habitat conservation and restoration can contribute to offsetting 3-7% of current fossil fuel emissions over two decades (Nellemann et al., 2009), such mitigation measures cannot act alone to counter escalating pressure on the global ocean from increased levels of atmospheric CO_2. As described in the previous chapter, the acidity of the ocean has already increased by 30% since the onset of the industrial revolution—this rate of change is approximately 10 times faster than ever experienced by marine ecosystems for the past 65 million years. If current rates of CO_2 emissions continue, oceanic pH could decrease further 150-200% by the end of this century. We cannot afford the rapid changes to ocean chemistry that over only a few centuries stand to dangerously impact marine organisms, food webs, ecosystems, and biogeochemistry. Global emissions of CO_2 must be substantially curbed.

Waters are already becoming increasingly inhospitable to species in certain areas of upwelling, estuaries, polar waters, coastal waters, and tropical coral reefs. Many of these same areas are important for food web processes that support key fisheries, and thus food security in some areas may also be threatened by ocean acidification. Until we implement strong and effective mitigation measures to curb CO_2 emissions, it is essential that the resilience of marine ecosystems be increased,[3] especially in vulnerable and productive

[2] According to Nellemann et al., "Out of all the biological carbon (or green carbon) captured in the world, over half (55%) is captured by marine living organisms—not on land—hence, it is called blue carbon."

[3] Though it is difficult to measure the resilience of an ecosystem, a wealth of information exists in the form of case studies, research and international guidance on agreed upon standards for fostering ecosystem resilience (e.g., see Walker and Salt, 2006; Hofmann and Gaines, 2008; Ferwerda, 2012).

areas, such that these ecosystems may be better positioned to withstand increased stress from ocean acidification.

Policy Recommendations

- To address the urgent threat of ocean acidification, it is imperative that action is taken at the global level to enact immediate and stringent reductions in CO_2 emissions.
- At the regional and national levels, actions should focus on fostering ecosystem resilience by maintaining and increasing biodiversity (Walker and Salt, 2006) and by conserving a diverse set of habitats.
- Particular attention should be given to increasing resilience of those areas that naturally exhibit a lower pH and carbonate ion concentration than the global average (e.g., high-latitude oceans (Orr et al., 2005; Steinacher et al., 2009) or upwelling zones off the west coasts of the continents (Feely et al., 2008)), and to especially vulnerable habitats (e.g., reef-building corals) that are key areas for the provision of ecosystem services that society depends upon.
- Emissions reduction is the only global means of reducing ocean acidification. To assist in mitigation efforts, however, the integration of blue carbon activities should be pursued at national and international levels, including under the framework of the United Nations Framework Convention on Climate Change (UNFCCC), should be supported by policy and financing processes, and should be included in other carbon finance mechanisms. Though blue carbon habitats represent only a small percent of the global ocean carbon sink, they are nonetheless important. To facilitate the inclusion of blue carbon in the valuation of ecosystem services, countries should work to develop a network of blue carbon demonstration projects.

Ocean Warming

The ocean has absorbed 80% of heat added to the climate system over the past 200 years (IPCC, 2007). Ocean warming has serious implications for marine ecosystems and resources and for coastal populations around the globe. There are two primary concerns regarding the impacts of ocean warming, other than sea-level rise: physical aspects related to changes in extreme weather events and biological consequences of ocean warming, including range shifts in fish populations and coral bleaching.

Regarding the physical aspects of ocean warming, observations and models point to increased intensity of precipitation that results in excessive flooding and erosion in some areas and drought in others. Though not solely the cause of ocean warming, the impacts of more water evaporating from the ocean and lakes due to a warming earth will be borne upon food security, infrastructure, and human health. More directly linked to ocean warming, there is also strong evidence supporting a future with fewer but more intense tropical storms, with one study showing the percentage of the strongest storms (category 4 and 5) increasing from less than 20% in 1970-1974 to about 35% in 2000-2004 across all major ocean basins (Webster et al., 2005), while another study projects a near double increase in the occurrence of the strongest storms over the Atlantic by 2100 (Bender et al., 2010). Coupled with higher projected sea levels, more intense tropical cyclones stand to severely damage densely populated coastal areas with dire economic consequences to coastal settlements and infrastructure, especially for some of the world's most vulnerable coastal populations that do not have the capacity to respond to such events.

The accelerated increase in ocean temperature is also impacting species distribution and food webs underlying globally significant fisheries, with range shifts of species moving toward higher latitudes (poleward) and into deeper waters. The anticipated increase in the level of stratification and changes in the thermocline may also reduce upward nutrient transport and thus the productivity of surface waters (Bindoff et al., 2007). These are key concerns—shifts in the distribution of fish, along with combined pressures already placed on marine living resources, will adversely impact food security for many coastal nations and further threaten increasingly stressed resources, which may lead to increased mortality in developing nations due to hunger and related diseases.

Coral reefs are also under threat from ocean warming, which will lead to massive bleaching and death of these valuable and highly productive ecosystems. Not only do reefs serve as nurseries for many fish species vital to commercial fishing industries, but coral reefs serve as buffers against storms, protecting many island and coastal States. For example, a 2004 study estimated that in the Caribbean region the total value of shoreline protective services provided by reefs was between US$740 million and $2.2 billion per year. Depending on how an adjacent coast was developed, this equated to protective benefits on the order of US$2000 to $1,000,000 per kilometer of coastline (Burke and Maidens, 2004).

Ocean warming is expected to have significant effects on fragile resources and economies. A primary concern is that many marine resources are already under severe human stress, in particular through overfishing. Coral reefs are also impacted on a daily basis from coastal development and growing tourism pressures. While these economic activities are vital and necessary, maintaining the resilience of marine ecosystems and proper management of coastal and open-ocean fisheries remain priorities and are ever more critical in light of the anticipated impacts of ocean warming.

Policy Recommendations

- To prepare for increased flooding and erosion resulting from more intense precipitation patterns, action should be taken at national and local levels to improve zoning laws, advance watershed management and drainage (including stormwater infrastructure), adjust planting and harvesting times, and allow for alternative crops.
- At the global scale, efforts should focus on improved forecasting, early warning systems, and deployable and effective relief efforts.
- Management strategies for high risk, high-impact events must be given higher priority, e.g., a one in 20-year event today could be one in five by 2100. In this respect, coastal zone planning and smart growth in coastal areas are crucial. Building in flood-prone areas should be discouraged.
- Considering potential range shifts in fish species and impacts on the affordability and availability of ocean-based protein resources, particularly in least developed countries, all action should be taken to lessen current stresses on these resources including overfishing, overcapacity, destructive fishing practices and damage to habitat, illegal, unregulated, and unreported fishing (IUU) and bycatch (IUU fishing is fishing

that is either expressly illegal or occurs in absence of applicable standards; see FAO, 2011). Where stocks are becoming unreliable due to ocean warming, overfishing, or a combination of factors, food security from alternative sustainable sources should be encouraged.

- The resilience of particularly vulnerable marine ecosystems, such as coral reefs, to ocean warming should be fostered by addressing additional sources of anthropogenic stress, such as overfishing, destructive fishing practices, invasive species and pollution, including through the use of marine-protected areas (MPAs). Protection also should be afforded to those ecosystems and resources most tolerable to an increase in ocean temperatures.

Hypoxia in the Ocean

Eutrophication, or the overenrichment of waters with nutrients or organic matter, has been recognized as a serious issue threatening human and environmental health for over 50 years. Globally, hypoxic zones in the ocean, one of the most acute consequences of eutrophication, have been increasing in number, endangering those ecosystems that provide humans with food by limiting supplies of oxygen to organism growth and production. There are now over 500 known hypoxic zones in the global ocean (found primarily in coastal areas), with the number increasing fastest in the developing world, while few have been reversed (Diaz et al., 2010). A primary contributor to these "dead zones" is anthropogenic nitrogen, which enters the ocean through water pathways primarily from runoff from agricultural lands and from sewage/industrial discharges. The occurrence of naturally low oxygen areas, or oxygen minimum zones, is also expanding due to global climate change and rising sea temperatures, which further exacerbates hypoxia in the ocean.

The extent to which eutrophication, and specifically hypoxia, is impacting human health and the continued provision of ecosystem services in the marine environment is largely unknown, but we know that hypoxia and dead zones result in economic losses and can shift ecosystem services away from highly valuable fisheries to those of lower values at the regional and local levels. One study estimates that the socioeconomic impacts of excessive nitrogen in the European Union alone are around US$100 billion per year and are approaching US$200-300 billion per year globally, resulting in a meaningful drain on economic development (Sutton et al., 2011). Excess nitrogen in the environment is considered one of the nine "planetary boundaries" that has already been exceeded—the limit for artificial nitrogen fixation (i.e., fertilizer production) is 35 Tg/year, yet current rates are approximately 110 Tg/year, meaning that we need a 70% reduction in the production of fixed nitrogen (IOC et al., 2011). Further, whereas nitrogen is considered to be in limitless supply, the amount of available phosphorus is fixed and finite. Thus, in both respects, the recycling of nutrients becomes an obvious and necessary measure.

Future scenarios for hypoxia will largely depend on a combination of factors related to climate change, land-use change, and changing social demographics. Changes in freshwater flow, expanding human populations (especially along the coasts), and unsustainable agricultural practices will continue to impact water quality and increase nutrient loadings. Under a business as usual scenario, projections for dissolved inorganic nitrogen entering the ocean show an increase by an additional 50% by 2050 (IOC et al., 2011), with the overall expectation that hypoxia in the global ocean will worsen, including increases in occurrence, frequency, intensity, and duration.

Approaches that enable and encourage the recovery and reuse of nutrients from waste streams are fundamental to a sustainable path. Such approaches should pursue new public-private partnerships across key sectors of society, including agriculture and wastewater management, and, in the longer-term, safeguard society and global food security by diversifying sources of nutrients. There is no other environmental variable of such ecological importance as dissolved oxygen to coastal ecosystems that has changed so drastically in such a short timeframe. Given this, and the fact that eutrophication has long been recognized as a threat to socioeconomic and ecosystem health, acute and integrated action should be taken to prevent and remediate coastal hypoxia.

It is important to note that, while the majority of the ocean's dead zones have not been reversed, there have been successful examples where concerted management efforts have resulted in the restoration of ecosystem services—currently, about 60 systems can be classified as recovering due to management efforts (Diaz et al., 2010). These cases have much to lend the global community. However, addressing the increasing number of oxygen minimum zones will be more challenging. As these areas are related to warming trends, in addition to curbing nutrient runoff, an effective solution must rely on mitigating climate change.

Policy Recommendations

- Anthropogenic oxygen-deficient dead zones in the ocean should be addressed by reducing nutrients, and particularly nitrogen inputs from land by 50% or more. This effort should include improved and integrated management of agricultural fertilization and wastes, human wastes, wastes from food industries, and emissions of nitrogen oxides from traffic and shipping. The most effective use of policy, legal, and regulatory command and control measures should be utilized to achieve this, including nutrient emissions limits, more stringent regulations on nutrient removal from wastewater, requirements for nutrient recovery and recycling, and the adoption and implementation of good agricultural practices.
- Nutrient management institutional capacity should be strengthened at the local, national, regional, and global scales. At the global scale, the Global Programme of Action for the Protection of the Marine Environment from Land-Based Activities (UNEP GPA) should be strengthened and expanded to include further capacity building and mobilization of resources for investment in treatment of human wastes and wastewater, including nutrient protocols that are river-basin wide.
- Reduction in fertilizer use and dramatic increases in fertilizer use efficiency are needed. Achieving these will require a broad-based effort involving, e.g., developing better combined irrigation/fertilization techniques, improving weather forecasting tailored for timing efficient fertilizer application, and adjusting crop types and cropping techniques to minimize fertilizer use.

- Economic instruments should be investigated to facilitate the efficient use of nutrients, including negative incentives (e.g., taxes on fertilizer and/or nutrient emissions) and positive incentives (e.g., subsidies that encourage the recovery and recycling of nutrients from waste streams). A cap and trade on manufactured nutrient production could also be examined.[4]
- New public-private partnerships across key sectors should be promoted that develop nutrient reduction technology in partnership with governments and promote innovation for nutrient recovery and reuse, including through the creation of new sectors and jobs to address this.

Sea-Level Rise

Though relative sea-level rise will differ from global sea-level rise, i.e., variations in sea level in a particular location will depend on local conditions and in many instances will differ from the global average, overall, sea-level rise is expected to exacerbate inundation, storm surge, erosion, and other coastal hazards, thus threatening vital infrastructure, settlements, and facilities that support the livelihoods of coastal and island communities. Sea-level rise from climate change is the result of a number of mechanisms, including the thermal expansion of water as heat is increasingly absorbed by the global ocean, added freshwater to the ocean from melting glaciers and ice sheets, and changes in ocean and atmospheric circulations.

Precisely how much global and relative sea level will rise is unclear, particularly due to difficulties in estimating how rapidly some of these impacts will unfold. Chapter 5 notes that if small glaciers and ice caps were to melt sea level would rise by 0.6 m, but if the Greenland ice sheet melts sea level would rise by over 7 m, and further that the melting of all the ice on Antarctica would cause a 56.6 m rise (Steffen et al., 2010). Additional processes could accelerate the melting of major ice sheets. For example, some ice shelves

[4] For example, an innovative cap-and-trade program for water quality is moving forward in the Ohio River Basin. The project, designed to establish a framework for interstate trading, will first undergo a pilot phase from 2012 to 2014, which will test different trading mechanisms in advance of regulatory drivers (EPRI, 2012).

effectively hold back even larger glaciers. If those ice shelves were to melt and flow into the ocean, these larger glaciers would more readily transfer melt and ice into the ocean. To date, we know that melting glaciers have been one of the largest contributors to sea-level rise. We also know that Greenland has witnessed an increasing net loss of mass by a factor of seven between the mid-1990s and mid-2000s, and that Antarctica witnessed almost double the ice loss in the same decade, primarily in West Antarctica and the Peninsula (Steffen et al., 2010). Although we have advanced our ability to quantify the sources of sea-level rise, it remains difficult to predict how these dynamic processes and feedback loops will continue to influence events.

Nevertheless, sea level has increased by about 25 cm since 1850, and we can confirm that this rate is accelerating: for the period 1961-2003, the rate of global mean sea-level rise was 1.8 mm/year, while from 2003 to 2007 it was 2.5 mm/year (Bindoff et al., 2007). Regarding future projections of sea-level rise, there will always be a range of possibilities dependent upon a number of factors and assumptions, and in particular upon the extent and rapidity with which glaciers, ice caps and ice sheets contribute to this rise. Regardless, it is important to note that the current rate of flow and melt into the global ocean is not appropriately incorporated into the most recent projections of the IPCC, leading many to believe that the range of possibilities in the AR4 report is at best underestimated, and that improvements in climate models are needed to better account for such interactions.

Although sea-level rise will impact coastal areas to varying extents, there are many areas that are particularly vulnerable, including coastal zones with high population densities, low elevations, high rates of land subsidence, and limited capacity. A meter rise in sea level would wreak havoc on these and other coastal areas. In particular, many of the most vulnerable areas are in the developing world, along the coasts of Africa, in South and Southeast Asia, and in Indonesia and other island nations (McGranahan et al., 2007). Many of these nations and peoples are the least responsible for greenhouse gas emissions, yet are the populations who will be most affected by the impacts of climate change, including, in some areas, the disappearance of some countries as islands become uninhabitable or heightened insecurity in others as high population, low-lying areas that are key to the functioning of the State are decimated by rising seas. Although the monetary costs of relocating a population may pale in comparison to those required to address the underlying threat of climate change through strong mitigation and adaptation measures, the social costs of a loss of national identity, of losing an entire peoples' culture and history, remain unaccounted for and should be considered.

The health effects of rising sea levels similarly remain largely unaddressed. Rising sea levels in some areas will impact coastal freshwater supplies, release toxins from coastal landfills, and increase the outbreak of disease, for example, by damaging sewers and other coastal infrastructure. Increased salination in water intakes, coastal areas, and low-lying islands will turn drinking water saline and increase treatment costs, and will further affect crops and harm aquatic animals and plants. These impacts should be given greater attention.

Policy Recommendations

- The international community should recognize that the gains that could be realized by accounting for a 1- to 2 m rise in seal level at present are even greater in terms of cost avoidance. The benefits of protection increase substantially with time, and substantial costs in terms of population displacement and land loss can be avoided through implementing protective measures now. Innovative financial instruments and funding mechanisms also must be mobilized to address this urgent need.
- States should prepare for sea level to rise at least 1 m by the end of the century by implementing measures for adaptation infrastructure and coastal defenses in vulnerable coastal areas, including through the relocation of coastal populations where necessary and possible.
- Adaptation measures should include a range of "hard" approaches, such as dikes, seawalls, reinforced structures, and ecological engineering, and "soft" approaches, including management, policy, and institutional considerations. An adaptation framework must recognize that impacts will occur at different intensities at different times, with impacts such as coastal erosion or habitat degradation often less immediate than inundation and flooding, but of no less concern.

- With urgency, the capacity of coastal and island areas to predict, understand, and respond to the risks posed by rising sea levels must be formalized and strengthened.
- International mechanisms through which low-lying island nations and coastal communities can be fairly compensated for the loss of their territory and cultural heritage should be prepared.

Pollution

Marine pollution, in its many forms, alters the physical, chemical, and biological characteristics of the ocean and coastal areas, negatively impacting the health of biodiversity and ecosystems. The three primary sources of marine pollution stem from direct discharge of effluents and solid waste on land and at sea, runoff from rivers, and atmospheric deposition. Overall, the most significant of these threats comes not from actions at sea but on land, with land-based activities contributing about 80% of all pollution entering the ocean. Globally, agriculture, sewage, urban runoff, industrial effluent, and fossil fuel combustion are the most common sources of nutrient pollution, with considerable regional variation in the relative importance of each (Selman et al., 2008). The impacts of toxic chemicals, such as DDT and PCB, and those of solid wastes (in particular plastics) have also gained much attention; however, impacts from invasive species, thermal energy, and noise pollution are disconcerting as well.

Due to a lack of data and long-term observations, as well as complex interactions that occur in the environment, it is difficult to estimate the amounts and impacts of specific pollutants, including how they interfere with ecosystem processes, especially at sea-land and sea-atmosphere boundaries. Nevertheless, enough is known about the sources, pathways, and threats of pollutants to advance actions to address harmful effects. We have long known some of the negative consequences of chemical and metal contamination in the marine environment, which in many instances has resulted in toxic effects in wildlife and humans. Humans are especially susceptible to toxicity through the consumption of contaminated seafood, leading to health effects including sensitization, allergy, neurological development deficits, and disturbance of the hormone system.

In prioritizing efforts to address these many and varied concerns, it is critical to understand the persistence, bioaccumulation potential, toxicity, radioactivity, and transboundary significance of a substance. For some chemicals, these criteria have helped to remediate impacts—at the global-, regional-, and national-level bans and restrictions on some of the most harmful chemical and metal contaminants have considerably addressed concentrations in the marine environment. However, other chemicals, for which impacts are unknown, have been introduced and are used in many areas without restriction. Further, there is very limited knowledge on the combined effects of these substances over time, and because ecosystems and wildlife are exposed to numerous substances at the same time, it is very difficult to directly link observed changes to a single contaminant. It is likewise difficult and complicated to scale up the effects of a contaminant on a single organism to the species or ecosystem level, though there is evidence that chemical contaminants affect ecosystem biodiversity and may increase susceptibility to other stressors, including disease (e.g., Khan, 2007).

Noise pollution also can have negative impacts in the marine environment, especially for marine animals that use sound for feeding, migrating, breeding, etc. Though the effects of noise in the global ocean on fish and mammals are not well understood, and there are many variables to further investigate, certain limits for lethal and sublethal effects are known. For a number of species where information is available, these limits should not be breached, especially for endangered species or populations of fish and whales.

Regarding oil pollution, there are natural releases of petroleum to the marine environment. However, direct anthropogenic releases from drilling operations and transportation, as well as indirect discharges from atmospheric deposition and municipal/industrial runoff, are of large concern. Recently, plastic and other debris that makes its way to the ocean has also been shown to accumulate and impact marine resources and ecosystems—plastic debris can be found in every region of the ocean, from the poles to the equator, at the surface and in the deep sea. While it is difficult to calculate the distribution of waste, recent studies and observations confirm that debris is subject to transport by ocean currents and tends to accumulate in a limited number of convergence zones, or gyres.

The debris causes a wide range of impacts, including lethal and sub-lethal effects on biodiversity, entanglement, chemical contamination and the alteration of community structures (Boerger et al., 2010). Various radionuclides entering the marine environment also are spread relatively widely through ocean current transport, as well as through the food chain. While global mean doses from anthropogenic radionuclides are thought to be well below acceptable levels with respect to human health, there is considerable regional variation, including some hot spot areas, which should be granted particular attention, as should for the long-term behavior of radionuclides in the marine environment.

Marine pollution has a relatively small impact on the global ocean as a whole, and thus when considering pelagic areas, the issue may appear to be diluted. However, these impacts can become more severe when considered at a more localized scale, particularly in coastal zones where pollution tends to be more concentrated and the risks substantially higher, as well as in pelagic areas, e.g., the gyres, or in the water column/along the seafloor. Most disconcerting with regard to marine pollution is the fact that we know so little of the long-term and compounding influences of varied types of pollution, especially regarding chemical and metal contamination, where only a miniscule percent of those entering the ocean are well investigated. Thresholds to contamination and pollution are not adequately anticipated or understood, including how a breach of these thresholds may impact ecosystems, human health, and the economic resources we depend upon. However, it is important to emphasize that while more detailed assessment is needed to properly allocate funding to address this issue, in many instances we have enough information to act.

Policy Recommendations

- The effects of climate change on the distribution of pollutants and contaminants are unknown, including how shifting climate zones, increased temperatures, and altered precipitation patterns will influence the behaviors, pathways, and effects of pollutants at the global and regional scales. Due to a lack of knowledge on concentration pathways and combined effects of pollutants, especially in light of climate change impacts, a precautionary approach to the issue of marine pollution should be pursued.

- A regional approach should also be pursued to the issue of marine pollution to assist in identifying research gaps, including gaps that enable assessment of the magnitude of the problem, e.g., the OSPAR Commission and resulting Quality Status Reports, among others. The capacity of regional institutions should be strengthened to identify priority hazardous substances and continue their abatement at the source, including emissions of polycyclic aromatic hydrocarbons from fossil fuels.
- At the global level, UNEP GPA should be strengthened to address pollution, including through capacity-building and increased investment for treatment of human wastes and wastewater. The international community should further support a global ban on persistent organic pollutants and worldwide control of mercury emission sources within the UN framework. Regarding noise pollution, global leaders should consider the EU Marine Strategy Framework Directive as a starting point on thresholds for noise levels that can be considered acceptable. A worldwide reduction in the production of new plastic and an increase in recycling should also be pursued to reduce the quantity of plastic debris entering the marine environment. In particular, a global public-private partnership could focus on the reduction of environmental impacts associated with single-use plastics packaging (STAP, 2011).
- Future coastal planning must consider the redistribution of sedimentation as a consequence of storms and/or flooding, as well as land-use change, including the potential release of contaminants as they become dispersed and oxygenated. Pollution impacts must further be considered in the broader context of marine ecosystem resilience.
- As action is pursued to address marine pollution, research should endeavor to better understand the effects of chemical contamination, including the combined effects of chemicals and synergies between various types of pollutants, as well as how these chemicals may behave in a changing climate.

Overuse of Marine Resources

Fishing is perhaps the most long-standing and common use of ocean resources and is the activity that most heavily exploits marine areas, with some type of fishing activity occurring in every region of the ocean. The United Nations Food and Agriculture Organization (FAO) estimates that fish provide over 3 billion people with 15% of their total animal protein intake. This particular food resource is even more critical in developing countries, SIDS, and coastal areas, where up to 90% of animal protein is provided by fish (FAO, 2010). The fishing industry is also a key economic sector in many countries—approximately 260 million people have full- and part-time jobs directly and indirectly related to global marine fisheries sectors (Teh and Sumaila, 2011). In developing countries, exports of fish are especially critical, contributing higher value than agricultural commodities in many areas (FAO, 2010).

Healthy fish stocks are a matter of food security for the globe—it is estimated that by 2050, global fish catch and aquaculture will have to grow by an additional 75 million tons per year to meet the animal protein needs of a burgeoning population (Rice and Garcia, 2011). However, global decline in marine fish catches began in the early 1990s, and today, with total catch exceeding 100 million tons per year (including discards, bycatch, and IUU fishing), the FAO estimates that 87% of stocks are fully exploited, overexploited, depleted, or recovering from depletion (SOFIA, 2012). Harmful subsidies continue to contribute to overcapacity of the fishing fleet (Sumaila et al., 2010), while inconsistencies in regional fishery management organizations (RFMOs) persist, leading to poor regulation and IUU fishing in many areas (Agnew et al., 2009). The World Bank (2009) estimated that by managing global fisheries ineffectively, we lose economic benefits on the order of US$50 billion annually. In another study, Dyck and Sumaila (2010) estimated that the total impact of marine fisheries on the current global economy could be up to US$240 billion a year.

To meet some of the challenges fisheries face in providing a steady and sustainable food source aquaculture is continuing to grow, representing the most rapidly growing animal food producing sector. In 1950, the global contribution of fish from aquaculture was less than 1 million tons, growing to 32.4 million tons in 2000 and to 52.5 million tons in 2008. In 2010, aquaculture provided almost half (45.6%) of fish consumed around the globe (FAO, 2010) and is expected to exceed 60% by 2020 (Nierenberg and Spoden, 2012). However, globally, aquaculture does not have an overarching framework, and the recent development and intensification of aquaculture have revealed a broad spectrum of associated environmental issues including domestication, introduced marine species, capture of wild stocks for aquaculture needs, unsustainable feed ingredients, excessive organic matter in the effluents and waste management, and effects on local flora and fauna, all of which put at risk the ability of aquaculture to mature into a sustainable economic sector and a stable contributor to food security (Naylor et al., 2000).

As aquaculture grows and confronts these challenges, global fisheries too must address long-standing threats to the sustainability of stocks and marine resources. In addition to overfishing and IUU fishing, destructive fishing practices damage marine habitats, causing further stress to ocean ecosystems. For example, bottom trawling, which involves dragging large, heavy nets along the sea floor, effectively decimates benthic habitats. In the deep ocean, huge reservoirs of marine biodiversity exist, for example, surrounding seamounts, hydrothermal vents, methane seeps, and deepwater corals (Rex and Etter, 2010; German et al., 2011). Organisms that thrive in these environments evolved under extreme environmental parameters and offer genetic material highly valuable for scientific discovery and commercialization. The extraction of marine genetic resources stands to greatly benefit humanity from the worldwide availability of products stemming these resources, and the contributions these products can make in fundamental areas such as improved public health, cheaper food, and new scientific knowledge, with numerous applications in pharmaceutical, biotechnological, and cosmetic fields. While the development of commercially viable products derived from marine genetic resources has, thus far, proven to be an expensive, risky, complex, and lengthy undertaking, as technology improves and access becomes more feasible, it is possible that these valuable resources also could be threatened by overexploitation.

The loss of marine biodiversity due in particular to overfishing (Pereira et al., 2010) will be exacerbated by climate change impacts. Global warming is already leading to shifts in the range and distribution of some species and is expected to further disrupt migration patterns, leading to uncertainty in the ability of coastal countries to rely on key fisheries to support their livelihoods, as well as higher risks that invasive species may cause environmental harm and destruction (e.g., Sumaila et al., 2011). Pelagic fisheries also are expected to be displaced from intense coastal upwelling zones to less productive open-ocean upwelling areas. These impacts threaten the well-being of vulnerable coastal populations. For example, along the Benguela Current research has reported a major shift of pelagic sardine and mackerel fisheries biomass south to eastward, which is likely due to climate change (Hampton and Willemse, 2012). This is concerning—the failure of some seasonally migrating species, such as sardines, to return to the area could result in the closure of South African and Namibian fish-processing facilities, which may not be able to operate at economically viable levels.

It is true that management efforts are resulting in the recovery of some fish stocks. However, in many areas of the world where infrastructure is lacking for rigorous fisheries science, monitoring, management, and enforcement, marine ecosystems and species continue to be overexploited. As shown in this book, there is a substantial net economic benefit to be gained from the recovery and rebuilding of global fish stocks, which would generate between three and seven times the mean cost of fisheries building reform and would have untold value for ecological health and socioeconomic well-being.

Policy Recommendations

- With urgency, fishing capacity and effort managed by national authorities and RFMOs should be reduced, and harmful and perverse subsidies that contribute to overfishing and excess capacity should be eliminated. As shown in UNEP's (2011) *Green Economy* report, efforts that address overcapacity should focus on large-scale vessels (e.g., through vessel buyback programs, compensation, retraining programs, etc.), among other actions to help the fisheries sector adjust to lower fishing capacity.
- The mandates of RFMOs should be reviewed and modernized, including through periodic performance reviews by an independent third party, and with monitoring of the implementation of recommendations made by reviews, including a focus on implementing an ecosystem approach to fisheries.

- States and RFMOs should confront IUU fishing by fully implementing the International Plan of Action to Prevent, Deter, and Eliminate Illegal, Unreported, and Unregulated Fishing (IPOA-IUU). States also should become party to the United Nations Fish Stocks Agreement and the FAO Agreement on Port State Measures to Prevent, Deter, and Eliminate Illegal, Unreported, and Unregulated Fishing. The use of destructive fishing practices should further be prohibited.
- Flag State control of vessels and nationals (both individuals and companies) should be improved through, *inter alia*, better compliance with, and the strengthening of, monitoring, control, and surveillance measures, including increased information-sharing, vessel-monitoring systems and observer programs, and increased participation in the International Monitoring, Control, and Surveillance Network.
- For aquaculture, ecosystem approaches should be advanced that integrate decisions on site selection and management, the selection of species and stocks suitable for local environments, the types and sources of feed, the use of veterinary medicines, discharges and emissions that can reach the marine environment, and impacts on wild species and other aspects of the environment.
- Research should be supported that explores the impacts of climate change on fisheries, including through the development of models for impacts on fish stocks and local fishing communities, and the integration of climate change projections and considerations into fisheries and aquaculture strategies at national and regional levels.
- Considering that some of the most serious impacts will be borne in developing countries, including the potential loss of fisheries due to climate change impacts, efforts should be accelerated to improve the design and implementation of ocean use agreements in the exclusive economic zones of developing countries to ensure local benefits, social equity, resource conservation, and public transparency. Capacity building and technology transfer should also be pursued to improve the capacity of developing States to achieve sustainable fisheries and aquaculture and to effectively participate in RFMOs.

- Existing international arrangements and mechanisms to manage and conserve marine biodiversity should be strengthened, and in areas beyond national jurisdiction (ABNJ) gaps in existing international instruments should be addressed.

MULTIPLE STRESSORS: PUTTING THE PIECES TOGETHER

The overarching lesson emanating from this research is that sustainable management of ocean resources and space cannot be achieved without careful consideration of the multiple stressors acting on marine ecosystems, how these stressors currently and may in the future interact, and how this may impact the efficacy of management strategies. We also must consider that many of the major impacts the ocean is facing originate on land or in the atmosphere. Currently, very few management efforts address multiple stressors in such a holistic manner. Rather, management, even that which begins on a basis of integrated coastal and ocean management and an ecosystem approach, continues to proceed largely on an issue-specific or sectoral basis. Those intersectoral management approaches that do exist, such as marine spatial planning,[5] may convey a clear picture of conflicting uses of an ocean area, but it is not clear that such approaches adequately focus on the *stressors* acting on (and potentially interacting in) an area. Approaches that focus on the cumulative effects of human activities within a marine ecosystem can paint a substantially different picture than approaches that focus on how an ocean or coastal area is being used. Without defining management targets at the ecosystem level we will continually undermine the resilience of marine ecosystems to withstand additional stress, and increasingly put at risk coastal communities around the globe.

It is critically important to better understand how these stressors interact. There are some studies, for instance, that show pollution

[5] According to UNESCO, marine spatial planning is "a public process of analyzing and allocating the spatial and temporal distribution of human activities in marine areas to achieve ecological, economic and social objectives that have been specified through a political process" (Ehler and Douvere, 2009).

can undermine the resilience of an ecosystem to climate change by decreasing the heat tolerance of organisms in an ocean area (e.g., Negri and Hoogenboom, 2011), yet these types of studies are, to date, sparse. Further, while spatial information is often available for specific locations where individual impacts are occurring, this information is rarely synthesized to convey a clear picture of where multiple impacts are occurring globally, regionally, or locally. The joint perspective of the combined effects of two or more stressors acting at the same time is largely unexamined. To this end, research should focus on advancing understanding of where multiple impacts are occurring in a holistic sense as well as on developing and utilizing ecosystem models to foster our understanding of ecosystem dynamics and the complex ways in which stressors may interact (Edwards et al., 2011). Research should also seek to conduct mesocosm experiments that use controlled environments to examine responses to multiple stressors at the ecosystem level. While these types of experiments are understandably difficult to conduct in the dynamic, ever-changing ocean environment, there are some successful examples upon which further research could be built (Russel et al., 2009). Coupled with laboratory experiments that utilize an ecosystem approach, mesocosm experiments would enhance our understanding of the interactions of multiple stressors acting on a marine ecosystem over time, including effects on keystone systems, critical processes, and tipping points in restructuring ecosystems. In conducting such research, it should be considered that these models and experiments are themselves complex. Whatever research limitations may exist, they are likely not readily transparent to resource managers, who similarly may have limited experience with understanding output data (Littell et al., 2011). Nevertheless, with adequate thought on how the user community (i.e., decision makers) will utilize research results, these models and experiments can go a long way toward reducing uncertainty.

However, and perhaps most importantly, while more research is critical to enable more effective management of ocean resources in light of multiple impacts, there are some provisions that can and must be enacted immediately to caution against the dangerous and catastrophic change marine ecosystems stand to undergo. A lack of perfect understanding must not delay the implementation of proven techniques that foster ecosystem resilience and strengthen the effectiveness of governance arrangements.

Regarding ecosystem processes, techniques that should be considered in a broader management context include, *inter alia*, scaling up efforts to implement and effectively manage MPAs.[6] The design of MPAs can be crafted to enhance the resilience of marine ecosystems, in order to better withstand climate change and other impacts (e.g., Game et al., 2008; Mumby et al., 2011). As detailed above, enhancing the resilience of marine ecosystems is critical to their ability to recover from and withstand additional stress. For example, impacts from marine pollution may undermine the resilience of ecosystems to accommodate impacts from several other single threats to the ocean, such as ocean acidification or commercial fishing. Management objectives for MPAs, such as those included in Table 12.1, should be pursued that can enhance ecosystem resilience to multiple stressors, including climate change (adapted from McLeod et al., 2009 and MPA FAC, 2010).

An additional tool that should be utilized to enhance ecosystem resilience is environmental impact assessment (EIA). If done correctly, EIAs can be useful planning tools to assess the potential impact, both positive and negative, that a proposed project or effort may have on an ecosystem and its resources, while also allowing for consideration of social and economic aspects (e.g., FAO, 2011). This decision-making tool, which is widely used throughout the world, can help to prioritize management efforts to foster resilience by setting principles, standards, and guidelines that require consideration of multiple stressors, with a view to prevent significant adverse impacts to key resources.

Whichever approaches are pursued, governance efforts must be integrated across and between institutions and instruments to provide for more effective management. As detailed throughout this book, issues affecting the marine environment are many and varied. There also are myriad international, regional, national, and local institutions with jurisdiction to address specific issues in managing

[6] An MPA is an area of "intertidal or subtidal terrain, together with its overlying waters and associated flora, fauna, historical, and cultural features, which has been reserved by law or other effective means to protect part or all of the enclosed environment" (Kelleher, 1999).

TABLE 12.1 MPA Design and Management Strategies to Enhance Ecosystem Resilience

Reduce additional stressors: Reduce anthropogenic stressors on marine ecosystems in order to increase their capacity to resist and recover from additional climate change impacts.

Protect the least exposed: Protect ecosystems that are least exposed to climate change by siting MPAs where impacts are expected to be ameliorated by local conditions.

Protect the least vulnerable: Protect ecosystems that are least vulnerable to climate change by siting MPAs where organisms are expected to be naturally more resistant or adaptable.

Protect critical areas: Protect especially valuable resources that are at risk, whether vulnerable or not, by siting MPAs to contain those resources. This could include areas that are biologically or ecologically significant, as well as areas of key economic or social importance.

Ensure diversity and size of populations: Ensure replenishment, viability, and genetic diversity of populations by designing MPAs and MPA networks to protect sufficiently large population sizes.

Account for range shifts: Site, design, and adapt MPAs and MPA networks to anticipate potential habitat or species range shifts in response to climate change.

Ensure connectivity: Ensure connectivity by siting and designing MPAs to create ecologically connected and functional networks with "corridors" or "stepping stones" that facilitate range shifts of populations and the movement of individuals and genes in response to climate change.

Spread risk: Protect a range of habitats within MPA networks, and replicate sites within those habitat types, to spread the risk of catastrophic loss due to more extreme impacts (i.e., reduce the chances habitat types will all be affected by the same impact).

ocean resources and space. However, at the ecosystem level, efforts are often largely uncoordinated and there remains a need to account for cumulative and potentially negatively synergistic effects of impacts. This is not to imply that individual efforts should cease, but rather that there is a need to scale up individual management efforts toward cross-issue, intersectoral, transboundary integration, with an overarching focus on enhancing ecosystem resilience to multiple stressors. While such efforts will be complex, it is critical that we acknowledge the entire institutional framework as an entity and, more importantly, work within it to strengthen holistic management in a wider context by strengthening its discrete parts, their streamlined interactions, and their ability to leverage resources and capabilities (Oberthur and Stokke, 2011).

Initiatives that explicitly adopt such network regime approaches to ocean governance, particularly focused on strengthening the entire governance framework at the regional level, are emerging and merit consideration. In the Wider Caribbean Region, arguably one of the most geopolitically complex regions of the world, it has become apparent that conventional approaches to ocean governance, typically in the form of commissions founded on legally binding agreements, cannot address the complexity of the region in terms of its size, wealth, and power, and where capacities and interests differ substantially. Rather, the region is pursuing innovative efforts toward reforming regional ocean management arrangements based on assessment of the fragmentation, gaps, and overlaps that occur among the organizations and arrangements, or regional ocean governance complex (CERMES, 2011). The approach uses the Large Marine Ecosystem Governance Framework as a model for regional ocean governance and seeks to identify interventions that would enhance or build key parts of the framework (Fanning et al., 2009). Reforms focus on disaggregated arrangements whereby the participation of States and organizations in ocean governance better aligns to appropriate levels of capacity and commitment for their specific level of development (Mahon et al., 2010). Notably, though substantial evidence exists that similar approaches are necessary, the operationalization of network governance will require data that demonstrate these approaches provide value added to the current array of management initiatives, as well as methods for periodic assessment and the improvement of efforts through an adaptive management process (Young, 2011). In this regard, the Global Environment Facility Transboundary Waters Assessment Programme has made considerable progress in developing methods for governance assessment in transboundary water systems, which could be looked to when constructing models for other areas and programs.[7] Further, any

[7] To view the GEF TWAP methods for governance assessments as related to oceans, see IOC-UNESCO (2011), Volumes 5 and 6. To view the conceptual background behind the assessments, see Mahon et al. (2011).

governance approach should take great care to ensure an effective framework exists for translating both biological and social sciences into management decisions, policy formulation, and broader governance issues (Mee and Adeel, 2012).

Finally, when considering multiple stressors, it is imperative to recognize that the isolated impacts of climate change will be exacerbated by additional effects, e.g., rising sea levels will cause an increase in hazardous water levels, which will be further increased in the event of a severe storm. Similarly, rising temperatures in the ocean can prevent oxygen-rich surface waters from circulating to lower depths, resulting in an increase in hypoxic events. Overall, multiple stressors in many areas will be negatively synergistic, and therefore, urgent action is needed to caution against these unknowns. Although at the global level political decisions and action necessary to combat the deleterious impacts of climate change in the ocean may be stalled, there are successful examples at the local or sub-State level where action has been taken to mitigate impacts from global concerns. In light of the current impasse in the climate negotiations and the absence of concerted efforts at the global scale, it is ever more essential that management decisions are enacted at the regional and local levels to build ecosystem resilience and provide for more effective governance, thus ensuring, to the highest degree possible, the sustained provision of marine resources and services.

The security of coastal and island populations is already on a risk trajectory due to increasing ocean acidification and warming, increasing intensity of extreme weather events, escalating incidences of hypoxia, pollution from multiple sources, and depleted living marine resources. The interactions and feedback mechanisms of these impacts, coupled with increasing scarcity of freshwater in many areas due to the influence of the greenhouse effect on the hydrological cycle, and combined with coastal population growth trends, threaten many States and, in particular, developing countries. Together, these effects may result in growing tides of environmental refugees either permanently—in the case of SIDS—or seasonally, as millions more are displaced by "natural" disasters. The scope of human security risks aggravated by anthropogenic climate and environmental change underscores the need for the global community to better safeguard socioeconomic well-being by beginning to manage for multiple stressors, and urgently so.

IMPLICATIONS OF VALUING OCEAN DAMAGES AND PLANNING FOR SURPRISE

Thus far, recommendations have focused on taking action on individual issues, but more importantly have emphasized a pressing need to develop management strategies for multiple stressors. However, given today's complex economic, social, environmental, legal, and regulatory challenges, a key concern is how ocean and coastal issues should be prioritized within the broader picture of ensuring economic well-being, and social and environmental security. We know that the economic and social values of coastal areas and the ocean are, for the most part, insufficiently researched, documented, and disseminated, resulting in diluted political will at the national, regional, and global levels to safeguard these resources. Yet in an era of global financial recession, with trade-offs necessarily occurring as we strive for fiscal stability and sustainable societies, where does protection of ocean resources and space fall on the spectrum of urgent action needed to secure the three pillars of sustainable development, i.e., economic development, social development, and environmental protection? In the broader context, what is the value of preventing further damage to the ocean, and what does this value imply for policies related to natural resource and disaster risk management?

As detailed in Chapter 10, best estimates for five categories of damages are developed under two emission scenarios (Table 10.3), for which: (a) it is possible to assign meaningful monetary values and (b) realistic policy decisions can be taken and enacted now and in the future. Under the low emissions, low-impact scenario, damages across these five issues are around US$612 billion per year, equivalent to 0.11% of world GDP in 2100. Under the high emissions, high-impact scenario, damages rise to US$1980 billion, equivalent to 0.37% of global GDP. The difference between the two, or the amount that can be saved by achieving the low emissions rather than the high emissions scenario is US$1367 billion, which is well over a trillion dollars per year by 2100, equivalent to 0.25% of world GDP.

Compared to the global economy these figures certainly matter, particularly when they are combined with the impacts of the same scenarios on agriculture, water supplies, human health, and other land-based sectors. Trillion dollar estimates of future damages, however, are not alone sufficient to foster political will, mobilize action, and catapult a shift toward implementing measures to mitigate and adapt to global environmental change; toward the sustainable management of the ocean and coasts. The deeper implication of this economic analysis is the need for better risk models, viewing the threats to the ocean through the lens of insurance rather than cost-benefit analysis. This is applicable on two levels: both creative use of insurance mechanisms to cope with extreme weather events and ensuing local disasters, and the development of long-term policies that function as insurance against catastrophic and irreversible global risks.

We need to rethink how we assess economic risks—comparison of long-run average impacts to regional or global economies is inadequate and misleading and does not account for different types of risk, e.g., the probability of an event actually occurring versus the probability of the type of impact that event would have. The early manifestations of climate change, which are already upon us, include more and bigger extreme weather events, such as an increase in the highest-intensity, most damaging hurricanes. The impacts of these events on the most exposed and vulnerable coastal communities, i.e., developing nations and SIDS, are greater than global averages suggest and may overwhelm local resources for adaptation and recovery.

Our traditional cost-benefit approaches often do not capture the effects of events that are in fact the result of several factors acting together. As one example, coral reefs are negatively impacted by ocean warming, acidification, overfishing, and excessive tourism. In today's world, all of these factors are acting simultaneously. We do not yet have the data or tools needed to properly assess the conditional probability of all of these threats acting together on specific reef communities and even less the exact nature of the impacts on specific reef systems. What we can say is that we expect the combined impacts of all of these factors acting together on reef systems to be greater than the sum of individual threats. Another example is the combined effect that the poleward movement of fish communities (due to ocean warming) and ocean acidification will have on the marine food chain in major ocean basins. Poleward migration of fish communities will displace or replace ones present today in polar ocean areas, while acidification may eliminate an entire class of marine organisms (e.g., those that form aragonite) with unknown—but undoubtedly very large—impacts on the marine food chain.

These examples illuminate some of the landscape in the area between *uncertainty* and *surprise* as discussed in Chapter 9. In these cases, proper risk assessment is not possible, as we know neither the probabilities of the various interactions occurring in a synergistic fashion nor the probabilities of the outcomes of these interactions. Nonetheless, we need to develop the capability of making subjective (but still informed) assessments of these probabilities while also building the capability to adapt to surprises into our management strategies.

The time to act is now if we are to avoid, manage, and transfer potentially devastating loss and damage related to global environmental change. While fleshing out the elements of such strategies is beyond the scope of this work, we can recommend increasing the use of one kind of risk/uncertainty framework, namely, insurance models, which can better assess economic risks associated with global environmental change. We must move from cost-based models to insurance-based models that account for combined, conditional probabilities of the effects of multiple stressors acting on the marine environment, including low probability but high-impact events. It is imperative that we adopt *ex ante* risk management models and tools without delay, with a focus on developing scenario models and insurance tools linked with disaster risk reduction and risk-transfer approaches. There are efforts that are developing new risk financing instruments that could serve as a basis for discussion, e.g., the Caribbean Catastrophe Risk Insurance Facility, a risk pooling facility designed to limit the financial impact of extreme events to Caribbean governments by providing expedited short-term liquidity.

Yet, beyond the existing threats of extreme events and local losses lies the greater unknown: the ultimate risk of reaching tipping points at which global environmental change and, in particular, those related to climate change, results in—perhaps quite rapidly—catastrophic and irreversible damages to the environment. It is, at this time, impossible to know what the probabilities are for reaching some critical and potentially disastrous tipping points with respect to marine resources, ecosystems, and processes; when these might occur; and how quickly they might advance. All too real, with incalculable economic impacts, are threats such as the extinction of all coral reef ecosystems, complete collapse of major fisheries, and/or the many meters of sea-level rise that would result from the total loss of the Greenland ice sheet. Entire ecosystems, ways of life for island and coastal populations, and ensembles of essential natural services are at risk. Analyzing such risks in a conventional cost-benefit framework is equally as impossible: both the monetary value of these existential threats and the probability of their occurrence remain unknown. While it is perhaps unknowable precisely when we will reach catastrophic tipping points, we do know that they become steadily more likely every year that emissions and climate change continue unabated.

Of course, the insurance framework does not literally apply to global threats; there is no interplanetary insurance company that will cover our destruction of the only planet we have. What is needed is the development of global policies that will "self-insure" humanity against tipping points and catastrophic losses by controlling global environmental change and degradation at a level that makes disasters less, rather than more, likely in the future. Again, detailing those policies and scenarios is outside the scope of this book, but this importunate concept cannot be overlooked as we consider the value of protecting the ocean environment.

Finally, it is critical to recognize that coastal and inland communities with ocean-based livelihoods are disproportionately at risk from the impacts of the threats we have examined, individually and holistically, and especially from sea-level rise, extreme weather events, the migration of important marine species, and ocean acidification. Specifically, coastal populations in developing countries and SIDS are extremely vulnerable—most developing countries are located in regions that are expected to be most affected by climate change, yet due to a lack of capacity, they cannot finance adaptation measures themselves. Further, there is no insurance market in many of these regions to cover at least part of the risks, such that, following a natural catastrophe, the entire economy of a nation can be set back many years.

In particular, SIDS—characterized by their small size, large ocean space, isolated locations, fragile economies that do not benefit from economies of scale, and vulnerability to externalities—are especially at risk. Sea-level rise in many areas will not only lead to saltwater intrusion of already scarce freshwater resources but will also lead to coastal erosion and the total disappearance of some beaches and low-lying islands. In Kiribati, for example, a 0.5 m rise in sea level and a reduction in rainfall of 25% would reduce the freshwater lens (floating freshwater store) by 65% (World Bank, 2000). The same increase in sea level would result in the loss of over one-third of the beaches in the Caribbean (UNDP, 2007/2008). More intense storms and hurricanes put SIDS at further increased risk, as does anticipated increases in floods—in 2080, flood risk is expected to be in the order of 200 times greater than at present for Pacific atoll countries (Nicholls et al., 1999).

In some cases, due to sea-level rise, the very existence of States and cultures is threatened. These kinds of threats are impossible to properly rationalize in purely economic terms. In the Maldives, a 1-m increase in sea level would result in the complete disappearance of the nation. The existence of Kiribati is at risk in this sense as well—President Tong of Kiribati is working with other countries to plan for the gradual evacuation of his nation of 110,360 people and has confirmed deals with Australia and New Zealand to train a small number of I-Kiribati to perform jobs those countries have trouble filling. He is also trying to find other options for Kiribati's citizens as he pursues his goal of relocating 1000 I-Kiribati per year over the next 20 years, stating "We want to move our people with as much dignity as possible. For us it is not a matter of economics, it is a matter of survival" (Tong, 2004).

The direct and indirect effects of these events on the economy, infrastructure, human well-being, and overall survival of these States, even given more modest impacts, for example, a 0.5 m sea-level rise as opposed to a 2 m rise, merit special consideration. This becomes even more important when one considers multiple stressors acting on already fragile coastal ecosystems and infrastructure. As detailed throughout this book, the benefits of protection can increase substantially over time. If a course of inaction is allowed, governments and the international community will face substantial costs in terms of population displacement, land loss, and the loss of ecosystem services that can be avoided through concerted action toward proactive and protective measures now.

PACIFIC REGION CASE STUDY—IMPLICATIONS FOR REGIONAL OCEAN GOVERNANCE

The challenges borne from multiple stressors acting on valuable marine resources and the need for expeditious action become incontrovertible when examining the Pacific Region. The Pacific Ocean is the largest ocean on Earth, borders the coastline of 50 countries or territories, harbors each of the major marine habitats, and is home to some of the highest density centers, or "hot spots," of biodiversity around the globe, most notably the coral triangle in the Indo-Pacific, which is recognized as a center of marine biodiversity for the world. As seen in the case study, the Pacific Ocean further provides hundreds of billions of dollars worth in ecosystem services each year, including over half of reported global fisheries landings at an estimated value of US$50 billion annually. Like other ocean areas, however, the Pacific suffers from overexploitation (with fisheries resources generally characterized as fully or overexploited), pollution and habitat destruction, invasive species, climate change, and ocean acidification, all of which threaten the sustained provision of these vital services to humankind. This is particularly disquieting when considering the replete dependence of many Pacific communities on coastal resources and space, as well as the potential interactions and impacts of the varied stressors, which are not yet well understood.

As shown in the case study, robust fisheries are critical to the health of Pacific ecosystems and peoples. Yet in many areas, including SIDS, they are overexploited—one study showed that over 50% of the fish stocks among SIDS in Oceania and Asia are likely to be overfished, resulting in a loss of fisheries production of about 55-70% (Sumaila and Cheung, 2009). A number of factors impacting the status of stocks have been identified, including, *inter alia*, technological advances, subsidies provided to the fisheries sector, IUU fishing, and the failure to account for and address the long-term costs of fishing.

Climate change and ocean acidification will only complicate the challenges fisheries management regimes face. Ocean warming is projected to affect the distribution of all exploited marine fishes and invertebrates in the Pacific Ocean by 2050, with shifts generally occurring toward higher latitudes. These shifts are expected to lead to high rates of species invasion in some high-latitude areas (e.g., North Pacific) and high rates of local extinction in others (e.g., tropical Pacific; Cheung et al., 2008). These changes, accompanied by acidifying waters, coral bleaching in many areas, and anticipated shifts in primary production, will likely alter potential fisheries catch throughout the region. While locally there may be winners and losers under various scenarios, the cumulative impacts may result in an overall decline in fisheries landings (Ainsworth et al., 2011). This becomes more likely under management regimes as they exist today, as by and large they lack the capability to address concurrent threats acting in synergistic ways.

There are proactive measures that can be enacted throughout the region to reduce negative consequences and rebuild fisheries in light of multiple stressors. For example, targeted establishment of MPAs could help key stocks recover, the elimination of harmful subsidies could address overcapacity at the State level, and as the case study argues, dividing fish stocks into domestic and international could foster a more coordinated approach for the latter. However, overfishing, IUU fishing, and threats to stocks are but one set of issues for the complex region. Pollution, coastal development, and other human pressures abound throughout the Pacific Region, and until the tendency toward sectoral approaches and shortsightedness is addressed and corrected, we will remain incapable of providing for true ecosystem-based management that can address multiple stressors, with a view to protect the interests of future generations.

Peoples and governments so reliant on marine resources and services, as is evident in the Pacific, cannot afford to suffer the consequences that will be brought by multiple stressors acting in a negatively synergistic fashion. A new norm should be immediately adopted that embraces an ecosystem-wide and holistic approach toward decision making, considers the broader institutional framework in a region, and works to leverage the resources and capabilities of all relevant entities within an overarching context—one that seeks to enhance ecosystem resilience to multiple stressors and thus incorporates a long-term view toward sustainable development and national security. With decisive action lacking at the global level, the duty falls to regional, State, and local authorities to reduce escalating risks to vulnerable coastal communities by prioritizing to build resilience today.

THE NEXT ERA OF GLOBAL OCEAN GOVERNANCE: PATHS TO SUSTAINABILITY

We must acknowledge that, even if emissions were stabilized at current levels, the world is already committed to additional amounts of ocean warming, acidification, hypoxia, and increased impacts to marine resources and services worldwide. The urgency for action has never been greater, and action can be taken now, despite the current impasse in a binding global agreement to drastically reduce greenhouse gas emissions and especially CO_2. As this book has shown, it is essential to economic, social, and environmental health and well-being that all efforts possible be pursued to counter some of the widespread and drastic global impacts that are assuredly approaching.

To forge a path to sustainability, we must address ocean governance gaps, secure sustainable financing for ocean and coastal areas, and further attempt to capture and incorporate the value of marine ecosystem services into our markets. Ultimately, a shift in thinking must occur regarding how human actions are managed with respect to marine resources and space. We cannot afford to manage actions for one place, one issue, one sector, one point in time, but must embrace multidimensional management schemes that consider the broader context, address multiple stressors, and allow for the optimization of more than one goal at a time. It is equally imperative that we adopt and implement *ex ante* risk management models and tools without delay, with a focus on developing scenario models and insurance tools linked with disaster risk reduction and risk-transfer approaches.

Address Global Ocean Governance Issues

Today, it is overtly clear that many global goals and targets related to the ocean and coasts are not being met. Though there has been progress in many areas, existing mechanisms at global and regional levels to address single or cross-sectoral issues are inadequate. The current institutional framework needs an infusion of commitment toward assuring the application of best practices and providing for innovation.

At the highest level, major gaps in the international ocean regime could be addressed effectively at the level of the United Nations through the designation of a UN High Commissioner for Oceans. This position would lead a high-level coordinating mechanism, reporting directly to the UN Secretary General. This effort also could examine practical relationships between the 1982 United Nations Convention on the Law of the Sea (UNCLOS), the premier international ocean-governing agreement, and other ocean-related legal instruments, e.g., 1992 Convention on Biological Diversity, 1979 Convention on Migratory Species, etc., as well as the strengthening of regional ocean governance regimes with the necessary capacity (i.e., technical, financial, human) to fulfill their mandates and ensure the effective implementation of existing commitments. Interagency coordination on ocean and coastal issues could also be advanced, as well as State-level participation in and implementation of existing multilateral agreements, including through capacity building and through the integration of ocean and coastal issues into national sustainable development strategies and frameworks.

However, while frameworks for ecosystem-based management and integrated ocean and coastal management must be established at the national level, actions cannot focus solely on the 200 nautical mile exclusive economic zones of States. Governance of the 64% of the ocean in ABNJ, currently absent a governing regime, must also be addressed. An emerging regime should pursue the adoption of shared principles; mandates for prior environmental assessment (e.g., EIAs); the establishment of MPAs in ABNJ; a mechanism for equitable benefit sharing of marine genetic resources in ABNJ; a process for addressing further governance gaps; and full implementation of UNGA resolution 61/105, which commits nations that authorize their vessels to engage in bottom fishing on the high seas to take a series of actions, as well as UNGA resolution 64/72, which places particular emphasis on conducting impact assessments for bottom fisheries on the high seas. An implementing agreement under UNCLOS could enable MPAs to be established in the high seas, provide a modality for carrying out EIAs and strategic environmental assessments in the high seas, provide a mechanism for access and benefit sharing of marine genetic resources in ABNJ, and implement necessary institutional and governance changes to ensure the protection of marine biodiversity in ABNJ.

Secure Sustainable Financing for the Ocean and Adaptation

To provide sustainable financing for the strengthening of global ocean governance initiatives, existing institutions must be fortified with specific resources and direction. The United Nations Economic and Social Council, UNFCCC, United Nations Commission on Sustainable Development, UNEP, and the World Bank, among others, must improve their ability to address the sustainable management of the ocean and coasts by developing a global financing strategy for these issues. To this end, the creation of a United Nations Secretary General Ocean Budget report could address financing needs and sources.

In particular, financing measures need to be undertaken to enhance capacity in developing countries and SIDS. Where possible, financial resources from existing and future global adaptation sources should be leveraged, intraregional and interregional (south-south) cooperation should be enhanced, and new/nontraditional resources should be explored. Linkages should also be built and strengthened among ocean-related processes in SIDS regions and the work of the Alliance of Small Island States under the UNFCCC, including advocating for effective coverage of island and ocean issues with a view to influence the IPCC and UNFCCC. To feed into the work of the IPCC, marine research related to climate change should be promoted throughout SIDS and developing countries, with a focus on providing ocean-related technology transfer, research, and capacity building.

The above actions are but a few necessary measures to provide for sustained ocean resources. Ultimately, however, sustainable financing for the ocean and adaptation must include the implementation of risk reduction and risk-transfer strategies, such as parametric insurance mechanisms at the national and regional levels. Governments should be motivated and facilitated to engage in serious and maintained dialog on the avoidance of loss and damage, with relevant professional risk management capacities, including insurance. Under current financing mechanisms, we cannot expect to meet the financial burdens that will be imposed upon us through global environmental change, especially with mounting uncertainty and an increasing likelihood of surprise. We must caution against an uncertain future by accounting for possible outcomes and associated economic risks today.

The "Blue Economy": The Importance of the Ocean and Coasts

As we have seen, a healthy ocean provides tremendous economic, social, and environmental benefits that directly support livelihoods around the globe and underpin life-sustaining processes for the planet. As governments, the private sector, NGOs, and other groups came together for the United Nations Conference on Sustainable Development (Rio+20) to "shape how we can reduce poverty, advance social equity and ensure environmental protection on an ever more crowded planet to get to the future we want,"[8] many organizations and countries promoted the "blue economy" that provides for a reduction of stressors and maintenance and/or restoration of the structure and function of marine ecosystems; recognizes the contributions that ocean products and services provide to society and the economy; and seeks to support sustainable ocean sectors and livelihoods.

Ocean-related sectors including fishing and aquaculture, ports and shipping, tourism, traditional and renewable energy, and other sectors should be examined as to their sustainability and how their frameworks and practices might be improved. Further, the international community should take a hard look at the contribution of marine ecosystem services, including biological services, regulating services, and cultural and aesthetic services. Because the values of these services are not easily captured in markets, those who have jurisdiction over these resources often do not consider their true values when choosing whether to use or clear a resource to produce goods that can be sold in the marketplace. This is a form of market failure that leads to excessive habitat and/or resource destruction. As a result, scientists, policymakers, and other concerned parties should pursue methods to change economic, social, and/or other incentives to correct the problem.

Governance and institutional deficiencies in the sustainable uses of the ocean, and particularly of the high seas, where no State has sole jurisdiction, have been major barriers toward progress in meeting global ocean and coastal-related goals. International cooperation, compliance and enforcement mechanisms, and governance mechanisms including international monitoring, accountability, and oversight are necessary to secure the conservation and sustainable use of the global ocean.

[8] See http://www.uncsd2012.org/rio20/about.html.

CONCLUDING RECOMMENDATIONS

Throughout the book and this chapter are numerous recommendations for better valuing ocean ecosystems and ensuring the sustainability of marine environments. Each chapter and issue should be given focused attention and weight, as each must be addressed if marine ecosystems and services are to be sustained. However, the overarching point is that focus on one issue concerning one area along one point in time is no longer adequate. Rather, we need to make policy decisions based on optimizing a complex system. Across all issue areas, governance and management schemes should endeavor to:

1. *Account for multiple stressors*: Accounting for multiple stressors means that management should seek to optimize against more than one goal at a time, e.g., not just address a pollution problem, or ocean acidification problem, but approach global, regional, and local issues from a large-scale, strategic point of view, through multimanagement schemes. This will no doubt be complex and difficult, but there are examples in other realms where this has been successfully implemented (e.g., control system of an airplane) and guidance exists in network and complexity theory. Such schemes should have management options for emergent properties, should be aware of the suite of problems that might impact the resource of concern (both now and in the future), and should bridge the gaps between economists, ecologists, and natural and social scientists.
2. *Adopt and implement insurance-based risk models, and develop tools to facilitate market capture of ecosystem services*: Unlike traditional cost-based models, insurance-based risk models can account for combined conditional probabilities of the effects of multiple stressors acting on the marine environment, including low probability but high-impact events. Such models should be adopted and implemented with urgency, including the development of tools that may not be predictive but are diagnostic, portraying possible outcomes, and the associated economic risks involved. Further, ocean services should be fully

included in the valuation of ecosystem functions, even considering those services that cannot, at this time, be appropriately measured (e.g., nutrient cycling). These services should be more effectively integrated into economic assessments and policies. Innovative instruments and incentives should be developed to allow these services—such as storm protection provided by mangroves and coral reefs—to be captured and appreciated by markets.

3. *Adopt a precautionary approach in the context of multiple stressors*: While there are more precedents for the precautionary approach to be used with a single stressor, it can also be applied in the context of multiple stressors. The genius of Chapter 17 of Agenda 21 was the realization that the ocean can no longer be managed as it has been traditionally, sector-by-sector, use-by-use. Instead, as Agenda 21 put it, approaches must be adopted that are "integrated in content, and precautionary and anticipatory in ambit."
4. *Utilize MPAs and EIAs to support the precautionary approach in the context of multiple stressors*: As seen above, the use of MPAs can aid in fostering ecosystem resilience and addressing uncertainty, while both prior EIA and strategic assessments are fundamental elements of the precautionary approach. Standards for environmental assessment in ABNJ have been developed for bottom fisheries (the FAO Guidelines for the Management of Deep Sea Fisheries in the High Seas), for seabed mining (by the International Seabed Authority), and in the Antarctic (by the Antarctic Treaty's Committee for Environmental Protection). For example, the 2009 UNGA Sustainable Fisheries resolution explicitly calls for States and RFMOs to ensure such assessments are conducted prior to allowing vessels to engage in bottom fishing, and requires that areas should be closed to bottom fishing until conservation and management measures have been adopted to prevent significant adverse impacts. These and other measures could guide the application of EIAs and MPAs in the context of multiple stressors, both under areas of national jurisdiction and in areas beyond national jurisdiction.

5. *Strengthen holistic management through institutional frameworks, including by ensuring regional and local issues and actions are embedded in the larger picture*: Local managers must be able to use the information generated globally to augment local management decisions. This is a critical component, as actions we pursue locally may buy us time or even mitigate global impacts. Effort should be expended to better ensure that regional and local decisions are consistent with global impacts and efforts, and are supported and built upon to improve and eventually achieve increased resilience for an ocean area. Local and municipal governments are often best placed to address these issues directly through tangible and concrete changes, and local governments are crucial connectors and coalition builders that can integrate citizens, industry, policymakers, and scientists toward a common goal. Local governments should thus be strengthened to promote the need for intersectoral coordination, especially across land-based and water-based agencies that do not typically work together. These efforts should be embedded and leveraged within the larger institutional framework for an ocean area, with an overall focus on cross-issue, intersectoral, and transboundary integration.

Of course, each of the overarching and issue-specific recommendations will require strengthened and renewed political will that prioritizes long-term needs over short-term gains. This will be difficult to foster, particularly given the resource constraints we currently face. It will similarly be undoubtedly challenging to provide for effective governance of such burgeoning complexity, especially given the sectoral nature of most management frameworks, the lack of coordination across institutions and between and within the global, regional, and local scales, and the severe lack of capacity to address even one threat in many areas. Yet we can attain these goals. We can infuse decision makers, and the public, with a better understanding of what is needed, and why, and with them take concerted and sustained action. We can instill the sense of urgency that is so direly needed to catalyze a full response in the face of constrained resources, dissipating time, and oft-dismaying complexity. For it is only by accepting and embracing this complexity, and advancing multidimensional management schemes, that we will ever be able to achieve a rapid reduction to simplicity.

Crafting, implementing, and enforcing the provisions of multidimensional management schemes may very well be as complex and challenging as rocket science. But if, more than 30 years after they left Earth, NASA's twin Voyager probes can reach the edge of the solar system, navigating the complexity of outer space, so too can we address the complexity of the ecosystems and human interactions that affect our home. We must revolutionize risk management and ocean and coastal governance. We can provide for sustainable marine resources and processes so critical to our lives, but to do so, we must commit ourselves and our capacities, shift our thinking, and act in collaboration with urgency.

References

Agnew, D. J., J. Pearce, G. Pramod, T. Peatman, R. Watson, J. R. Beddington, and T. J. Pitcher (2009), Estimating the worldwide extent of illegal fishing, *Plos One*, **4**(2), 1–8.

Ainsworth, C. H., J. Samhouri, D. Busch, W. Cheung, J. Dunne, and T. Okey (2011), Potential impacts of climate change on Northeast Pacific marine foodwebs and fisheries, *ICES J. Mar. Sc.*, **68**, 1217–1229.

Bender, M. A., T. R. Knutson, R. E. Tuleya, J. J. Sirutis, G. A. Vecchi, S. T. Garner, and I. M. Held (2010), Modeled Impact of Anthropogenic Warming on the Frequency of Intense Atlantic Hurricanes, *Science*, **327**(5964), 454–458.

Bindoff, N. L., et al. (2007): Observations: Oceanic Climate Change and Sea Level. In: IPCC. (2007). *Climate Change 2007: The Physical Science Basis. Contribution of Working Group I to the Fourth Assessment Report of the Intergovernmental Panel on Climate Change* [Solomon S., D. Qin, M. Manning, Z. Chen, M. Marquis, K. B. Averyt, M. Tignor, H. L. Miller (eds.)]. Cambridge University Press, Cambridge, United Kingdom and New York, NY, 996.

Boerger, C. M., G. L. Lattin, S. Moore, and C. Moore (2010), Plastic ingestion by planktivorous fishes in the North Pacific Central Gyre, *Marine Policy Bulletin*, **60**, 2275–2278.

Burke, L., and J. Maidens (2004), *"Reefs at Risk in the Caribbean."* World Resources Institute, Washington, D.C.

Canadell, J. G., C. Le Quere, M. R. Raupach, C. B. Field, E. T. Buitenhuis, P. Ciais, T. J. Conway, N. P. Gillett, R. A. Houghton, and G. Marland (2007), Contributions to accelerating atmospheric CO2 growth from economic activity, carbon intensity, and efficiency of natural sinks, *Proceedings of the National Academy of Science*, **104**, 18866–18870.

Centre for Resource Management, Environmental Studies (CERMES) (2011), The emerging ocean governance regime in the Wider Caribbean Region. Policy

Perspectives CERMES, UWI, Cave Hill Barbados. Available: *http://cermes.cavehill.uwi.edu/PolicyPerspectives/CERMES_Policy_Perspectives_2011_11_01.pdf*.

Cheung, W., C. Close, V. Lam, R. Watson, and D. Pauly (2008), Application of macroecological theory to predict effects of climate change on global fisheries potential, *Mar Ecol.-Prog. Ser.*, **365**, 187–197.

Diaz, R., M. Selman, and C. Chique (2010), Global Eutrophic and Hypoxic Coastal Systems. World Resources Institute. Eutrophication and Hypoxia: Nutrient Pollution in Coastal Waters, http://www.wri.org/project/eutrophication.

Dyck, A., and R. Sumaila (2010), Economic impact of ocean fish populations, *Bioecon*, **12**, 227–243.

Edwards, H., I. Elliott, C. Eakin, A. Irikawa, J. Madin, M. McField, J. Morgan, R. Van Woesik, and P. Mumby (2011), How much time can herbivore protection buy for coral reefs under realistic regimes of hurricanes and coral bleaching? *Glob. Change Biol.*, **17**, 2033–2048. http://dx.doi.org/10.1111/j.1365-2486.2010.02366.x.

Ehler, C, and F. Douvere (2009), *Marine Spatial Planning: a step-by-step approach toward ecosystem-based management*. Intergovernmental Oceanographic Commission and Man and the Biosphere Programme. IOC Manual and Guides No. 53, ICAM Dossier No. 6. UNESCO, Paris.

Electric Power Research Institute (EPRI) (2012), Ohio River Basin Trading Project. Available: http://my.epri.com/portal/server.pt?open=512&objID=423&mode=2&in_hi_userid=230564&cached=true.

Fanning, L., R. Mahon, and P. McConney (2009), Focusing on living marine resource governance: the Caribbean Large Marine Ecosystem and Adjacent Areas Project, *Coast. Manage.*, **37**, 219–234.

Feely, R. A., C. L. Sabine, J. M. Hernandez-Ayon, D. Lanson, and B. Hales (2008), Evidence for upwelling of corrosive "acidified" water onto the continental shelf, *Science*, **320**, 1490–1492.

Ferwerda, W. (2012), *Nature Resilience, Organising Ecological Restoration by Partners in Business for Next Generations,* Rotterdam School of Management, Erasmus University, i.a.w. IUCN Commission on Ecosystem Management. Available: *http://data.iucn.org/dbtw-wpd/edocs/2012-068.pdf*.

Food and Agriculture Organization of the United Nations (FAO) (2001), International Plan of Action to Prevent, Deter and Eliminate Illegal, Unreported and Unregulated Fishing. Available: http://www.fao.org/docrep/003/y1224e/y1224e00.htm.

Food and Agriculture Organization of the United Nations (FAO) (2010), The state of world fisheries and aquaculture report 2010. Available: http://www.fao.org/docrep/013/i1820e/i1820e.pdf.

Food and Agriculture Organization of the United Nations (FAO) (2011), Environmental Impact Assessment: Guidelines for FAO Field Projects. Available: http://www.fao.org/docrep/014/am862e/am862e00.pdf.

Food and Agriculture Organization of the United Nations (FAO) (2012), The state of world fisheries and aquaculture report 2012 .Available: http://www.fao.org/docrep/016/i2727e/i2727e.pdf.

Game, E., E. McDonald-Madden, M. Puotinen, and H. Possingham (2008), Should We Protect the Strong or the Weak? Risk, Resilience, and the Selection of Marine Protected Areas, *Conserv. Biol.*, **22**, 1619–1629, http://dx.doi.org/10.1111/j.1523-1739.2008.01037.x.

German, C., E. Ramirez-Llodra, M. Baker, P. Tyler, and The ChEss Scientific Steering Committee. (2011), Deep-Water Chemosynthetic Ecosystem Research During the Census of Marine Life Decade and Beyond: A Proposed Deep-Ocean Road Map, *PLoS ONE*, **6**(8), e23259, http://dx.doi.org/10.1371/journal.pone.0023259.

Hampton, I., and N. Willemse (2012): Potential Effects of Climate Change and Variability on the Resources of the Benguela Large Marine Ecosystem. In: *Frontline Observations on Climate Change and Sustainability of Large Marine Ecosystems* [Sherman K. and G. McGovern (eds.)]. UNDP. Available: http://www.undp.org/content/dam/undp/library/Environment%20and%20Energy/Water%20and%20Ocean%20Governance/Frontline%20Observations%202012.pdf

Hofmann, G. E., and S. D. Gaines (2008), New tools to meet new challenges: emerging technologies for managing marine ecosystems for resilience, *BioScience*, **58**, 43–52.

Ibid.

Ibid.

Intergovernmental Panel on Climate Change (IPCC) (2007), *Climate Change 2007: Synthesis Report. Contribution of Working Groups I, II and III to the Fourth Assessment Report of the Intergovernmental Panel on Climate Change (Fourth Assessment Report: Synthesis Report)*, 8pp., IPCC, Geneva.

IOC-UNESCO. (2010), *Methodology for the GEF Transboundary Waters Assessment Programme*. Volume 6. Methodology for the Assessment of the Open Ocean, UNEP, vi + 71 pp; Volume 5. Methodology for the Assessment of Large Marine Ecosystems, UNEP, viii + 115 pp.

IOC/UNESCO, IMO, FAO, UNDP (2011), *A Blueprint for Ocean and Coastal Sustainability* IOC/UNESCO, Paris. Available: http://www.unesco.org/new/fileadmin/MULTIMEDIA/HQ/SC/pdf/interagency_blue_paper_ocean_rioPlus20.pdf.

Kelleher, G. (1999), *Guidelines for Marine Protected Areas*, IUCN, Gland. Switzerland and Cambridge, UK. xxiv +107pp. ISBN: 2-8317-0505-3

Khan, N. Y. (2007), Multiple stressors and ecosystem-based management in the Gulf, *Aquatic Ecosystem Health & Management*, **10**, 259–267.

Littell, J., D. McKenzie, B. Kerns, S. Cushman, and C. Shaw (2011), Managing uncertainty in climate driven ecological models to inform adaptation to climate change, *Ecosphere*, **2**(9), 102, http://dx.doi.org/10.1890/ES11-00114.1.

Mahon, R., P. McConney, K. Parsram, B. Simmons, M. Didier, L. Fanning, P. Goff, B. Haywood, and T. Shaw (2010), Ocean governance in the Wider Caribbean Region: Communication and coordination mechanisms by which states interact with regional organisations and projects. CERMES Technical Report No. 40. 84pp.

Mahon, L. Fanning, and P. McConney (2011): TWAP common governance assessment. In: *Volume 1. Methodology and Arrangements for the GEF Transboundary Waters Assessment Programme, United Nations Environment Programme, at 55-61* [Jeftic L., P. Glennie, L. Talaue-McManus and J. A. Thornton (eds.)]. Available: http://twap.iwlearn.org/publications/databases/volume-1-methodology-for-the-assessment-of-transboundary-aquifers-lake-basins-river-basins-large-marine-ecosystems-and-the-open-ocean/view.

McGranahan, G., D. Balk, and B. Anderson (2007), The rising tide: assessing the risks of climate change and human settlements in low elevation coastal zones, doi:http:// dx.doi.org/10.1177/0956247807076960 20072007; 19; 17 Environment and Urbanization. Available: http://eau.sagepub.com/cgi/content/abstract/19/1/17.

McLeod, E., R. Salm, A. Green, and J. Almany (2009), Designing marine protected area networks to address the impacts of climate change, *Frontiers in Ecology and the Environment*, **7**, 362–370.

Mee, L., and Z. Adeel (2012), *Science-Policy Bridges Over Troubled Waters-Making Science Deliver Greater Impacts in Shared Water Systems*, United Nations University Institute for Water, Environment and Health (UNU-INWEH), Hamilton, Canada.

Mumby, P., I. Elliott, C. Eakin, W. Skirving, C. Paris, H. Edwards, S. Enriquez, Iglesias-R. Prieto, L. Cherubin, and J. Stevens (2011), Reserve design for uncertain responses of coral reefs to climate change, *Ecol. Lett.*, **14**, 132–140.

Murray, B., A. Jenkins, S. Sifleet, L. Pendleton, and A. Baldera (2010), *Payments for Blue Carbon: Potential for Protecting Threatened Coastal Habitats*, Nicholas Institute for Environmental Policy Solutions, Duke University. Available: *http://nicholasinstitute.duke.edu/oceans/marinees/blue-carbon*.

Naylor, R. L., R. J. Goldburg, J. H. Primavera, N. Kautsky, M. C. M. Beveridge, J. Clay, C. Folke, J. Lubchenco, H. Mooney, and M. Troell (2000), Effect of aquaculture on world fish supplies, *Nature*, **405**, 1017–1024.

Negri, A. P., and M. O. Hoogenboom (2011), Water contamination reduces the tolerance of coral larvae to thermal stress, *PLoS ONE*, **6**, e19703, http://dx.doi.org/10.1371/journal.pone.0019703.

Nellemann, C., E. Corcoran, C. Duarte, L. Valdés, C. De Young, L. Fonseca, and G. Grimsditch (eds). (2009), Blue Carbon. A Rapid Response Assessment. United Nations Environment Programme, GRID-Arendal. Available: http://www.grida.no/files/publications/blue-carbon/BlueCarbon_screen.pdf.

Nicholls, R. J., F. Hoozemans, and M. Marchand (1999), Increasing Flood Risk and Wetland Losses Due to Global Sea-Level Rise: Regional and Global Analyses, *Global Environ. Chang.*, **9**, 69–87.

Nierenberg, D., and K. Spoden (2012), Aquaculture Tries to Fill World's Insatiable Appetite for Seafood. Vital Signs Online: Worldwatch Institute. Available: http://vitalsigns.worldwatch.org/vs-trend/aquaculture-tries-fill-world%E2%80%99s-insatiable-appetite-seafood.

Oberthur, S., and O. S. Stokke [eds]. (2011). *Managing institutional complexity: Regime interplay and global environmental change*, 352 pp., MIT Press, Cambridge.

Orr, J. C., V. J. Fabry, and O. Aumont, and 24 others. (2005), Anthropogenic ocean acidification over the twenty-first century and its impact on calcifying organisms, *Nature*, **437**, 681–686.

Pereira, H., P. W. Leadley, V. Proença, R. Alkemade, J. P. W. Scharlemann, J. F. Fernandez-Manjarrés, M. B. Araújo, P. Balvanera, R. Biggs, W. W. L. Cheung, L. Chini, H. D. Cooper, E. L. Gilman, S. Guénette, G. C. Hurtt, H. P. Huntington, G. M. Mace, T. Oberdorff, C. Revenga, R. Rodrigues, R. J. Scholes, U. R. Sumaila, and M. Walpole (2010), Scenariosfor Global Biodiversity in the 21st Century, *Science*, **330**, 1496–1501.

Rex, M., and R. Etter (2010), *"Deep-Sea Biodiversity: Pattern and Scale."*, 354 pp., Harvard University Press.

Rice, J., and S. Garcia (2011), Fisheries, food security, climate change, and biodiversity: characteristics of the sector and perspectives on emerging issues, *ICES J. Mar. Sci.*, http://dx.doi.org/10.1093/icesjms/fsr041.

Russel, B. D., J. -A. I. Thompson, L. J. Falkenberg, and S. D. Connell (2009), Synergistic effects of climate change and local stressors: CO2 and nutrient-driven change in subtidal rocky habitats, *Glob. Change Biol.*, **15**, 2153–2162.

Selman, M., S. Greenhalgh, R. Diaz, and Z. Sugg (2008), *Eutrophication and Hypoxia in Coastal Areas: A Global Assessment of the State of Knowledge*, World Resources Institute, Washington, DC.

Sigler, M. F., R. J. Foy, J. W. Short, M. Dalton, L. B. Eisner, T. P. Hurst, J. F. Morado, and R. P. Stone (2008), Forecast fish, shellfish and coral population responses to ocean acidification in the north Pacific Ocean and Bering Sea: An ocean acidification research plan for the Alaska Fisheries Science Center. AFSC Processed Rep. **2008–07**.

STAP (2011), *Marine Debris as a Global Environmental Problem: Introducing a solutions based framework focused on plastic. A STAP Information Document*, Global Environment Facility, Washington, DC.

Steffen, R. H. Thomas, E. Rignot, J. G. Cogley, M. B. Dyurgerov, S. C. B. Raper, P. Huybrechts, and E. Hana (2010): Cryospheric Contributions to Sea-Level Rise and Variability, in Understanding Sea-Level Rise and Variability [Church, J. A., P. L. Woodworth, T. Aarup and W. S. Wilson (eds.)] Wiley-Blackwell, U.K, 177–225.

Steinacher, M., F. Joos, T. L. Frölicher, G. -K. Plattner, and S. C. Doney (2009), Imminent ocean acidification in the Arctic projected with the NCAR global coupled carbon cycle-climate model, *Biogeosciences*, **6**, 515–533.

Sumaila, R., and W. Cheung (2009): Vulnerability and Sustainability of Marine Fish Stocks Worldwide: With Emphasis on Fish Stocks of the Commonwealth of Nations. In: *From hook to plate: The state of marine fisheries – A Commonwealth perspective* [Bourne, R., M. Collins (eds).] 195–210.

Sumaila, R., A. Khan, A. Dyke, R. Watson, G. Munro, P. Tydemers, and D. Pauly (2010), A bottom-up re-estimation of global fisheries subsidies, *Bioecon*, **12**, 201–225.

Sumaila, R., W. Cheung, V. Lam, D. Pauly, and S. Herrick (2011), Climate change impacts on the biophysics and economics of world fisheries. Review Article, Natureology.

Sutton, M. A., C. M. Howard, J. W. Erisman, G. Billen, A. Bleeker, P. Grennfelt, H. vans Grinsven, B. Grissetti (eds.) (2011), *The European Nitrogen Assessment*, 664 pp., Cambridge University Press.

Teh, and R. Sumaila (2011): Contribution of marine fisheries to worldwide employment. Fisheries Centre, University of British Columbia. In: *Fish and Fisheries*. Blackwell Publishing Ltd.

The Royal Society (2005), Ocean acidification due to increasing atmospheric carbon dioxide Policy document 12/0. ISBN 0 85403 617 Available: http://royalsociety.org/uploadedFiles/Royal_Society_Content/policy/publications/2005/9634.pdf.

Tong. (2004), Statement by His Excellency Anote Tong Beretitenti (President) of the Republic of Kiribati, at the 59th Session of the United Nations General Assembly, 28 September 2004.

UNDP. (2007/2008), Fighting Climate Change, Human Solidarity in a Divided World. Available: http://hdr.undp.org/en/reports/global/hdr2007-8/.

UNEP. (2011), Towards a Green Economy: Pathways to Sustainable Development and Poverty Eradication. Available: http://www.unep.org/greeneconomy/greeneconomyreport/tabid/29846/default.aspx.

U.S. Marine Protected Areas Federal Advisory Committee (MPA FAC) (2010), Climate Change in the Ocean: Implications and Recommendations for the National System of Marine Protected Areas. Available: http://www.mpa.gov/pdf/helpful-resources/mpafac_tor_doi_5-3-10-1.pdf.

Walker, B., and D. Salt (2006), *"Resilience Thinking: Sustaining Ecosystems and People in a Changing World."* Island Press, Washington, D.C., USA.

Webster, P. J., G. J. Holland, J. A. Curry, and H. R. Chang (2005), Changes in Tropical Cyclone Number, Duration and Intensity in a Warming Environment, *Science*, **309**, 1844–1846.

World Bank (2000), *Cities, Seas, and Storms: Managing Change in Pacific Island Economies,* The World Bank, Washington, D.C.

World Bank (2009), The Sunken Billions: The Economic Justification for Fisheries Reform. Available: http://siteresources.worldbank.org/EXTARD/Resources/336681-1224775570533/SunkenBillionsFinal.pdf.

World Bank, IUCN, ESAPWA (2010), Capturing and conserving natural coastal carbon: Building mitigation, advancing adaptation. Available: http://siteresources.worldbank.org/EXTCMM/Resources/coastal_booklet_final_nospread11-23-10.pdf.

Young, O. R. (2011), *If an Arctic Ocean treaty is not the solution, what is the alternative?* Polar Record 8p. Cambridge University Press.

Index

Note: Page numbers followed by *f* indicate figures and *t* indicate tables.

A

Absolute *vs.* RSLC
  measurements, 98, 99*f*
  observations, locations, 98–100
  tide gauges, 98
Arrow-Hurwicz analysis, 230

B

Biological effects, ocean warming
  availability, dissolved oxygen, 59–60
  climate change impacts, marine organisms, 58, 59*t*
  seasonal cycles, 57–58
  temperature-dependent respiration and metabolism, 60, 61*f*
  thermal limits and organisms distribution, 60–62
"Blue economy"
  description, 339
  governance and institutional deficiencies, 339
  ocean-related sectors, 339
Blueprints scenarios, 232

C

Calcification, ocean acidification
  pteropod shells, 26–27
  shells and skeletons, 26
  tropical coral reefs, 26
  upwelling, CO_2 levels, 27
Carbon dioxide (CO_2) emissions
  acidification, 195
  carbonate ion concentration, 195–196
  chemical and physical changes, 197*f*, 198
  fossil fuels, 195
  greenhouse gas, 196–198
  ocean's pH, 195–196
  Revelle factor, 195–196
  saturation state, 195–196, 197*f*
Chemical pollution
  baltic herring and salmon, 133
  bans and restrictions, 135
  bioaccumulation efficiency, 131
  climate change (*see* Climate change)
  fish, health benefits, 133
  human activities, 129
  human health, 132–133
  immune dysfunction, 132
  industrial emissions, 130
  inorganic compounds, 131
  LRTAP, 135
  mammals and predatory birds, 132
  Manila Bay, Philippines, 130
  ocean acidification, 134–135
  organic compounds, 131
  persistence, 130–131
  production and use, emissions, 129–130
  recreational activities, 133
  Stockholm Convention, 135
  toxic effects, 131–132
Climate change
  acidification, ocean, 291–293
  compelling evidence, 287
  coral reefs, 290
  fisheries, ocean environment
    catch volumes, high latitudes, 252
    climate-related variability, 253
    direct effects, warming, 253–254
    economic analysis, 253
    ENSO, 252
    food expenditures, 254
    mollusk fisheries, 253
    ocean acidification, 252
    phytoplankton growth, 253
    pole-ward migration, 252
    vulnerability, 252
  food web structures, 134
  high-latitude regions, 288*f*, 289–290
  human impacts, 293
  immune and reproduction systems, 134
  marine fish stocks
    affect, ocean changes, 179–180
    commercial species, 180

Climate change *(Continued)*
distributions, exploited species, 180
effectiveness, conservation plan, 181
global greenhouse gas emissions, 179–180
ocean primary production, 180
resource availability, 180
weather conditions, 181
NEMURO projects, 291
ocean warming, 287–289
persistent organic pollutants, 133–134
pollution, eutrophic/hypoxic zones, 293, 294*f*
primary production, 291
SEAPODYM, 289–290
sea surface temperature, 287, 288*f*
spawning habitat, *Thunnus obesus*, 289–290
SRES A1B scenario, 291, 292*f*
temperature, 133–134
tourisms, ocean environment
Caribbean islands, 255
commercial anticipation, 256
coral bleaching, 256
coral reefs, 256
current warm-weather destinations *vs.* gains, 254–255
description, 254
econometric analysis, international, 255
income elasticity, demand, 255
meta-analysis, 257
net benefits, 257
and recreation value, 257
wind fields and speed, 134
CMIP3. *See* Coupled Model Intercomparison Project Phase 3 (CMIP3)
Coupled Model Intercomparison Project Phase 3 (CMIP3), 46–47

E

Economic consequences, SLR
benefits, protection, 122
damage components, 120–121, 121*f*
fighting and fleeing, 120
levels, 120
protection costs, 121–122
purpose, 120
weakness, 122
Economic effects, hypoxia
fisheries landing data, 86, 87
mortality, oysters, 85–86
multiple stressors, 85
recreational fishing, 86
water quality and nutrient cycling, 87
Ecosystem approaches
aquaculture, 322
coastal and ocean management, 323
to fisheries, 321
mesocosm experiments, 323–324
Ecosystem resilience
biodiversity and diverse set of habitats, 306
effectiveness, governance arrangements, 324–325
EIA, 325
MPA design and management strategies, 326*t*
multiple stressors, 325
pollution impacts, 318
Ecosystem services
natural barriers, cyclone damages, 262
nutrient cycling, 248
ocean values, 247, 248*t*
significance, biodiversity, 249
substantial economic value, 248–249
TEEB Database, 249, 250*t*
tourism revenues, 256
Ecosystem valuation
climate economics and science literature, 243
oceans
atmospheric CO_2 concentrations, RCPs, 4, 5*f*
economic activities, 1
economic and earth system function value, 9
expert surveys, 9, 10*f*
feedback processes, 3
global-scale economic valuation, 7–8
global to regional aspects, 8
impacts and climate research communities, 4
knowledge, decision support, 12–13
matrix elements, 11–12
multiple stressors, 3
"plain vanilla", 3
policy decisions, 3
potential impacts, 5–6
RCP2.6 and RCP6 scenarios, 6
RCPs vs. SRES scenarios, 4–5, 6*f*
scope, 2
sea levels, 2
socioeconomic and demographic development, 3–4

5th IPCC Assessment Report, purposes, 4
ocean services, TEEB Database, 250*t*, 251
EIA. *See* Environmental impact assessment (EIA)
Environmental consequences, hypoxia
Drammensfjord, 81–82
elimination, benthic prey, 83–84, 84*f*
faunal response, 82
fish and shrimp, 83
"Jubilees", 81–82
mortality effects, 85
planetary boundary, 74*t*, 81
population-level effects, 81, 82*t*
secondary production, 83
stressors, 81
Environmental impact assessment (EIA), 325
Eutrophication, hypoxia
"dead zones", 69
definition, 69
Drammensfjord, 73
economic effects, 85–87
ecosystem function and services, 68, 69*t*
environmental consequences, 81–85
fertilization, marine systems, 70–71
fishes and invertebrates, 71–73
and global change, 87–90
global patterns, 73–78
harmful algal blooms, 71
land-use activities, 68–69
nitrogen cycles, land and ocean, 71, 72*f*
OMZs, 78–81
oxygen solubility, 69–70
restoration, 90–92
salmonid fishes, 70
Extreme weather
changes, atmospheric circulation, 47
CMIP3 models, 46–47
description, 46
flooding, greenhouse gas emissions, 49
global climate, 46
impacts and potential mitigation options, 47–49, 49*t*
RX1D and RX5D precipitation, 46–47, 48*f*
water vapor, 46

F

Fisheries
aquaculture, 173
commodities, 171–172
diet component and food security, 172–173
discount rate assumptions, 286
elimination, harmful subsidies, 295
fishing gear and vessel technology, 284
illegal/unreported catches, 171
importers, 171–172
IUU fishing (*see* Illegal, unregulated and unreported (IUU) fishing)
low-income, food-deficit countries, 171–172
manufactured goods, 171–172
ocean environment
A1B *vs.* continuation of year 2000 conditions, 267
and climate change (*see* Climate change)
high and low emissions, 267
price stability, 267
open-access, 283–284, 295
shortsightedness stems, 286
SIDC, 283
small-island developing states and coastal areas, 172–173
subsidies, 284, 285*f*
wheat/rice production, 173
World Bank and UN FAO, 172–173
Food webs and ecosystems
benthic community, 33–34
growth rates and toxicity, *Pseudo-nitzschia multiseries*, 33
mounting evidence, 34–35
Pisaster ochraceus and *Crossaster papposus*, 33
reef development, 33–34
zooplankton, 32–33

G

Glacier and ice sheet melting
atmospheric CO_2 concentrations, 102
description, 122
elevation, 104, 104*f*
freshwater, 102
ice volume, 102, 103*f*
mass loss, Greenland and Antarctica, 104, 105*t*
measurements, 103, 104, 104*f*
volume change and cumulative contribution RSL, 105, 106*f*

Global change and hypoxia
environment, society and economies consequences, 90
global warming, 88
influence, climate drivers, 88, 89*t*
land management, 89–90
predictions and observations, 87–88
regional scales, 87–88
wind patterns, 90
Global marine ecosystems
chemical pollution, 159–160
climate-change scenarios, 157–158
cocktail effects, 160
food quality and access, 159
fossil fuels, 158
population growth, 158
production, use and dissemination, 158–159
resilience, 160
stressors, pollutants, 157
waste dumping, shipping and sewage discharge, 159
Global ocean governance
"blue economy", 339
ex ante risk management models, 336
greenhouse gas emissions, 336
issues, 336–337
ocean and coastal areas, 336
secure sustainable financing, 338
Global patterns, hypoxia
development, 76, 77*f*
green revolution, 76
human population, 76–78
industrial revolution, 76
nitrogen deposition, 76–78, 78*f*
planetary boundary processes, 73–76, 74*t*
Global-scale economic valuation
categorization, 8
monetary *vs.* nonmonetary damages, 7
natural resources, 7
thresholds and discontinuities, 7–8
Global-scale stressors
CO_2 emissions (*see* Carbon dioxide (CO_2) emissions)
dissolved oxygen, 199
Earth System Model, 197*f*, 199–200
global warming (*see* Global warming)
NCAR CSM 1.4 model, 197*f*, 199
vulnerability, 200, 200*f*
Global warming
chemical and physical changes, 197*f*, 198
greenhouse gas, 196–198
magnitude, sea-ice loss, 198–199
NCAR CSM 1.4 model, 197*f*, 198–199
nutrients supply, 196–198
ocean deoxygenation, 196–198
upper-ocean stratification, 197*f*, 199

H

Hypoxia
dissolved oxygen, 201
eutrophication (*see* Eutrophication, hypoxia)
nutrient enrichment, 201
in ocean
classification, management efforts, 310–311
coastal, 310
"dead zones", 309
eutrophication, 309
future scenarios, 310
human health and ecosystem services, 309–310
"planetary boundaries", 309–310
policy recommendations, 311–312
waste streams, 310
and pollutants, stressors, 200–201
population density, 201–202
water quality, 201–202

I

Illegal, unregulated and unreported (IUU) fishing
disadvantages, market, 284–285, 286*t*
economic models, 295–296
estimation, 284–285, 285*f*
issue, 296
organizations, 296
stock assessments, 284–285
Insurance-based risk models, 340
IPCC's scenarios, greenhouse gas emissions, 231
IUU fishing. *See* Illegal, unregulated and unreported (IUU) fishing

L

Land storage
description, 105–106
flow from groundwater mining, 106
human modifications, 105–106
terrestrial water storage, 106, 107*f*
Long-Range Transboundary Air Pollution (LRTAP), 135
LRTAP. *See* Long-Range Transboundary Air Pollution (LRTAP)

M

Marine debris
description, 143
"ghost fishing", 143–144
plastic (*see* Plastic debris)
rubber, clothing and paper, 143–144, 144*f*
wood, metals and glass, 143–144, 144*f*
Marine fish stocks
aboriginal hunting, 175–176
biotic resources, 173–174
capacity-enhancing subsidies, 178
catch per unit effort, 177
climate change (*see* Climate change, marine fish stocks)
in coastal waters, 177–178
"colonial" phase, 175–176
commercial exploitations, 175
development, sonar, 177
economic losses
global fisheries, 184, 185*t*
profits and wages, rebuilding, 185–186
resource rent, 184–185
fisheries (*see* Fisheries)
freshwater/migratory species, 174
global economic loss, overfishing (*see* Overfishing, global economic loss)
oyster reefs, estuaries, 176
phase shifts, ecosystem engineer species, 178
predator populations, 178
prey species, California, 173–174
recovery plans, 178–179
regional economies, 175
social norms and community institutions, 179
South Pacific Islands, 174
steam trawlers, 177
stock status and fishing mortality, 178–179
technological advances, 176
terrestrial species, 174
Marine pollution
chemical and metal contaminants, 316
chemicals, 129–135
definition, 128
description, 128, 315
global marine ecosystems, 157–160
lethal and sublethal effects, 316–317
noise, 152–157, 316
oils, 135–142, 316–317
pathways and impacts, 128, 129*f*
in pelagic areas, 317
plastic debris, 316–317
policy recommendations, 317–318
radioactive waste, 150–152
sea-land and sea-atmosphere boundaries, 315
sewage, 315
solid substances, 143–150
toxic chemicals, 127–128, 315
wildlife and humans, 315
Marine protected areas (MPAs), 294, 309
Marine resources
benefit humanity, 320
FAO, 318–319
fishing, 318–319
healthy fish stocks, 319
loss of biodiversity, 320–321
management efforts, recovery, 321
migrating species, sardines, 320–321
pelagic fisheries, 320–321
policy recommendations, 321–323
RFMOs, 319
steady and sustainable food source aquaculture, 319–320
and stocks, 320
Maximum sustainable yield (MSY)
practical and policy reasons, 182
stock assessment, 182
taxon-EEZ pair, 182–183
MDR. *See* Mean damage ratio (MDR)
Mean damage ratio (MDR), 56, 56*f*
Mine tailings, 146–148, 147*f*
MPAs. *See* Marine protected areas (MPAs)
MSY. *See* Maximum sustainable yield (MSY)
Multiple stressors
categorization, 194
coastal and island populations, 328
cumulative effects, human activities, 323
description, 193–195
ecosystem dynamics and complex ways, 323–324
efficacy, management strategies, 323
EIAs, 325
enhance ecosystem resilience, MPA, 325, 326*t*
global-scale, 195–200
governance arrangements, 324–325
governance efforts, 325–327
human impacts, 194, 194*t*
human security risks, 328
hypoxia (*see* Hypoxia)

Multiple stressors *(Continued)*
intensity and spatial distribution, 193–194
leverage resources and capabilities, 325–327
marine ecosystem, 323–324
MPAs, 325
network regime approach, 327–328
operationalization, network governance, 327–328
oxygen-rich surface waters, 328
pollution (*see* Pollution)
synergistic effects (*see* Synergistic effects, stressors)

N

Noise pollution
fish and mammals, 153–154, 154*f*
impacts, marine mammal and fish assemblages, 156
invertebrates, 155–156
marine animals, 153
naval sonars, 154–155
ship, 152–153
sound travels, 152
sublethal effects, anthropogenic noise, 155
threats and pressures, 156–157
underwater (*see* Underwater noise pollution)
variety, activities, 152–153

O

Ocean acidification
biodiversity and ecosystems, 303–304
buffering capacity, 304
chemical-physical process, air-sea gas, 304–305
CO_2 emissions
activity, hydrogen ions, 17
carbonate ions, 16
coastal shelf seas, 24–25, 25*f*
description, 35–36
dissolved inorganic carbon, 15–16
food webs and ecosystems (*see* Food webs and ecosystems)
H^+ concentrations, 16
industrial revolution, 15
lower pH and carbonate ion concentrations, 21–24, 23*f*
ocean warming and deoxygenation, 24
pH stabilization levels, 21, 22*f*
physiological processes and behavior, 25–32
saturation state, $CaCO_3$, 16–17
time and space scales, 17–21
upwelling, waters, 23*f*, 24
food web processes, 305–306
fossil fuel emissions, 305
marine ecosystems, 304–305
policy recommendations, 306
vegetated habitats, 304–305
vulnerable and productive areas, 305–306
Ocean carbon sink
absorption, 269–270
emission scenarios, 270
shrinking
A1B scenario with "E1" stabilization, 263
anthropogenic carbon emissions, 262
atmospheric concentration, 263
climate change, 262–263
11 coupled climate-carbon cycle models, 263–264
physical and biological effects, warming, 264–265
physical processes, 264
Ocean environment
aggregate economic valuation, 246
Atlantic cod and bluefin tuna stocks, 246
biomes, ecological services, 251
carbon emissions, 246–247
catastrophic risk, economic analysis, 244
categories, services, 244
determination, price/dignity, 245
economic valuation studies, 247
ecosystem valuation, TEEB, 251
ecosystem values, 247, 248*t*
fisheries (*see* Fisheries, ocean environment)
high and low impact climate scenarios, 244
marginal values, 248–249
meaningful prices, 245
nutrient cycling and cultural services, 248–249
ocean carbon sink, 262–265, 269–270
open ocean and coastal zones, 248
policy decisions and economic calculations, 247
reefs and coastal areas, 250–251
scenario definitions, 266

selected climate impacts, 265, 265*t*
SLR (*see* Sea-level rise (SLR))
storms (*see* Storms, ocean environment)
TEEB, 249
tourisms (*see* Tourisms, ocean environment)
tropical storms, 251
valuable functions, 243
valuation, selected ecosystem services, 249, 250*t*
warming and SLR, 246–247
whale-watching trips, 245–246
Ocean warming
biological effects, 57–62
CO_2 emissions, human activities, 62
coral reefs, 307–308
description, 62–63, 306–307
estimation, 45
flooding and erosion, 307
fragile resources and economies, 308
intense tropical cyclones, 307
physical effects, 46–57
policy recommendations, 308–309
species distribution and food webs, 307
weather/hurricane intensity, 45–46
Oil pollution
alkanes and cycloalkanes, 135–136
aquaculture installations, 142
causes, shipping, 137, 138*t*
cleanup operations, 142
degradation process, 139
droplets, 141
evaporation and dispersion, 137–138
fisheries, 142
heavy oils, 137–138
littoral zone, 141
marine environment, petroleum hydrocarbons, 136, 137*t*
oxidative phosphorylation, 139
petroleum/crude oil, 135–136
plankton organisms, 140–141
seabirds, 140
sedimentation process, 138–139
tanker accidents, 136–137
temperatures, 139–140
whales and dolphins, 142
OMZs. *See* Oxygen minimum zones (OMZs)
Open-access fisheries, 283–284, 295
Overfishing
global economic loss
catch losses, 182
cost, overusing, 182
database, ex-vessel fish prices, 182–183
description, 181
environmental resources, 184
log-linear relationship, 182
MSY, 182
prone to overestimate depletion, 183
resource rent, 183–184
UNEP's, 183
value, net gains, 184
landed values and catches, 282–283, 282*f*
SIDC, 283
skates, rays and large-bodied croakers, 283
Oxygen minimum zones (OMZs)
demersal and pelagic fisheries, 80
description, 70–71
expansion, 80–81
mid-water layers, 78–79
OML, 78–79
upwelling regions, 79–80
world distribution, shelf areas, 79, 79*f*

P

Pacific Ocean
bordering, countries, 277, 278*f*
capture fisheries, 281, 281*f*
climate change, 287–293
coral reefs, seagrass, mangroves, seamounts and estuaries, 278–281, 279*f*
description, 277–278
MPAs, 294
overfishing, 282–283
stressors, 296
Paleocene-Eocene Thermal Maximum (PETM), 18–19, 18*f*
Pentagon planning
climate change risks, 233
decade-long droughts, 233
description, 233
impacts, abrupt change, 234
scarce and necessary resources, 233
"threat multiplier", volatile regions, 234
uncertainty, climate change, 234
PETM. *See* Paleocene-Eocene Thermal Maximum (PETM)
Physical effects, ocean warming
description, 45
extreme weather, 46–49
tropical cyclones, 49–57

Physiological processes and behavior, ocean acidification
calcification, 26–27
CO_2 concentrations, 30
coral recruitment, 30
hypercapnia, respiration, energetics and growth, 30–32
Ophiothrix fragilis, 29–30
primary production and microbial loop, 27–28
sea urchin gametes, 29–30
sensory cues, 28–29
Plastic debris
description, 143–144, 144*f*
marine animals, 144
microplastics, 144–145
packaging, 144
production, carbon emissions, 145–146
variability, abundance, 144–145
Pollution
cocktail and cumulative effects, noise, 203
impacts, 202
marine environment, 202
persistent chemicals, 202–203
solid waste and oil spills, 203
Precautionary approach
issues, marine pollution, 317
MPAs and EIAs, 341
of multiple stressors, 341

R

Radioactive waste
behavior, radionuclides, 152
collective doses, sea food, 151
estimation, 151–152
radionuclides, 150–151
shell fish and algae, 151
RCPs. *See* Representative concentration pathways (RCPs)
Rebuilding fisheries
catch losses, overfishing, 182
profits and wages, 185–186
real cost to society, 185–186
value, net gains, 184
Regional fishery management organizations (RFMOs), 319
Regional ocean governance
climate change and ocean acidification, 334–335
multiple stressors, 335
Pacific Ocean, 334
peoples and governments, 335
SIDS, 334
Representative concentration pathways (RCPs)
anoxia, ecosystem function, 12
atmospheric CO_2 concentrations, 4, 5*f*
RCP2.6 and RCP6 scenarios, 6
vs. SRES scenarios, CO_2 emissions, 4–5, 6*f*
Restoration, hypoxia
Baltic Sea coastal regions, 91–92
nutrient management, 91
physical, chemical and biological processes, 91
socioeconomic commitments, nutrients, 91
stressors, 91
RFMOs. *See* Regional fishery management organizations (RFMOs)

S

Scenarios analysis
Blueprints, 232
IPCC's, 231
and planning exercises, 231
RCP, 231
Scramble, 232
shell, 231–232
SRES, 231
Scramble scenarios, 232
Sea-level rise (SLR)
absolute *vs.* RSLC, 98–100
adaptation costs, 259
analysis, global costs, 258
assessments, economic impact, 257–258
changes, land storage (*see* Land storage)
changes, mean sea level, 97–98
from climate change, 312
on coastal areas, 115–116
coastal zones, 313–314
description, 267–268, 312
developing countries, 258
direct measurements, tide gauges, 109–110
dynamic processes and feedback loops, 312–313
economic consequences (*see* Economic consequences, SLR)
glacier and ice sheet melting, 102–105
glaciers and ice caps, 312–313
greenhouse gas emissions, 313–314
health effects, 314

hypothetical beach community, 258–259
past sea-level changes, 109
policy recommendations, 314–315
projections
description, 111–114, 113*f*
glaciers and ice sheets, 114
intentions, 114
post-IPCC AR4, 115
semiempirical methods, 114–115, 115*f*
projections, IPCC, 313
"residual" fraction, 108, 108*f*
satellite measurements
information, 110
mean sea level, changes, 110, 110*f*
rates, 111, 112*f*
scientific advances, 97
on society, 97
sources, 107, 107*t*
take-home messages, 122
thermal expansion/density changes (*see* Thermal expansion)
types, 116, 117*t*
vulnerability (*see* Vulnerability, SLR)
SEAPODYM. *See* Spatial ecosystem and population dynamic model (SEAPODYM)
Shell scenarios approach
Blueprints scenario, 232
decision pathways, 232
description, 231–232
Scramble scenario, 232
SIDC. *See* Small islands developing countries (SIDC)
SLR. *See* Sea-level rise (SLR)
Small islands developing countries (SIDC), 283
Solid substances, pollution
description, 143
hard structures
artificial reefs, 149, 149*f*
climate change, 149
oil and gas rigs, 148
quantity, 150
marine debris, 143–146
mine tailings, 146–148, 147*f*
sedimentation, 146, 148
Spatial ecosystem and population dynamic model (SEAPODYM), 289–290
Special Report on Emissions Scenarios (SRES), 231
SRES. *See* Special Report on Emissions Scenarios (SRES)
Storms, ocean environment
A1B, RCP6, 268
weather
A1B climate scenario, 261
correlation, 259–260
cyclone damages, 262
FUND model, 261
historical Atlantic hurricane record, 259–260
intense storms, 262
intensification, tropical cyclones, 259
tropical cyclones, 260–261
U.S. hurricane losses, socioeconomic changes, 260
wind speeds, 268
Surprise ("unknown unknowns")
catastrophic and irreversible damages, 331–332
planning
catastrophic and irreversible damages, 331–332
catastrophic tipping points, 331–332
coastal populations, 332
coral reefs, 330–331
cost-based to insurance-based models, 331
cost-benefit analysis, 329–330
economic risks, 330
emission scenarios, 329
global financial recession, 329
kinds, threats, 333
for multiple stressors, 329
poleward migration, fish communities, 330–331
"self-insure" humanity, 332
SIDS, 332–333
and uncertainty, 331
preparation
abrupt changes, 223–224
analogy, private insurance, 227–228
atmospheric aerosol loading and chemical pollution, 236
boundaries, characteristics, 236–237
climate scientists, 235
collapse, North Atlantic cod stocks, 224, 224*f*
cost-benefit analysis, 227
decisions making in dark, 228–230
global oceans, 223
gradual changes, 225–226
insurance *vs.* personal catastrophes, 226–227

Surprise ("unknown unknowns") *(Continued)*
IPCC's scenarios, 231
life insurance, 226
links, planetary boundaries, 237, 238, 238*f*
marine issues, 238–239
marine threats and uncertainties, 225
meteorite insurance, 227
ocean acidification, 224, 237
oceanic sediment measurements, 224–225
ocean-related climate impacts, 223
ominous scientific findings, 239
pentagon planning (*see* Pentagon planning)
"planetary boundaries" analysis, 235–236, 236*f*
planning, marine, 239
prediction, 225
private and policy sectors, 237
"safe minimum standards", 235
scenario analysis and planning exercises, 231
shell game (*see* Shell game)
"tolerable windows", 235
uncertainty, 226
U.S. fire departments, 226
Sustainable ocean resources
anthropogenic CO_2 release, 301
global ocean governance (*see* Global ocean governance)
human-caused environmental change, 302
hypoxia in ocean, 309–312
multiple stressors (*see* Multiple stressors)
ocean acidification (*see* Ocean acidification)
ocean and coastal areas, 301
ocean damages and surprise planning (*see* Surprise ("unknown unknowns"))
ocean warming (*see* Ocean warming)
overuse, marine resources (*see* Marine resources)
pollution (*see* Marine pollution)
regional ocean governance (*see* Regional ocean governance)
SIDS, 302
SLR (*see* Sea-level rise (SLR))
threats to global ocean, 302–303
vulnerable coastal communities, 302
Synergistic effects, stressors
Acropora intermedia and *Porites lobata*, 207, 208*f*
Acropora tenuis and *Acropora millepora*, 207–209, 209*f*
additive and antagonistic responses, 203–204, 204*f*
bleaching, coral reefs, 207
calcifying algae, 211–212
Cancer pagurus, 206–207
chaperone *hsp70*, 205–206, 206*f*
community-level response, 205
coral bleaching, 210
crustaceans, 210–211
herbicides, 209–210
hypoxia, sediments and pollutants, 207–209
impacts, anthropogenic activities, 205
jumbo squid *(Dosidicus gigas)*, 213
local/regional impacts, 210–211, 213–214, 215*f*
nutrients, 211, 212*f*
pCO_2 and temperature, 206–207
pH and O_2 content, fisheries, 213, 214*f*
population dynamics and ecosystem traits, 204–205
Porolithon onkodes, 207, 208*f*
reduction, nutrients, 214–215
temperatures and Cu concentration, 209–210
trophic levels, 204–205

T

TEEB. *See* The Economics of Ecosystems and Biodiversity (TEEB)
The Economics of Ecosystems and Biodiversity (TEEB)
coral reefs/coastal wetlands, 250–251
UNEP, 249
valuation, selected ocean ecosystem services, 250*t*
Thermal expansion
CCSM3, GISS, MICRO3.2 and MRI-CGCM2.3.2 models, 101–102
comparison, observations, 100–101, 101*f*
heat content, 100, 100*f*
measurements, 102
observations and model calculations, 102
volcanic eruptions, 101
Time and space scales, ocean acidification
burning, fossil fuel, 17
carbon perturbation, 17
catastrophic impact events, 19

ocean pH and carbonate ion concentration, 19–21, 20*f*
PETM, 18–19, 18*f*
surface mixed layer, 17–18
Tourisms, ocean environment
Caribbean and Pacific islands, 268–269
and climate change (*see* Climate change)
coral reef-based, 268–269
demands, 269
income elasticity, 269
Tropical cyclones
Accumulated Cyclone Energy index, 54
Atlantic hurricanes, 53–54, 54*f*
economic damage, 57
factors, 50
frequency, 51–52, 51*f*, 52*f*
life cycles, 49–50
map, 50, 50*f*
MDR, 56, 56*f*
projections, 55
Saffir-Simpson hurricanes, 53, 53*f*
US damage costs, 57, 58*f*
variability, formation and life cycle, 54–55
wind speeds, 56, 56*t*

U
Underwater noise pollution
fish and mammals, 153–154
impacts, individuals and populations, 157
invertebrates, 155–156
population, fish and whales, 156
ship, 152–153

UNEP. *See* United Nations Environment Program (UNEP)
UNFCCC. *See* United Nations Framework Convention on Climate Change (UNFCCC)
United Nations Environment Program (UNEP), 249
United Nations Framework Convention on Climate Change (UNFCCC), 306

V
Vulnerability, SLR
aquaculture, 119–120
coastal locations and islands, 118
health effects, 119–120
limitations, 118–119
populations, small island states, 118
regions, flooding, 116–118, 118*f*
socioeconomic sectors, impacts affect, 119, 119*t*

W
World Resources Institute eutrophication project, 91–92

Z
Zooplanktons
excretion, fecal pellets, 138–139
ocean acidification, 141
structure, 141

Printed and bound by CPI Group (UK) Ltd, Croydon, CR0 4YY

08/05/2025

01864866-0001